MICROEMULSIONS
AND RELATED SYSTEMS

SURFACTANT SCIENCE SERIES

CONSULTING EDITORS

MARTIN J. SCHICK
Consultant
New York, New York

FREDERICK M. FOWKES
Department of Chemistry
Lehigh University
Bethlehem, Pennsylvania

OTHER VOLUMES IN PREPARATION

Library of Congress Cataloging-in-Publication Data

Bourrel, Maurice.
   Microemulsions and related systems : formulation, solvency, and
physical properties / Maurice Bourrel, Robert S. Schechter.
         p.   cm. -- (Surfactant science series ; v. 30)
   Includes index.
   ISBN 0-8247-7951-7
   1. Emulsions.   I. Schechter, Robert Samuel, [date].   II. Title.
III. Series.
TP156.E6S34 1988
660.2'9451--dc19   88-3624

Cover art: *Sky and Water,* *M. C. Escher (1938).* *Copyright 1988*
   by M. C. Escher heirs, Cordon Art, Baarn, The Netherlands.

MARCEL DEKKER, INC.
270 Madison Avenue, New York, New York 10016

Current printing (last digit):
10 9 8 7 6 5 4 3 2 1

PRINTED IN THE UNITED STATES OF AMERICA

MICROEMULSIONS AND RELATED SYSTEMS

Formulation, Solvency, and Physical Properties

Maurice Bourrel

Groupement de Recherches de Lacq
Elf-Aquitaine
Artix, France

Robert S. Schechter

Departments of Chemical
and Petroleum Engineering
The University of Texas at Austin
Austin, Texas

MARCEL DEKKER, INC.   **New York and Basel**

Sky and Water, M. C. Escher (1938).

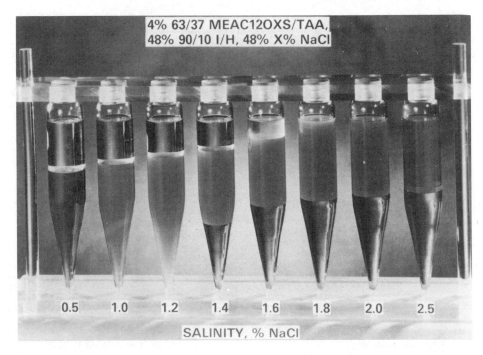

4% 63/37 MEAC12OXS/TAA, 48% 90/10 I/H, 48% X% NaCl

| 0.5 | 1.0 | 1.2 | 1.4 | 1.6 | 1.8 | 2.0 | 2.5 |

SALINITY, % NaCl

Microemulsion Phase Behavior, R. N. Healy, R. L. Reed and D. G. Stenmark (1976).

Human mortality is most certainly the origin of our fascination with continuity, eternity, and infinity. Escher has captured the essence of these concerns in his woodcarving <u>Sky and Water</u> (opposite upper). The similarity of this representation with the continuity of states found in micellar solutions (opposite lower) is striking.

This is the principal subject treated in this monograph and it is entirely appropriate, perhaps even mandatory, that this book be dedicated to our children, Antoine, Geoffrey, Lawrence, Richard, and Sophie, who represent the succeeding generation.

Preface

This book is written for those scientists and engineers who wish to
understand the relationship between the molecular structure of amphiphilic compounds and their ability to cause water and apolar liquids
(oils) to mix forming a single-phase, isotropic solution. Although
knowledge of this subject has expanded enormously over the past
fifteen years, primarily because the use of micellar solutions for the
enhanced recovery of oil from subterranean formations seems an attractive possibility, a wide spectrum of other applications has emerged,
and many others are under active development. Thus, it became evident to us that a book unifying the results of this flurry of research
would be of considerable interest and help to surface and colloid
chemists interested in selecting surfactant molecules for various applications. We hope that this monograph responds to this need. We
are certainly persuaded that the R-ratio first proposed by P. A.
Winsor and greatly extended and applied here is an intrinsically valuable concept without which it would be difficult, if not impossible, to
reconcile many of the results. Future research may very well advance
the theories outlined in this book, but we believe that they will spring
from the foundations set down in it. The R-ratio or some equivalent
concept is, we believe, indispensable in understanding the behavior
of micellar solutions.

For the past twelve years the authors have been engaged in re-
search that has had as its central goal the development of improved
surfactants for enhanced oil recovery. Early on in this effort Solvent
Properties of Amphiphilic Compounds (Butterworths, 1954) by P. A.
Winsor came to our attention and the value of the approach taken by
Winsor to help us to achieve our goals became evident.

His primary concept is that the R-ratio of cohesive energies, stemming from the interaction of the interfacial layer with oil divided by the energies resulting from the interaction with water, determines the curvature of the interface. Thus, if R > 1, the interface tends to increase its area of contact with oil while decreasing its area of contact with water. Similarly, a balanced interfacial layer is represented by R = 1.

This concept proved to be so useful in the conduct of our research that we first proposed that the original Winsor book be revised and reprinted in a form closely resembling the 1954 version. However, once we began to enumerate those aspects of the original book requiring updating to incorporate the astonishing quantity of research reported since its original publication, it became evident that a simple revision would not do justice to the great spectrum of new ideas which have emerged over the past thirty years.

This book is new. It follows the general outline of Solvent Properties of Amphiphilic Compounds but retains little of the original content. The R-ratio as initially proposed has been extended beyond the limits originally visualized by Winsor and is the main focal point of the monograph. We have corresponded with Winsor, and his remarks, questions, and corrections stimulated considerable thought and certainly helped to improve the presentation. We are grateful for his assistance.

The terminology used in this monograph primarily conforms to that originally used by Winsor, although we employ interchangeably the terms micellar solution and microemulsion to describe the solutions he called micellar solutions. The term microemulsion has been used by a number of authors and especially by Schulman and his co-workers. It is thought that these solutions containing significant amounts of both oil and water together with an amphiphile differ from those that consist primarily of water and an amphiphile together with trace quantities of oil. The oil in these latter solutions is termed solubilized. Thus, some authors who describe certain solutions as microemulsions perceive that there is a fundamental difference between micellar solutions and microemulsions and indeed there may very well be differences; however, there is no distinction that can be sharply drawn. Thus, it seems counterproductive to attempt to define one. Indeed, to make a sharp distinction would obscure one of the primary features of this book, namely, the unity of the sequence of states that exist between micelles and inverted micelles.

The applications for surfactants are pervasive. They are used in the formulation of pharmaceuticals, agricultural chemicals, foods, cleaning compounds, cosmetics, lubricants, oil recovery agents, and a myriad of other formulations. These applications have prompted much of the research that has been reported over the years; however, the ability of solutions containing surfactants to self-organize has also been a major stimulus. This remarkable property of amphiphiles, which tends to result in interfacial regions that are bilayers and closely resemble biological membranes, has long been a source of scientific curiosity. The tendency for self-aggregation is one of the features of amphiphilic systems stressed in this monograph.

Chapters 2 and 3 are devoted to a discussion of micelles and inverted micelles, respectively. These are the beginning and ending states of the sequence of isotropic states which are the primary subject of this monograph and therefore their properties are of considerable interest. The main theme of the discussion is the unity of these states so that their properties evolve smoothly and continuously toward those described in Chapter 4. At no point can one make a sharp distinction between the states appearing in the sequence. For this reason the characteristics of micelles and inverted micelles are both interesting and relevant.

The formulation of micellar solutions containing significant quantities of both oil and water is a matter of considerable importance and is an aspect emphasized in this monograph. Chapter 4 describes the physical properties of these solutions and demonstrates their relationship to the phase behavior of mixtures of oil, water, and amphiphiles. This connection is important because many characteristics of micellar solutions can be inferred based on a knowledge of phase behavior which can often be obtained from simple experiments.

Having considered the importance of the phase behavior, it becomes important to be able to change it systematically. Chapter 5 focuses on this issue. It is seen that the qualitative trends can be predicted based on an understanding of the R-ratio. The relative sensitivity of the phase behavior of micellar solutions to changes in the formulation variables is established in Chapter 6. The procedure used to fix these quantitative relationships is to make a change in one of the formulation variables and to compensate for the effect of this change by altering a second formulation variable. By this procedure the relative importance of a given variable can be established.

In addition to influencing the phase behavior, the formulation variables also control the solvency or effectiveness of an amphiphilic compound. The R-ratio provides guidance for improving the solvency. The hypothesis is stated in Chapter 7 and is supported by extensive evidence. Its relationship to solvency is one of the most important characteristics of the R-ratio.

In Chapter 8 the thermodynamic implications related to the existence of a natural bending of the interfacial region are developed. It is seen that the existence of the natural bending is symmetry breaking; that is, the thermodynamic models imply a degree of symmetry not consistent with observation unless a natural bending is imposed. Thus, the R-ratio is a concept needed to reconcile thermodynamic prediction with observation.

Because of the wide range of applications, many different researchers have found micellar solutions to be a subject of considerable interest and have over the years added to our understanding of these systems. However, the contributions of a few researchers are prominently used because they directly support the concepts developed in the monograph. The work of K. Shinoda and his co-workers and R. L. Reed and his co-workers is extensively cited. The authors would like to acknowledge these early efforts and their impact in helping to formulate the authors' own views.

Finally, a number of other able investigators deserve special mention not only for their significant research but also because the authors have greatly benefitted from extensive discussions with them. In particular, W. H. Wade has been a coauthor with one or both of us of many of the papers cited. His pioneering ideas on modeling the oil structure provided the basis for many subsequent developments. S. J. Salter, C. Chambu, and J. Biais are acknowledged for their helpful discussions.

We also want to express our appreciation to our wives, Marie-Jo and Mary Ethel, for their patience and understanding during the preparation of this monograph. We are especially indebted to Mary Ethel, who undertook the difficult task of helping to edit and type the manuscript. In this effort Mrs. Joye Johnson provided considerable assistance to help complete the project. Finally, one of us (M. B.) wishes to acknowledge the management of Elf-Aquitaine, who encouraged the preparation of this monograph.

M. Bourrel

R. S. Schechter

Contents

Acknowledgments

Sky and Water, M. C. Escher (1938). Copyright 1988 by M. C. Escher heirs, Cordon Art, Baarn, The Netherlands.

Figures 2.22, 4.11, 4.19, 4.25, 4.30, 4.31, 4.32, 4.35, 4.36, 4.40, 4.41, 5.11, 5.12, 5.20, 5.28, 5.29, 5.34, 6.1, 6.2, 6.3, 6.4, 6.5, 6.6, 6.7, 6.8, 6.12, 6.31, 7.1, 7.4, 7.7, 7.8, 7.9, 7.10, 7.11, 7.14, 7.17, 7.20, 7.21, 7.23, 7.25, 7.26, 7.28, 7.30, 7.31, 8.17, 8.18: Reproduced by permission. Copyright 1976, 1977, 1978, 1979, 1982, 1983, 1984, 1985, 1986, 1987. Society of Petroleum Engineers of AIME.

Figures 2.1, 2.2, 2.5, 2.6, 2.7, 2.12, 2.13, 2.15, 3.5, 6.14, 7.16, 7.33: Reproduced by permission. Copyright 1947, 1948, 1951, 1963, 1964, 1977, 1978, 1979, 1984, 1986. American Chemical Society.

Figure 2.27: Reprinted by permission from Nature, Vol. 222, pp. 1159. Copyright © 1969, Macmillan Magazines Ltd.

Table 2.10: Reproduced with permission of the copyright owner, The American Pharmaceutical Association.

MICROEMULSIONS
AND RELATED SYSTEMS

1

The R-Ratio

I. AMPHIPHILIC COMPOUNDS

Amphiphilic compounds, or more briefly, amphiphiles, are characterized by possessing in the same molecule two distinct groups which
differ greatly in their solubility relationships [1]. Such compounds
were termed "amphipathic" by Hartley [2] to denote the presence of
a "lyophilic" group having an affinity (sympathy) for the solvent
and a second group, sometimes referred to as "lyophobic," which is
antipathetic to the solvent. Whenever water is the solvent, the two
groups are often designated as hydrophilic and hydrophobic portions,
or more commonly and perhaps more pictorially as the "head" and the
"tail," respectively. The terms "lipophile," "hydrophobe," and "amphiphile's tail," therefore, are all used in the literature to describe
the same part of the amphiphilic compound, and these terms may be
regarded as being interchangeable. Similarly, "hydrophile" and
"amphiphile's head" both refer to the same part of the amphiphile
and are also interchangeable.

   Amphiphilic compounds tend to concentrate as a monolayer at a
water interface, with the tendency increasing as both the lipophilic
and hydrophilic character of the amphiphile become more pronounced.
At the interface, amphiphilic compounds are arranged so that the
lipophile is removed from the water while the hydrophile remains in
contact with the aqueous solution. This molecular orientation is consistent with a number of experimental observations, including the
sharp reduction in surface or interfacial tension attending the addition of small quantities of amphiphile to an aqueous phase. Because
of this pronounced tendency for amphiphiles to accumulate as a monolayer at an interface, and there to reduce the surface tension, they

1

are sometimes called surface-active agents or, more simply, surfactants. The terms "surfactant," "amphiphile," or "amphiphilic compound" are used interchangeably throughout this book.

Within the definition of amphiphilic compounds are included compounds that range in character from predominantly lipophilic to predominantly hydrophilic. This is exemplified by the following series: $C_nH_{2n+1}X$, where n has values from 1 upward and where X = $-N^+(CH_3)_3$, $-NH_2$, $-OH$, $-CO_2H$, $-CO_2^-$, $-SO_4^-$, and so on. As shown by distribution experiments with liquid members, the higher-molecular-weight end of these series are oil[†]-miscible and water-immiscible compounds, while at the lower end are those which are water miscible and oil immiscible. Compounds of intermediate molecular weights show marked miscibility with both organic solvents and water. These are the compounds more typically amphiphilic.

It is evident from experience that the range of molecular weight where high oil solubility is lost and where high water solubility is gained differs greatly in the different series. Thus for the normal alcohols, amines, and acids it lies at about C_6 to C_4. For the polyethylene oxide condensation products, it varies according to the length of the polyethylene oxide chain. For n-alkylated ions (e.g., RSO_4^-, RNH_3^+) it is above C_{14}.

It must be emphasized, however, that with the crystalline higher n-alkyl ionic compounds, a low solubility in both hydrocarbon and water at room temperature results from another factor—their high tendency to crystallize. Since the solubility of a substance is determined by the equality of its chemical potential in both the solid and the liquid phases, conclusions concerning the intermolecular relationships in liquid phases can be drawn from comparisons of solubilities only if the solubilities refer to comparable crystalline phases.

From these brief considerations it is evident that the selection of a chemical structure with lyophobic and lyophilic regions suitable for a particular application will vary with the nature of the solvent and the conditions of use. To achieve high surface activity, the amphiphile must be appropriately balanced. One of the primary goals of this book is to elucidate those factors that influence the lyophobic/lyophilic balance and to establish methods for quantifying them. Different tests, such as emulsion stability, partitioning between oil and water, interfacial tension, solubilization, and so on, designed to reveal the character of an amphiphile may at first sight seem disconnected and unrelated, not subject to interpretation or in other cases, apparently contradictory. The particular observation may, for example, depend on surfactant concentration, be dramatically altered

---

[†]The term "oil" will be used loosely to include organic liquids insoluble in water. In the present context it refers more particularly to liquid hydrocarbons.

in the presence of small concentrations of impurity, or be quite sensitive to temperature. These phenomena can generally be interpreted and complex behavior predicted based on a simple, yet profound concept introduced by Winsor [1]. This concept, which derives from a consideration of the cohesive energies acting within the interfacial region, is expanded and given semiquantitative interpretation in this book. Its thermodynamic origins are explored. In the hands of an experimentalist interested in selecting or synthesizing amphiphilic compounds for practical application, it is a powerful, indispensable tool. To the theorist interested in classifying or understanding complex behavior, its predictions are crucial. The formulation of the concept is here called the R-ratio. To appreciate fully the meaning of the R-ratio, it is necessary to have some understanding of the molecular forces that act on amphiphilic compounds positioned at an interface. The origins of these molecular forces are reviewed briefly in the following paragraphs.

II. MOLECULAR INTERACTIONS

A. Van der Waals Forces

Three types of forces acting between molecules contribute to the Van der Waals forces. They are all attractive, and remarkably, each varies inversely with the distance between molecules taken to the sixth power, even though they originate from very different mechanisms [3–5].

1. Forces Between Polar Molecules

These are known as Keesom forces, or orientation forces. The interaction energy between two permanent dipoles μ_i^o and μ_j^o is obtained by considering the coulombic forces between the four charges. The interaction energy depends on the distance r between dipoles and on their relative orientation, which is of course affected by temperature as well as the electric field. Applying Boltzmann statistics, Keesom [3] derived the potential resulting from dipole-dipole interaction[†]:

$$u_{ij}^K = -\frac{2}{3}\frac{\mu_i^{o2}\mu_j^{o2}}{r^6 kT} \tag{1.1}$$

[†]It has, however, been argued by J. S. Rowlinson that a proper average of the dipole orientations would give $-1/3$ instead of $-2/3$ for the numerical coefficient in Eq. (1.1) [6].

where μ_i^o and μ_j^o are the dipole moments and k is Boltzmann's constant. The important features of this potential are that it is attractive, it decreases when the temperature increases (as one would expect), and it is short range, since it varies like r^{-6}. In condensed phases, when steric and other factors, such as hydrogen bonds, prevent free molecular rotation, Boltzmann statistics do not apply and Eq. (1.1) is in error.

2. Forces Between Permanent and Induced Dipoles

When a nonpolar molecule i having no permanent dipole is subjected to the electric field of a polar molecule j, the electrons of i are displaced from their equilibrium positions and a dipole is induced. The resultant force between the permanent dipole and the induced dipole is always attractive.

The mean potential energy was first calculated by Debye [4]:

$$u_{ij}^D = -\frac{\alpha_i \mu_j^{o2}}{r^6} \tag{1.2}$$

where μ_j^o is the dipole moment of molecule j and α_i is the polarizability of molecule i. The characteristic features of this potential are that it is short range and does not depend on temperature.

3. Forces Between Nonpolar Molecules

The deviations from ideal gas behavior of a nonpolar molecule such as argon are indications of the existence of types of forces other than those described above. London [5] showed first that so-called nonpolar molecules are, in fact, nonpolar only when viewed over a period of time. At a given instant, however, the oscillations of the electrons about the nucleus result in distortion of the electron arrangement sufficient to create a temporary dipole moment. This dipole moment is rapidly changing in magnitude and direction and produces an electric field which then induces dipoles in the surrounding molecules.

Using quantum mechanics, the energy due to the London forces, the dispersion forces, is calculated for two spherically symmetric molecules i and j as

$$u_{ij}^L = -\frac{3\alpha_i \alpha_j}{2r^6} \frac{I_i I_j}{I_i + I_j} \tag{1.3}$$

where I_i and I_j are the ionization energies. The important features of this latter potential are that it is attractive, short range, and independent of temperature.

If the two molecules are of the same species,

$$u_{ii}^{L} = -\frac{3\alpha_i^2}{4r^6} I_i \tag{1.4}$$

Accordingly,

$$|u_{ij}^{L}| = \frac{2\sqrt{I_i I_j}}{I_i + I_j} \sqrt{u_{ii}^{L} u_{jj}^{L}} \tag{1.5}$$

In many cases, the ionization energies of the different molecules are not much different, and hence $2\sqrt{I_i I_j}/(I_i + I_j) \approx 1$. Thus it follows that

$$|u_{ij}^{L}| = \sqrt{u_{ii}^{L} u_{jj}^{L}} \tag{1.6}$$

Equation (1.6) provides a theoretical basis for the frequently applied "geometric-mean rule," which is often used in equations of state for gas mixtures and in theories of liquid solutions. The restrictive conditions imposed in its derivation must, however, be kept in mind.

The three types of forces discussed above are often considered to be the components of the Van der Waals forces. The order of magnitude of Van der Waals interactions is 2 kcal/mol for moderately polar or nonpolar molecules (1.75 kcal/mol for benzene, 1.7 kcal/mol for CCl_4, 1 kcal/mol for CO_2). To show the relative magnitude of dipole, induction, and dispersion forces in some representative cases, potential energies can be written in the form $u_{ii} = -B/r^6$ and B can be calculated separately for the contributions due to the dipole, induction, and dispersion effects with the aid of Eqs. (1.1), (1.2), and (1.4).

The values of B in erg-cm^6 $\times$ 10^{60} are, respectively, for HCl, 24.1:6.14:107; for cyclohexane, 0:0:1460; and for water, 203:10.8:38.1 [6]. In the latter case, however, specific interactions are operative. These are discussed next. The computed values of B indicate that the contribution of induction forces is small.

B. Specific (Chemical) Forces: Hydrogen Bonding

A good summary has been given by Prausnitz [6]. The main difference between a physical and a chemical force lies in the criterion of saturation: chemical forces are saturated but physical forces are not [7]. The covalent bond between two hydrogen atoms is an example of "satisfied" (or saturated) chemical attractive forces. On the other hand, the purely physical force between two argon atoms, for example, is not similarly saturated. Two argon atoms which are attracted to form a doublet still have a tendency to attract a third argon atom, then a fourth, and so on.

A variety of specific or chemical forces, such as electron donor-acceptor interactions, acidic solute-basic solvent (or vice versa) interactions, and hydrogen bonds cannot be expressed by simple interaction energies as in the previous cases of physical interactions. Chemical effects in solution are generally classified in terms of asso-ciation or solvation. The former refers to the tendency of some molecules to form extended structures, especially through hydrogen bonding. The latter refers to the tendency of molecules of different species to form complexes.

Detailed discussion of hydrogen bonding can be found elsewhere. It appears that two sufficiently negative atoms X and Y (which may be identical) may, under suitable conditions, be bonded with hydrogen according to $X-H\cdots Y$. Consequently, molecules containing hydrogen attached to an electronegative atom (as in acids, alcohols, and amines) show strong tendencies to associate with each other and to solvate with other molecules possessing electronegative atoms.

C. Hydrophobic Interactions

Compounds immiscible with water, such as hydrocarbons, have been called hydrophobic, a term that tends to convey the concept of a repulsive force between the water and the immiscible compound. Actually, as pointed out by Hartley [2], there is no conceivable mechanism for this, and the reasons for water-hydrocarbon immiscibility now appears to be rooted primarily, although certainly not entirely, in entropic effects (see Section 2.V). This mechanism was first discussed in detail by Frank and Evans [8] and more recently by Kauzmann [9] and Tanford [10]. The accepted view, which is consistent with a large number of observations, is that when a nonpolar solute is dissolved in water, very few of the hydrogen bonds that link water molecules together are disrupted. There is often little heat associated with the solution of hydrocarbon in water. Although hydrogen bonds are not disrupted, the presence of the nonpolar solute does tend to distort the water structure locally and does restrict the motion of the water molecules. Thus a large entropy increase in the water molecules is associated with removal of the apolar solute from solution.

The entropy increase is primarily responsible for the surface activity of amphiphilic compounds, which, as noted previously, tend to become oriented at a water interface so that the lipophile part of the amphiphile is removed as nearly as possible subject to other constraints from the aqueous phase. The hydrophobic interaction is also the primary free energy driving force for the formation of micelles and thus is of considerable importance. It is discussed in greater length in Section 2.V, where measurements quantifying the entropic change associated with the transfer of apolar compounds from an aqueous phase are cited.

D. Electrostatic Forces Involving Ions

The well-known interaction potential between two point charges e_i and e_j separated by a distance r in a dielectric medium is given by

$$u_{ij}^E = \frac{e_i e_j}{\varepsilon r} \tag{1.7}$$

where ε is the dielectric constant of the medium. Since u_{ij}^E decreases as r^{-1}, this potential is long range compared with the other intermolecular potentials.

E. Repulsion Forces

At very small distances of separation between the molecules, the electron clouds overlap and the forces between molecules become strongly repulsive rather than attractive. The repulsive potential given by a high inverse power law as suggested by Mie [11] has the form

$$u = \frac{\Omega}{r^n} \tag{1.8}$$

where Ω is a positive empirical constant and n is a positive number.

Using n = 12, Lennard-Jones [12] combined Eqs. (1.4) and (1.8) to give the total potential energy as

$$u_{total} = 4\xi \left[\left(\frac{\sigma}{r} \right)^{12} - \left(\frac{\sigma}{r} \right)^6 \right] \tag{1.9}$$

where $-\xi$ is the minimum potential energy and σ is the distance where $u_{total} = 0$. For many properties, reasonable predictions are obtained with $u = \infty$ for $r \leq \sigma$ and $u = 0$ for $r > \sigma$, that is, by assuming the particles to be hard spheres.

F. Long-Range Van der Waals Forces

The interaction between two macroscopic objects, such as spheres or plates, requires the summation of the elementary attraction between isolated molecules. Calculations were carried out first by de Boer [13] and Hamaker [14], assuming the additivity of the molecule-molecule dispersion forces and that the distance between the two bodies is large compared to the molecular dimension. The addition was achieved by integration over volume elements in the two bodies,

which has the effect of changing the dependence on r and of introducing the number of molecules per cubic centimeter in each body.

For example, the long-range Van der Waals energy of attraction
between two parallel flat plates of thickness d separated by a distance r in a vacuum is given by [15]

$$u = -\frac{H_m}{12\pi}\left[\frac{1}{r^2} + \frac{1}{(r + 2d)^2} - \frac{2}{(r + d)^2}\right]$$ (1.10)

H_m is known as the Hamaker constant and is equal to $\pi^2 n^2 B$, where
B is derived from the Van der Waals forces (see Section 1.II.A) and
depends on the nature of the molecules (i.e., their polarity) and n
is the number of molecules per cubic centimeter in the plates.
If the two plates are separated by a liquid rather than a vacuum,
H_m must be replaced by an effective Hamaker constant:

$$H_m = H_{11} + H_{22} - 2H_{12}$$ (1.11)

where H_{11} and H_{22} are the Hamaker constants for the plates and
the liquid medium, respectively, and H_{12} is the Hamaker constant
for the interaction between phases 1 and 2. To the approximation
given by Eq. (1.6), $H_{12} = \sqrt{H_{11}H_{22}}$ and thus [16]

$$H_m = (\sqrt{H_{11}} - \sqrt{H_{22}})^2$$ (1.12)

The additivity of molecular forces as well as the procedure of integration over volume elements is questionable when the distance between interacting bodies is of the order of the intermolecular distances, which is often the case. An alternative, more general approach has been developed by Lifshitz [17] [see also 18−21].

III. MICELLES

A. Formation of Amphiphilic Aggregates

The physical properties of aqueous solutions of amphiphilic compounds
often undergo a more or less abrupt change over a narrow concentration range as shown by Fig. 2.1. It is generally accepted that this
rapid change in the property/concentration curve is due to the formation of aggregates of amphiphiles or micelles in solution. The nu-
cleation of micelles is spontaneous and the remarkable feature is that
quite often a large number of amphiphiles, more than one hundred,
aggregate together, forming a structure that can vary between spherical and cylindrical depending on the molecular architecture of the
amphiphilic compound, the solution composition, and the temperature.

Aggregates containing fewer amphiphiles are unstable and do not exist in significant concentrations. The various contributions to the free energy of micellization and the reasons why large aggregates rather than smaller ones such as dimers, trimers, and so on, are often the most prevalent aggregates are discussed in Chap. 2.

The particular arrangement of the amphiphilic compounds in a micelle is to be visualized as being such that the hydrophiles are in contact with water, whereas the lipophiles are collected together within the interiors of micelles to create small regions from which water is essentially excluded. Although certainly not precise, it is useful to visualize the interior of a micelle to be liquidlike. Evidence supporting the view is presented in Chap. 2. The micellar structures designated as S_1 in Fig. 1.1 are intended to represent the aggregates described here. Note that in this drawing the lipophilic portion of an amphiphilic compound is represented by a single, solid line. This is the tail. The hydrophilic part, the head, is represented as a circle attached to the tail. Shown also is oil included within the interior of the micelles and in contact with the tail of the amphiphile. Since the interior of a micelle is liquidlike, certainly apolar compounds will dissolve preferentially within this region.

Aggregates of amphiphilic compounds can also form in apolar solvents. These are called inverted micelles. In the absence of water, the free-energy contributions both supporting and opposing the formation of inverted micelles are all smaller than those which exist in aqueous solutions. Thus in apolar solvents the aggregation numbers tend to be much smaller, often less than 20 amphiphilic compounds per inverted micelle, and therefore, changes in the property/concentration slope are much less abrupt, weakening, if not entirely eliminating, the concept of a critical micelle concentration (CMC). Furthermore, the degree of aggregation tends to be sensitive to impurities, especially water, and the distribution of aggregate sizes tends to be relatively broad, with dimers, trimers, and so on, often exhibiting a significant lifetime. A discussion of inverted micelles is given in Chap. 3.

Although inverted micelles tend to consist of small numbers of amphiphiles, they do promote the solubility of water in apolar solvents. Water can be visualized as residing within the interior of the inverted micelle as depicted by Fig. 1.1. In this case the S_2 structure is composed of a water core in contact with the hydrophile and an oil exterior in contact with the lipophile.

B. Fluctuations

It is worth noting that micellar aggregates have a finite mean lifetime (see Chap. 2). The structures are mobile, the interfaces are flexible, and there is a rapid exchange of molecules between the neighboring region and the aggregate. At higher concentrations

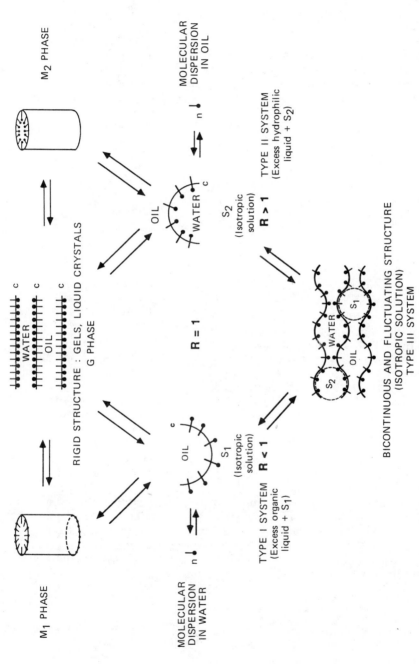

FIGURE 1.1 Intermicellar equilibrium and associated phase changes shown by certain series of amphiphilic solutions (see Chap. 4). Phase symbols are defined in Table 1.1. R = Ratio of the dispersing tendencies on lipophilic and hydrophilic faces of C respectively.

more extended structures form, but these are subjected to local fluctuations in both form and composition. Micelles should not, as
pointed out by Winsor [22], be regarded as persistent entities having well-defined geometrical shapes. Thus both the S_1 and S_2 forms
shown in Fig. 1.1 represent isotropic solutions containing aggregates
which are dynamic entities uniformly distributed and rapidly exchanging molecules with the surrounding continuous region.

C. Solubilization

The increased solubility of a compound associated with the formation
of micelles or inverted micelles has been termed "solubilization" by
McBain [23]. Depending on the structure of the amphiphile, the
type of solvent, and the nature of the compound to be solubilized
(the "solubilizate"), the mechanism enhancing solubility will vary.
The solubilizate may simply be incorporated into the central core of
the micelle or inverted micelle or it may be associated only with the
amphiphile within the interfacial layer. It may also partition between
the continuous phase, the interior of the micelle (or inverted micelle),
and the interfacial region. Thus the interfacial region may not consist entirely of the primary amphiphilic compound. Molecules that
exhibit a substantial presence within the interfacial layers are sometimes called cosolvents or cosurfactants. The distinction between
surfactants, cosolvents, or cosurfactants is often difficult to make
and for the most part is entirely unnecessary. These terms are
used in this book since they have been referred to in the literature,
but care will be taken here to ensure that the chemical compound so
designated is clearly identified.

In connection with the solvent properties of aqueous solutions
of members of homologous series of amphiphilic salts, a distinction
is often drawn between "solubilization" and "hydrotropy." With the
amphiphilic salts of short chain length (e.g., C_3-C_8), the solubilization effect becomes marked only with their rather concentrated
aqueous solutions. This has been termed "hydrotropy" [24]. This
distinction is artificial and misleading [1,25] (Chap. 7), since an
entirely gradual change in solubilizing (i.e., hydrotropic) properties
is observed on studying a complete homologous (C_3-C_{11}) series of
amphiphiles [26]. According to Hartley [25], the difference between
a system where alcohol or acetone is the cosolvent and one where a
paraffin-chain salt is the cosolvent is quantitative only. It is not
really possible to make a distinction in qualitative behavior. The
terms "hydrotropy" and "solubilization" have, however, a certain
customary significance apart from any mechanistic implications and
will therefore be used in this book as convenient terms.

D. Transition from S_1 to S_2 Micelles

By changing the proportions of oil relative to water or any one of a
myriad of other variables (see Chap. 5) a micellar solution can progressively be converted from a S_1 system to a S_2 system. Two possible paths for this transition are indicated by Fig. 1.1. The upper
path shows the transformation taking place by passing through a
series of well-defined liquid-crystalline mesophases. This path and
the corresponding liquid-crystalline structures are defined in Section
1.IV.

The lower path denotes a transition through a series of isotropic
states. It is possible to pass from S_1 to S_2 observing gradual
changes in the properties of the micellar solution. The microstructure of the intermediate states, especially those containing substantial amounts of oil and water, as well as amphiphilic compound, is not
fully understood. In Fig. 1.1 these intermediate states are depicted
simply as somewhat extended micelles which are highly fluid, lacking
in any long-range order and rapidly fluctuating. Since the solutions
are isotropic, many of the techniques used to establish the structure
of the intermediate liquid crystals are not useful in elucidating the
underlying structure of these intermediate states. Scriven [27] has
proposed that for some of the intermediate states, the interface separating oil from water may be bicontinuous, that is, a surface of constant mean curvature which nowhere closes upon itself. Both oil and
water are, in Scriven's view, continuous. Although this hypothesis
is quite appealing, its verification has proven difficult because the
interfaces are extremely fluid and are frequently disrupted by thermal
fluctuations. Whatever the microstructure existing within the micellar
solution lying along the lower path, experiments indicate that most
physical properties, such as electrical conductivity, change gradually
in going from S_1 to S_2. Therefore, there are no precipitous changes
in the microstructure at discrete compositions. Changes are gradual
and progressive.

Because the microstructure of the sequence of states progressing
from S_1 to S_2 is difficult to study, these systems were largely ignored
until recent years, when it has been recognized [28,29] that the isotropic solutions between S_1 and S_2 are agents which can significantly
enhance the recovery of oil from subterranean reservoirs. The micel-
lar/polymer process is one of the most flexible of the oil recovery
processes [30-32]. It is capable of application in a wide variety of
oil reservoirs. This potential application has ignited a surge of interest in both industrial and academic research laboratories. Substantial
progress has been made in understanding the nature of the transformation and the properties of the intermediate micellar states.

It is the sequence of isotropic states that is of primary interest
and is the focus of this book. These systems exhibit quite remarkable
properties which are noted particularly in Chap. 4. Once having

established their properties, it is of interest to be able to create them. This is discussed in Chaps. 5 and 6. In Chap. 5 those variables that can influence the behavior of the micellar solutions are described. These are termed formulation variables, and Chap. 5 considers the influence of one of the variables holding all of the others constant. To derive some insight into the relative importance of the formulation variables, a strategy is employed whereby two of them are changed at once in such a way that the variation caused by the change of one of the variables is precisely cancelled by a compensating change in a second formulation variable. These compensating changes are compared in Chap. 6.

The question of effectiveness of an amphiphile in causing water and oil to be mixed together to form a single-phase isotropic solution is addressed in Chap. 7. The essential issue is resolved based on consideration of the R-ratio, which can serve as a guide in the structuring of more effective surfactant molecules.

Chapter 8 treats the thermodynamic aspects of these interesting intermediate states. The main thrust of the treatment is to justify the need to incorporate an intrinsic curvature of the interfacial region, which is here called the R-ratio. This natural curvature and its relationship to the state of the micellar solution intermediate between S_1 and S_2 is in fact the central issue treated in this book.

In Section 1.IV the sequence of liquid-crystalline forms encountered in progressing from S_1 to S_2 is described briefly. Since previous review articles and monographs have dealt at length with these interesting fluids, they are mentioned in subsequent discussion only when doing so will contribute to the understanding of the isotropic states. Those states lying along the lower path are the focus of this book.

IV. LIQUID-CRYSTAL MESOPHASES

Depending on a large number of factors, including the structure of the amphiphile, the temperature, and the presence or absence of certain additives, some or all of the intermediate liquid-crystalline phases may appear [33–36]. There are a large variety of mesophases, but generally they appear in a given sequence, such as shown by Fig. 1.1 and listed in Table 1.1. It should be stressed that many other mesophases, both anisotropic and isotropic, may appear. Those listed here are believed to be the most frequent.

The G phase (or "neat phase") shown in Fig. 1.1 is the most widely occurring liquid crystal. It has a lamellar structure in which there are alternate extended aqueous and oil lamellae separated by an intervening lamella of amphiphile. The oil in a simple binary aqueous solution of amphiphile is composed of juxtaposed lipophiles.

TABLE 1.1   Designation of Some of the Common Lyotropic Phases

		Basic structure	Designations
1.	Micellar solutions (optically isotropic)	More or less spherical, swollen micelles containing solubilized, organic liquids	S_1
2.	Middle phase, normal (anisotropic)	Indefinitely long, mutually parallel rods in hexagonal array. The rods consist of more or less radially arranged amphiphiles. The hydrophiles are in contact with the surrounding continuous aqueous phase.	M_1
3.	Neat phase (anisotropic)	Coherent double layers of amphiphilic molecules with the hydrophiles in the interfaces with intervening layers of water.	G
4.	Middle phase, reversed (anisotropic)	Indefinitely long mutually parallel rods in hexagonal array. The lipophiles are arranged so that the surrounding continuous organic phase is in contact with the lipophile.	M_2
5.	Inverted micellar solutions (optically isotropic)	More or less spherical inverted micelles containing solubilized water.	S_2

Sources: Refs. 33 and 34.

This structure is within certain limits capable of dimensional changes, with variation in both temperature and concentration providing it with a certain degree of stability.

It is possible, however, to break down the G phase so that amorphous, isotropic S_1 and S_2 phases result; however, in some cases further liquid-crystal structures have been observed. These consist of amphiphilic molecules micellized as cylindrical fibers of indefinite length and within local domains arrayed in a regular two-dimensional hexagonal configuration. As described in Table 1.1 and shown in Fig. 1.1, these hexagonal arrangements can occur in both micelles and inverted micelles. These two phases are denoted as M_1 and M_2.

In addition to those intermediate liquid-crystalline phases indicated in Fig. 1.1 or Table 1.1, other mesophases have been identified. Some of these appear as almost isotropic and are thought to consist of structures in more or less cubic array.

V. THE R-RATIO

A. Relationship to Cohesive Energies

In micellar solutions such as those represented by compositions appearing along the lower path in Fig. 1.1, three distinct regions can be recognized: an aqueous region, W, an oil or organic region, O, and an amphiphilic or bridging region C.[†] The O and W regions both coexist within a single-phase isotropic solution and are therefore quite small, being submicroscopic in size, but on the other hand are much larger than the dimensions of the molecules composing them. When the scale of the O and W regions is large compared to the size of a molecule, it is useful to consider the interfacial region as a zone having a definite composition separating essentially bulk-phase water from bulk-phase oil. Although this is an exceedingly simple picture, it will often be invoked successfully to explain seemingly complex issues.

The interfacial zone has a finite thickness and hence will contain, in addition to the amphiphilic compounds, some oil and some water. This thickness can be considered to be similar to an interfacial region separating oil and water, for example, and regarded as that distance over which the pressure is anisotropic [38]. This definition of the thickness is discussed in some detail in Chap. 8, and some of the concepts introduced here without proof are developed further there.

Within the C layer certain cohesive interaction energies exist which govern the stability of this layer. These interaction are depicted schematically in Fig. 1.2. The cohesive energy A_{xy} between molecules x and y is regarded here as the negative of the interaction energy potential discussed in Section 1.II, which is in some sense averaged over the positions of x and y, taking into account changes in their distances of separation due to the thermal fluctuations. Furthermore, A_{xy} is considered to be calculated on the basis of a unit area and includes only the interaction between those molecules

---

[†]Winsor [37] used the symbols $\overline{W}$, $\overline{O}$, and $\overline{C}$ to emphazise that each region may contain a number of components. Here the bars are omitted, but it must be understood that W, O, and C are not necessarily composed of a single component.

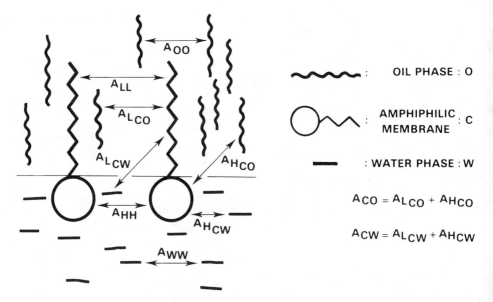

FIGURE 1.2   Interaction energies in the amphiphilic membrane at the water/oil interface.

residing within the C layer. The cohesive energy as used here is therefore positive whenever interaction between molecules is attractive.

In Fig. 1.2 A_{OO} is depicted as the cohesive energy per unit area between oil molecules residing in the anisotropic interfacial C layer. There are two important points to be made. Since the cohesive energy is expressed on a per unit area basis, then A_{OO} is given by $(1/2)(\Gamma_O - 1)(\Gamma_O)\varepsilon_O a^*_{OO}$, where Γ_O is the number of oil molecules per unit of area in the C layer, ε_O is the fraction of pairs of molecules that are interacting, and a^*_{OO} is the mean cohesive energy per interacting pair of oil molecules. Since the number of molecules per unit area is large, the quantity $(\Gamma_O - 1)$ is essentially Γ_O.

The cohesive energy can be considered to be composed of two additive contributions as follows:

$$A_{OO} = A_{LOO} + A_{HOO} \tag{1.13}$$

where A_{LOO} represents the interaction between the nonpolar portions of the two oil molecules which are typically the London forces discussed in Section 1.II.A and A_{HOO} represents the interaction arising from forces of a specific or polar charcter (see Section 1.II.B),

especially those due to hydrogen bonding or of an ionic character.
For the oil molecules A_{Loo} is the dominant term and in the discussion
to follow, A_{Hoo} is ignored. To an approximation the polarizability
of a molecule can be obtained by summing the polarizabilities of the
constituent groups [39]. For a series of alkanes it might be expected
that

$$A_{oo} = \frac{1}{2} \varepsilon_o \Gamma_o^2 a_{oo}^* = a(ACN)^2 \qquad (1.14)$$

where ACN, the alkane carbon number, is the number of carbon atoms
in the alkane, and a is a constant. Equation (1.14) arises because
the molecular interaction as expressed by Eq. (1.4) is proportional
to polarizability squared.

As depicted in Fig. 1.2, the lipophilic portions of the amphiphilic
compound interacts with the oil, with other lipophiles, and with the
water. Since the lipophiles are often nonpolar hydrocarbons, the cohesive energy per unit area is given by

$$A_{\ell\ell} = \frac{1}{2} \varepsilon_s \Gamma_s^2 a_{\ell\ell}^* = b(n)^2 \qquad (1.15)$$

where Γ_s is the number of amphiphiles per unit of area, $a_{\ell\ell}^*$ is the
cohesive interaction energy of two lipophiles, and ε_s is the fraction
of amphiphiles interacting with one another. Here n represents the
number of carbon atoms in the lipophile.

Similarly,

$$A_{Loo} = \varepsilon_{so} \Gamma_s \Gamma_o a_{co}^* = Cn(ACN) \qquad (1.16)$$

where ε_{so} is the fraction of the oil and amphiphile pairs which are
interacting, a_{co}^* is the mean cohesive energy, and C is a constant.
Note that $a_{co}^* = \sqrt{a_{oo}^* a_{\ell\ell}^*}$ provided that the geometric mean defined by
Eq. (1.6) applies. For all cases studied in this book, this approxima-
tion is accurate.

This term is normally to be added to the contribution A_{Hco}, which,
however, is genrally much smaller and is ignored. Their sum will be
denoted simply as A_{co}.

The other cohesive energies per unit area depicted by Fig. 1.2
relate the interaction between water molecules, A_{ww}, the amphiphile
with water, A_{Hcw}, and the interaction between hydrophiles, A_{hh}.
The contribution A_{Hcw} will normally be added to A_{Lcw}, which is usually much smaller, and their sum will generally be denoted as A_{cw}.

The cohesive energy A_{co} evidently promotes the miscibility of
the amphiphilic compound with the oil region, whereas A_{cw} promotes
the miscibility of it with the aqueous region. On the other hand,
both A_{oo} and $A_{\ell\ell}$ oppose miscibility with the oil. Similarly, A_{hh} and
A_{ww} oppose miscibility with water.

Clearly, a condition for the stability of a micellar solution is that the solvent affinities of the C region not be too unevenly divided between the O and W regions. If a strong affinity exists for one phase or the other, the solution will break down into two phases, with the amphiphilic species confined largely to the organic phase or to the aqueous phase, according to whether the preferential solvent affinity of C is exerted toward O or W, respectively.

Where the difference in the solvent interactions of C with O and W is sufficiently small to permit the existence of a stable interface, the ratio of these interactions remains of great importance in determining the properties of the solution formed. this is due to the influence of the ratio, in conjunction with thermal motion, on the average form or curvature of the C layer, which can be regarded as having fluid character when the micellar solution is isotropic, that is, corresponds to one of the states along the lower path of Fig. 1.1. The variation of the dispersing tendencies at the O and W faces of the C region may be expressed qualitatively, as proposed initially by Winsor [1], by

$$R = \frac{A_{co}}{A_{cw}} \tag{1.17}$$

This ratio does not take into account the fact that segregation of the components either as separate phases or as regions of higher local concentration will be promoted by A_{oo}, A_{ww}, A_{hh}, and $A_{\ell\ell}$. Thus R as defined by Eq. (1.17) does not include all of the cohesive energies represented in Fig. 1.2. To account for the structure of the oil, Winsor [37] considered the difference between A_{co} and A_{oo} and proposed an extended version of the original R-ratio: $(A_{co} - A_{oo})/(A_{cw} - A_{ww})$. Since there is a priori no reason for neglecting the interactions A_{hh} and $A_{\ell\ell}$ between the surfactant molecules, which tends to promote their separation as precipitated liquid- or solid-crystalline phases, they should also be taken into account [40,41]. Thus the following extended definition is obtained and is the one applied throughout the book:

$$R = \frac{A_{co} - A_{oo} - A_{\ell\ell}}{A_{cw} - A_{ww} - A_{hh}} \tag{1.18}$$

In many cases (e.g., with paraffins), A_{Hco} is negligible; similarly, A_{Lcw} can be neglected and A_{co} and A_{cw} can be approximated, respectively by A_{Lco} and A_{Hcw}.

For ionic amphiphiles the term A_{hh} is generally negative since in aqueous solutions the hydrophiles tend to dissociate to yield the amphiphilic ion plus a counterion. Thus in the C region the charges are not equally distributed since the counterions will tend to diffuse

from the amphiphiles because of thermal fluctuations. Electric fields
will then develop which tend to oppose charge separation. On the
average, at least some of the amphiphiles will not be closely associated with a counterion. Those headgroups not completely neutralized by a counterion will tend to repel one another, leading to a
negative A_{hh}. (Recall that attractive interactions lead to positive
cohesive energies.)

The number of amphiphiles not closely associated with a counterion
will certainly decrease if additional counterions are added to the solution. This in turn will result in an increased A_{hh} (less negative).
A calculation of the electrostatic free energy of a micelle is presented
in Chap. 2 and the role of the added electrolyte is made explicit in
Chap. 5.

B. Relationship to Micellar Structure

All the terms in Eq. (1.18) will be dependent on the chemical nature
of the components of the system, on their relative concentrations,
and on the temperature. Because of thermal fluctuations in concentration, R will also be subject to fluctuation with time from point to
point within the solution. The extent of these fluctuations will depend on a number of factors, many of which are included in the definition of the R-ratio as given by Eq. (1.18) and are discussed in
some detail in Chap. 5. The condition for the maximum mutual solubility of C, O, and W is that the mean value of R be unity. Where
R = 1, no tendency for the liquid C layer to be preferentially convex
toward either O or W exists and its optimum form will be planar.
Under these conditions there would seem to be nothing to limit the
extension of the C layer. It must be remembered, however, that,
as has been noted, R may vary from point to point because of the
thermal fluctuations in concentration within the solution, so that, in
practice, unlimited extension of the micelle may not occur, even in
the range of relative concentrations where the mean value of R is
unity.

In certain circumstances, however, the thermal fluctuations are
not sufficient to disrupt long-range order, and a stable lamellar
(for R = 1) arrangement of lipophilic, amphiphilic and hydrophilic
layers may persist. Thus, as shown in Fig. 1.1, lamellar phases or
isotropic micellar phases are both possible for the condition R = 1.
Which of the two types will be observed will depend on a number of
factors, but certainly the cohesive energies and the temperature will
be the critical factors.

Where the mean value of R is <1 or >1, the affinity of the C
layer is no longer the same for the O and W regions. The C layer
will tend to become convex toward the W region for R < 1, since
this inequality implies that the relative miscibility of the C layer
with the W region has increased and/or that with O region has de-

creased.   The tendency of the C region to spread preferentially into
the W region, that is, become convex toward W, will be assisted by
A_{cw} and A_{hh} (when it is negative) and resisted by A_{ww}.   This tendency for the interface to curve can also be visualized by imagining
that when R $\ll$ 1, the tendency will be to maximize the interfacial
area of contact with W while minimizing it with the O region.

It should be reemphasized that the concept of intermicellar equilibrium shown in Fig. 1.1 does not mean that only S_1, S_2, or G micellar forms are present at equilibrium, but rather that a fluctuating
micellar form may be regarded as resolved into S_1 and S_2 forms in
equilibrium with G micelles.   In the isotropic solutions the concentration of G micelle is probably small and its development incomplete.
It is to be looked upon as intermediate between S_1 and S_2 forms.
When the concentration and development of the G micelle reach certain critical values, separation of a corresponding liquid-crystalline
G phase occurs.

Similarly, the tendency of the C region to expand preferentially
into the O region, that is, to become convex toward O, will be assisted by A_{co} and resisted by both A_{oo} and $A_{\ell\ell}$.   Thus the R-ratio
defines the tendency for the interface to assume a certain shape.
This tendency may in some sense be considered as representative of
an intrinsic curvature, that is, the curvature the interface would
assume in the absence of all other constraints.

C. Conjugate Phases

Figure 1.1 delineates the variety of micellar and liquid crystalline
states anticipated depending on the value of R.   Since along the
lower path the succession of states between S_1 and S_2 are all isotropic, it is difficult to tell by simple observation to which intermediate state a particular micellar solution may correspond.   The
micellar states as well as the liquid-crystalline mesophases often appear together with conjugate phases, depending on the proportions
of the constituents present in the system, and by observing these
conjugate phases it is often qualitatively possible to ascertain the
state of a micellar solution (see Section 4.II).   The arrows in Fig.
1.1 denote reversibility and indicate phases that may coexist.   For
example, the G phase can appear together with M_1 or M_2.   Similarly,
as shown by the arrows, G can also be found conjugate to S_1 or S_2
or perhaps both simultaneously.

An important observation is that a molecular dispersion of amphiphile in water and of amphiphile in oil (the two extreme states shown
in Fig. 1.1) can sometimes both exist in equilibrium with a micellar
phase.   Since the micellar phase is often intermediate in density
between the oleic and aqueous phases, it appears in the middle.
This micellar solution is therefore sometimes called a "middle phase."
Perhaps even a better and more descriptive name is "surfactant

phase." Both of these terms have been used in the literature. The appearance of three mutually saturated isotropic phases is important because it marks systems for which R is unity or near unity. There are, in fact, other measures, such as electrical conductivity, which help establish the value of R for isotropic systems but are not as convenient as phase behavior for establishing the state of a micellar system. The transition from S_1 to S_2 is a gradual one and is not interrupted by any sudden or even notable transitions. The great importance of three-phase (type III) systems, for which an amphiphile-rich phase containing substantial amounts of oil and water is in equilibrium with both oil and water phases containing disperse amphiphile, will shortly become apparent. Fascinating and important properties can be ascribed to systems exhibiting type III behavior (see Chap. 4).

Type I systems are those for which S_1 solutions are in equilibrium with an oil phase containing dispersed amphiphile. For these systems, R is generally less than unity since the tendency is higher for the C layer to disperse in water than in oil. Type II systems represent those for which R is greater than unity. A typical pair of conjugate phases would be S_2 in equilibrium with water containing dispersed amphiphile. Thus phase behavior is seen to be a qualitative measure of the R-ratio. This important observation serves as the basis for many of the arguments given in this book.

In some cases more than three conjugate phases may appear and one or more may be anisotropic and therefore liquid crystalline. In complex systems, care should be taken to ensure that all phases are at equilibrium and that emulsions or precipitates that will ultimately disappear are not counted as conjugate phases. Many of these non-equilibrium phases may persist for extended periods of time and present difficulties in identifying the true equilibrium phase behavior.

D. Working Hypotheses

1. Optimal Systems

As noted above, those systems exhibiting isotropic behavior for R = 1 have been found to be both theoretically and practically important. Such systems are called "optimal systems" throughout this book. Since R is not experimentally measured, a criterion is needed to identify the optimal state. Type III systems, that is, those which exhibit three mutually saturated conjugate phases, closely correspond to R = 1, and this fact can be used in the development of a working criterion. Various approaches are discussed at length in Chap. 4. For example, a commonly used criterion is that an optimal system is one for which the micellar phase contains equal volumes of oil and water [42]. The value R = 1 can be assigned to these systems.

2. Solubilizing Property

According to Winsor [37], "the cosolvent effect of C will be higher, the greater the values of both A_{co} and A_{cw}, and the more nearly they are equal." Originally, Winsor observed that "there is a sense in which one solubilizing agent may correctly be considered as possessing more 'solvent power' than another, viz. in regard to the range of organic liquid mixtures included [between type I and type II systems] in comparative experiments.... This range, at a given concentration, will depend, other things being equal, on the sum A_{co} and A_{cw} and will increase with both the hydrophilic and lipophilic solvent power of the amphiphile."

A convenient way of defining the solubilizing power or solvency of an amphiphile and the one primarily used here (see especially Section 7.I) is simply to consider the amount of oil and water which can be dissolved per unit of mass of amphiphile in the micellar solution at the optimal state. This amount of oil per unit mass of amphiphile is called the solubilization of oil. Similarly, the solubilization of water is defined as an amount, usually volume, per unit mass of amphiphile. These solubilizations are simultaneously maximized when R = 1 and when the numerator and hence the denominator of the R-ratio are as large as possible (Chap.7).

In regard to the relationship with the R-ratio, the interactions are considered per unit area of the interfacial C layer. In some cases it is convenient to visualize each term in Eq. (1.18) as referring to a single surfactant molecule and to let the number Γ_s of the surfactant molecules per unit area of interface appear explicitly in the numerator and denominator of R. The definition of R thus becomes

$$R = \frac{\Gamma_s(a_{co} - a_{oo} - a_{\ell\ell})}{\Gamma_s(a_{cw} - a_{ww} - a_{hh})} \tag{1.19}$$

where the a_{ij} are the cohesive energies expressed per mole of amphiphile in the C layer.

It is then postulated that, at optimum, R = 1 and the solubilization of oil or water varies in the same manner as the numerator (or denominator) of R. Improved solubilization at optimum will thus be obtained by increasing Γ_s or/and $(a_{co} - a_{oo} - a_{\ell\ell})$. This is the subject of Chap. 7. A similar view regarding the relationship of the size of the hydrophilic and lipophilic groups to the mutal solubilization of oil and water by nonionic surfactants has been expressed by Shinoda and Kunieda [43].

E. Thermodynamic Stability and the R-Ratio

1. Interfacial Curvature

The R-ratio as defined by Eq. (1.19) relates to the energies of inter-action between the C layer and the O and W regions. This formula-tion does not refer to entropic effects and clearly does not, therefore, provide a thermodynamic description of micellar solutions. It is rea-sonable to inquire just how the R-ratio does in fact relate to the complete free energy of a micellar solution.

This question is the subject of continuing research and a full and complete answer is not yet available. Considerable progress has, however, been recorded and much of this progress is detailed in Chap. 8. The important conclusion is that the interfacial C layer possesses an intrinsic curvature, that is, a shape the interface will assume if all other constraints are removed [44–48]. This intrinsic or natural curvature is determined by the interaction energies and is therefore akin to the R-ratio [45].

The actual curvature of the C layer of any micellar solution is determined by a number of thermodynamic variables, only one of which is its natural curvature. Thus the thermodynamic state dic-tated by the minimum of free energy will generally exhibit an inter-facial curvature different from the natural one. This difference will result in a positive contribution to the free energy by inclusion of a bending free energy which is proportional to the difference between natural and actual curvatures.

In Chap. 8 it is shown that if a bending energy or some equiv-alent term is not included as a part of the total system free energy, the thermodynamic analysis confers on micellar solutions a degree of symmetry that is not consistent with their observed behavior. Thus a bending energy is an essential concept. It must be included in one form or another as a part of every satisfactory thermodynamic analysis of the stability of micellar solutions. It is at this point that the R-ratio enters the analysis because this ratio defines the natural curvature of the C layer.

Gibbs [49] was the first to include a bending component in for-mulating the interfacial free energy. For a spherical interface, the form proposed by Gibbs can be written in the following form (see Section 8.II.A):

$$\frac{\partial f^S}{\partial r} = -\frac{H}{r^2} \tag{1.20}$$

where f^S is the interfacial free energy per unit of area and r is the radius of the spherical interface. In this formula, H is evidently a key quantity. It is the parameter that should be relatable to the R-ratio. As an example, the approximate form developed by Miller

and Neogi is modified and its consequences studied in Chap. 8.
The approximate equation given by Miller and Neogi [45] is

$$H = K_r \left(\frac{1}{r} - \frac{1}{R_N} \right) \qquad (1.21)$$

where R_N is the natural radius of curvature and K_r is a parameter
related to the flexibility of the interface. If r exceeds R_N, H is a
negative quantity and it is seen from Eq. (1.20) that f^s will decrease
when r is decreased. The interfacial free energy reaches a minimum
at $r = R_N$.

For small K_r (highly flexible interfaces), deviations between the
actual interfacial radius and the natural one do not result in large
contributions to the free energy of the micellar solution. As K_r increases, the penalty increases and for large values of K_r, the bending contribution to the total free energy becomes the overriding contribution. Such systems will exhibit rigid interfaces, which is characteristic of liquid-crystalline solutions.

R_N provides some of the same information embedded in the R-
ratio. The interfacial region is convex toward the O region for
$R < 1$ and toward the W region for $R > 1$. Thus balanced systems for
which $R = 1$ tend to have planar interfaces and this corresponds to
$R_N \rightarrow \infty$.

This brief discussion is intended to show that the R-ratio is a
real thermodynamic quantity even though a precise relationship connecting R to H is not yet available. It is important for the reader
to recognize the fundamental nature of the R-ratio even though much
of the subject matter included in this book is descriptive and qualitative. Although the primary application of the R-ratio now resides
in its use for predicting qualitative trends, it seems likely that more
quantitative uses will ultimately emerge. Here it is sufficient to
suggest that the free-energy contribution expressed by Eq. (1.20)
is entirely consistent with the Winsor concept and provides a thermodynamic basis in support of the hypotheses, which have been stated
here without full mathematical justification.

2. Solubilization

A great deal of emphasis will be given to the phase behavior of the
micellar solutions, that is, the simple observation of the number of
conjugate phases that exist when mixtures of oil, water, and amphiphile are allowed to equilibrate. For isotropic states, which are the
central focus here, the systems are classified as type I, II, or III,
depending on whether the micellar solution is S_1 in equilibrium with
an excess oil phase, S_2 in equilibrium with an excess water phase,
or a surfactant phase in equilibrium with both excess oil and water
phases. As noted, type III systems, especially those for which the

micellar phase (often called the middle-phase microemulsion) contains equal volumes of oil and water, are particularly significant here. These optimal systems provide the reference state against which all changes are measured. The R-ratio is for these balanced systems roughly unity, indicating that the natural curvature of the interface is planar.

One of the most important properties of micellar solutions is the solvency of the amphiphile, that is, the quantity of amphiphile required to cause oil and water to mix. Clearly, the solvency is an issue of considerable practical importance. In Section 1.V.D it is stated that solubilization (solvency) is increased by adjusting the system so that $R = 1$ while simultaneously increasing the numerator and denominator of R so that they are as large as possible. In terms of Eq. (1.21) this procedure corresponds to $R_N \rightarrow \infty$ and K_r large. Detailed thermodynamic calculations provided in Section 8.II.F demonstrate that solvency is predicted to increase under conditions of increasing R_N and K_r.

Since solubilization does increase as R_N increases, two different systems can only be compared at comparable values of R_N to decide which system possesses the greatest solvency. According to Eq. (1.21), such a comparison amounts to determining which of the two systems has the largest K_r. However, as K_r is increased, the rigidity of the the interface increases correspondingly and the tendency to form liquid-crystalline phases rather than isotropic micellar solutions also becomes more pronounced. Thus it is not possible to improve solvency without limit. There is an upper bound for K_r beyond which liquid crystals dominate.

The trends found based on Eq. (1.21) are consistent with the hypotheses set forth by Winsor. Thus while Winsor's hypotheses are accepted here without mathematical proof, it is important to recognize that thermodynamic arguments demonstrate the need to introduce quantities that capture the essence of the R-ratio. The proof of Winsor's hypotheses stand essentially on the experimental evidence presented in this book. However, it seems certain that the R-ratio will evolve from a complete thermodynamic analysis of micellar solutions as a natural and important parameter.

VI. MICELLAR SOLUTIONS, MICROEMULSIONS, OR EMULSIONS

Mixtures of aqueous electrolytes, hydrocarbons, and amphiphilic compounds have been the subject of extensive research, especially those systems forming amorphous isotropic solutions, so-called microemulsions. Several books and reviews have treated this subject [50–56]. The term "microemulsion" was introduced by Hoar and Schulman [57] to describe transparent or translucent systems obtained by titration

of an ordinary emulsion having a milky appearance to clarity by
addition of a medium-chain alcohol such as pentanol or hexanol.
These alcohols were later referred to as cosurfactants or cosolvents.
There has been much debate about the use of the term "microemulsion" to describe such systems [58]. Some prefer the names "swollen
micellar solutions" or "solubilized micellar solutions" [1,59] to describe precisely the same systems as those called microemulsions by
Hoar and Schulman.

It is certain that Schulman and his coworkers envisioned fundamental differences between the state of matter that they called microemulsions and those systems which are called micellar solutions or
ordinary emulsions. In this book a sharp distinction is drawn between microemulsions or micellar solutions on the one hand and emulsions on the other hand. Emulsions are thermodynamically unstable.
The drops of dispersed phase are generally large, perhaps larger
than 0.1 μm, so that emulsions often take on a milky rather than
the transparent or translucent appearance usually associated with
micellar solutions. The drop size is not, however, the critical factor.
Emulsions are distinguished from microemulsions and micellar solutions
by the fact that the average drop size grows continuously with time.
When exposed to a body force proportional to the mass, such as a
gravitational field, emulsions will ultimately separate into two distinct
phases. The fact that the drop sizes increase continuously with
time is a manifestation of thermodynamic instability. Micellar solutions and microemulsions are, on the other hand, thought to be
thermodynamically stable. Their properties are time independent.
They are, furthermore, independent of the order of mixing, and
they return to their original state when subjected to a small disturb-
ance which is subsequently relaxed. This book deals primarily with
thermodynamically stable systems. Their states are defined by
specifying the concentrations of the various constituents, the pressure, and the temperature.

It is also clear that many investigators have perceived a difference between microemulsions and micelles containing solubilizate. The
authors of this book believe that in principle a difference may very
well exist; however, there is no operational method which is available
to distinguish this difference. All previous attempts to distinguish
between micellar solutions and microemulsions introduce into the definition a degree of arbitrariness which makes the distinction rather
artificial, especially in view of the continuity that one observes in
physical properties as solubilization is increased. One possible definition is to assert that a microemulsion is composed of bulk isotropic
(diagonal pressure tensor) O and W regions separated by an anisotropic C layer. A micelle (see Chap. 2) would not satisfy this definition because the interior of a micelle, the O region, is composed
entirely of the lipophilic portions of the amphiphile and is probably
not isotropic. Even a micelle containing a small quantity of solu-

bilizate would not be expected to have an isotropic interior. However, as the amounts of oil and water in the micellar solution both become large and approximately the same volume, it would be anticipated that essentially bulk isotropic O and W regions would at some point both exist. These systems can therefore be thought to correspond to those sytems called microemulsion by Schulman and his co-workers. Analyses of thermodynamic stability all invoke the concept that a microemulsion is composed of bulk phases separated by an interfacial region. Microemulsions may therefore, be thought to be distinct and separate from micellar solutions; however, since there does not exist any technique for distinguishing between these two states of matter, it seems unreasonable to try. Furthermore, often precisely the same system considered to be a microemulsion by some writers is called a micellar solution by others. This is, of course, simply a manifestation of the absence of a method of distinguishing one system from another.

In this book the authors have elected to call all thermodynamically stable transparent or translucent blends of oil, water, amphiphiles, and other additives either micellar solutions or microemulsions, without distinction.

REFERENCES

1. P. A. Winsor, Trans. Faraday Soc., 44:376 (1948).

2. G. S. Hartley, Paraffin-Chain Salts, Hermann & Cie, Paris (1936).

3. W. H. Keesom, Phys. Z., 22:126; 22:643 (1921); 23:225 (1922).

4. P. J. W. Debye, Phys. Z., 21:178 (1920).

5. F. London, Z. Phys., 63:245 (1930); Z. Phys. Chem., B11:222 (1930); Trans. Faraday Soc., 33:8 (1937).

6. J. M. Prausnitz, Molecular Thermodynamics of Fluid-Phase Equilibria, Prentice-Hall, Englewood Cliffs, N.J. (1969).

7. J. H. Hildebrand and R. L. Scott, Solubility of Nonelectrolytes, Reinhold, New York (1950).

8. H. S. Frank and M. W. Evans, J. Chem. Phys., 13:507 (1945).

9. W. Kauzmann, Adv. Protein Chem., 14:1 (1959).

10. C. Tanford, The Hydrophobic Effect: Formation of Micelles and Biological Membranes, 2nd ed., Wiley, New York (1980).

11. G. Mie, Ann. Phys., 11:657 (1903).

12. J. E. Lennard-Jones, Proc. R. Soc., A112:214 (1926).

13. J. H. de Boer, Trans. Faraday Soc., 32:10 (1936).

14. H. C. Hamaker, Physica, 4:1058 (1937).

15. J. T. G. Overbeek, in Colloid Science, Vol. 1 (H. R. Kruytt, ed.), Elsevier, Amsterdam (1952).

16. M. J. Vold, J. Colloid Sci., 16:1 (1961).

17. E. M. Lifshitz, Zh. Eksp. Teor. Fiz., 29:94 (1955).

18. J. Mahanty and B. W. Ninham, Dispersion Forces, Academic Press, New York (1976).

19. V. A. Parsegian, in Physical Chemistry: Enriching Topics from Colloid and Interface Science, (H. van Olphen and K. Mysels, eds.), Academic Press, New York (1985).

20. J. N. Israelachvili, Intermolecular and Surface Forces, Academic Press, New York (1985).

21. R. M. Pashley, J. Colloid Interface Sci., 80:153 (1981).

22. P. A. Winsor, Chem. Rev., 68:1 (1968).

23. J. W. McBain, Colloidal Science, D. C. Heath, Lexington, Mass. (1950).

24. C. Neuberg, Biochem. Z., 76:107 (1916).

25. G. S. Hartley, Rep. Prog. Appl. Chem., 45:33 (1948).

26. R. Durand, C. R. Acad. Sci., 223:898 (1946).

27. L. E. Scriven, in Micellization, Solubilization, and Microemulsions, Vol. 2 (K. L. Mittal, ed.), Plenum Press, New York, p. 877 (1977).

28. L. W. Holm and G. G. Bernard, "Secondary Recovery Waterflood Process," U.S. Patent No. 3,082,822 (1959).

29. W. B. Gogarty and R. W. Olson, "Use of Microemulsions in Miscible-Type Oil Recovery Procedure," U.S. Patent No. 3,254,714 (1962).

30. J. J. Taber, Soc. Pet. Eng. J., 9:3 (1969).

31. R. N. Healy and R. L. Reed, Soc. Pet. Eng. J., 14:491 (1974).

32. D. O. Shah and R. S. Schechter, Improved Oil Recovery by Surfactant and Polymer Flooding, Academic Press, New York (1977).

33. P. Ekwall, in Advances in Liquid Crystals, Vol. 1 (G. H. Brown, ed.), Academic Press, New York, p. 1, (1971).

34. R. G. Laughlin, in Surfactants, (T. F. Tadros, ed.), Academic Press, New York, p. 53 (1984).

35. V. Luzzatti and F. Husson, J, Cell. Biol., 12:207 (1962).

36. G. J. Tiddy, Phys. Rep., 57:1 (1980).

37. P. A. Winsor, Solvent Properties of Amphiphilic Compounds, Butterworth, London (1954).

38. J. S. Rowlinson and B. Widom, Molecular Theory of Capillarity, Oxford University Press, New York (1982).

39. J. O. Hischfelder, C. F. Curtiss, and R. B. Bird, Molecular Theory of Gases and Liquids, Wiley, New York (1954).

40. A. Beerbower and M. W. Hill, in McCutcheon's Detergents and Emulsifiers, Allured Publishers, Ridgewood, p. 223 (1971).

41. M. Bourrel and C. Chambu, Soc. Pet. Eng. J., 23:327 (1983).

42. R. N. Healy, R. L. Reed, and D. G. Stenmark, Soc. Pet. Eng. J., 16:147 (1976).

43. K. Shinoda and H. Kunieda, J. Colloid Interface Sci., 42:381 (1973).

44. M. Robbins, in Micellization, Solubilization, and Microemulsions, Vol. 2 (K. L. Mittal, ed.), Plenum Press, New York, p. 713 (1977).

45. C. A. Miller and P. Neogi, AIChE J., 26:212 (1980).

46. S. Mukherjee, C. A. Miller, and T. Fort, J. Colloid. Interface Sci., 91:223 (1983).

47. P. G. de Gennes and C. Taupin, J. Phys. Chem., 86:2294 (1982).

48. B. Widom, J. Chem. Phys., 81:1030 (1984).

49. J. W. Gibbs, Collected Works, Vol. 1, Yale University Press, New Haven, Conn., p. 55 (1948). See also F. P. Buff, J. Chem. Phys., 25:146 (1956) for a discussion of spherical interfaces.

50. L. M. Prince, Microemulsions: Theory and Practice, Academic Press, New York (1977).

51. M. Rosoff, in Progress in Surface and Membrane Science, Vol. 12 (D. A. Cadenhead and J. F. Danielli, eds.), Academic Press, New York, p. 405 (1978).

52. S. L. Holt, J. Disp. Sci. Tech., 1:423 (1980).

53. I. D. Robb, Microemulsions, Plenum press, New York (1982).

54. A. M. Bellocq, J. Biais, P. Bothorel, B. Clin, G. Fourche, P. Lalanne, B. Lemaire, B. Lemanceau, and D. Roux, Adv. Colloid Interface Sci., 20:107 (1984).

55. K. Shinoda and S. Friberg, Emulsions and Solubilization, Wiley, New York (1986).

56. S. Friberg and P. Bothorel, eds. Microemulsions: Structure and Dynamics, CRC Publishers, Cleveland, Ohio (1987).

57. T. P. Hoar and J. H. Schulman, Nature, 152:102 (1943).

58. K. Shinoda and S. Friberg, Adv. Colloid Interface Sci., 4:281 (1975).

59. S. R. Palit, V. A. Moghe, and B. Biswas, Trans. Faraday Soc., 55:463 (1959).

2
Aqueous Solutions Containing Amphiphiles

I. GENERAL

The principal subject discussed in the present chapter is aqueous solutions of amphiphiles which in a certain concentration range tend to form large aggregates known as micelles (from the Greek "mica," meaning "small bit" or "crumb"). In particular, the molecules of interest are composed of hydrophilic moieties which are ionic or which display a sufficient interaction with water to maintain solubility of those compounds having longer-chain lipophiles. Examples of representative molecules are cited in Table 2.1.

The primary factor contributing to the strong affinity for water exhibited by the ionic compounds is the entropy increase (increased randomness) resulting from the dissociation of the amphiphile into an amphiphilic ion and a counterion [1]. The amphiphile is designated as anionic if the amphiphilic ion bears a negative charge and is called cationic if it is positive. Examples of both anionic and cationic amphiphiles are listed in Table 2.1.

The water affinity of the nonionic compounds listed in Table 2.1 derives primarily from the strong association of the polyethylene oxide glycol group of the monoether with water. This association is attended by a considerable enthalpic contribution to the free energy of mixing. ΔH is -439 J/g mol of mixture when 0.96 mol of water is blended with 0.04 mol of diethylene glycol [2].

The strong hydrophilic affinity of the moieties given in Table 2.1 confers miscibility of the amphiphiles with water up to quite high molecular weights. With the higher n-alkane-1 or -2 salts (above about C_{14}), another factor, the increasing tendency to crystallize, limits solubility at room temperature, but concentrated isotropic solutions

TABLE 2.1   Representative Amphiphilic Compounds

Anionic

Sodium n-dodecyl sulfate	$CH_3(CH_2)_{11}OSO_3^-Na^+$
Potassium n-octanoate	$CH_3(CH_2)_7CO_2^-K^+$
Sodium dodecylbenzenesulfonate	$CH_3(CH_2)_{11}C_6H_4SO_3Na^+$

Sodium di-2-ethylhexylsulfo-
succinate (Aerosol OT)

$$\begin{array}{c} \qquad\qquad\qquad C_2H_5 \\ \qquad\qquad\qquad | \\ CH_3-(CH_2)_3-CHCH_2OCOCH_2 \\ \qquad\qquad\qquad\qquad\qquad\qquad\qquad | \\ CH_3-(CH_2)_3-CHCH_2OCOCHSO_3^-Na^+ \\ \qquad\qquad\qquad | \\ \qquad\qquad\qquad C_2H_5 \end{array}$$

Cationic

n-Hexadecyltrimethylammonium- bromide (CTAB)	$CH_3(CH_2)_{15}N(CH_3)_3^+Br^-$

Nonionic

n-Dodecyloctaethylene glycol monoether	$CH_3(CH_2)_{11}(OCH_2CH_2)_8OH$
n-Nonylphenol hexaethylene glycol monoether	$CH_3(CH_2)_8C_6H_4(OCH_2CH_2)_6OH$

---

of these salts may be obtained with rise in temperature (see Section 2.VIII.A). These solutions may often be supercooled without separation of a second liquid phase. With the longer-chain salts the solutions of higher concentration obtainable at elevated temperatures show one liquid-crystalline phase, possible more (Section 1.IV).

The crystals of the n-alkane-1 salts consist of alternate layers of a paraffin lattice and an ionic lattice [3]. To construct a similar crystal from, say tetradecane-4 sodium sulfate would be sterically difficult and would involve a considerably greater lateral separation of the ionic groups on the face of the paraffin layers. Accordingly, those n-alkane salts that contain the ionic group in a position numerically higher than 2- possess a much reduced tendency to crystallize and, in consequence, show a much higher solubility at normal temperature in both water and organic solvents.

Similar behavior is shown by branched-chain salts and by compounds in which the ionic group occupies a medial position in the molecule. With these salts however, the separation of liquid-crystalline aqueous phases occurs at a much lower concentration than for the 1-alkane salts. The concentration at which the liquid-crystalline phase separates is, moreover, reduced by an increase in branching or a rise in molecular weight.

Although the discussion in this chapter focuses primarily on those molecules that form large aggregates, lower amines, alcohols, or carboxylic acids may also be properly included in this section even though with these nonionized compounds complete miscibility with water is lost even at low molecular weight (C_4-C_5) because of the relatively low hydrophilic solvent affinity of the polar group. Nevertheless, the qualitative behavior of these low-molecular-weight compounds in aqueous solution is similar to that of the longer-chain compounds, and the two cannot be sharply distinguished. Thus even these low-molecular-weight compounds are considered here to be amphiphiles. In this book low-molecular-weight alcohols are often referred to as cosurfactants or cosolvents because they are often applied mixed with higher-molecular-weight compounds. Thus even though there is, in fact, no clear definition of the terms "cosurfactant" or "cosolvent," they are used here in some cases simply to designate a certain compound that would not be used as the primary amphiphile because of its poor solvency but is blended with other molecules to form the surfactant system.

II. STRUCTURE OF ISOTROPIC AQUEOUS SOLUTIONS OF AMPHIPHILES AT CONCENTRATIONS IN THE REGION OF THE CRITICAL CONCENTRATION FOR FORMATION OF MICELLES

A. Aggregation of Amphiphiles

When curves are drawn illustrating the variation with concentration of the physical properties of aqueous solutions of amphiphiles, it is often found that the curves show a series of fairly sharp breaks which, for a given molecule, occur at roughly the same concentration for all properties. Such a series of curves is illustrated in Fig. 2.1 for aqueous dodecane-1 sodium sulfate [4]. It is generally agreed that the initial sharp inflection in the property/concentration curve marks the commencement of the formation of aggregates, or micelles, of amphiphilic ions. Interestingly, at concentrations somewhat but not greatly less that than at the breakpoint, many physicochemical properties, such as electrical conductivity, self-diffusion, surface tension, turbidity, and others, indicate no appreciable aggregation of amphiphiles, whereas at concentrations just in excess of those at the initial sharp inflection, substantial aggregation is found. Micelles appear to form over a very narrow range of concentrations, prompting many experimentalists, as will be noted in a subsequent section, to define a critical micelle concentration (CMC) which sharply separates those concentrations at which it is assumed that no aggregates

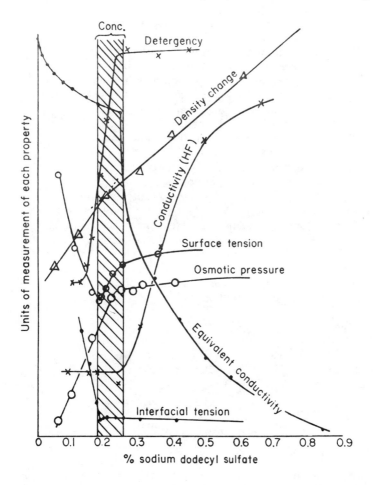

FIGURE 2.1   The influence of micelle formation on properties of solutions of dodecane-1 sodium sulfate at 20 or 25°C.   (After Ref. 4.)

exist (less than CMC) from those at which all surfactant in excess of the CMC exists as aggregates.   Even though micelles are known to form over a range of concentrations, the concept of a critical micelle concentration has proven to be a useful one.

The aggregates of amphiphilic compounds that form in water are visualized as being arranged so that the hydrophilic moiety is in contact with water while the lipophilic chains cluster together so as to minimize their contact with water.   The free energy change responsible for promoting this structural arrangement is a subject for subsequent discussion (Sections 2.V and 2.VI.A), but it should not be thought that a repulsive force between the hydrocarbon chain and water exists as might be implied by the name hydrophobic effect [5], which is often used to characterize the origin of the free-energy force driving micelle formation. Actually, the force between water and hydrocarbon is probably dispersive and therefore slightly attractive (see Section 1.II). It appears that the free-energy change associated with the removal of the lipophilic chains from water is primarily an entropic contribution. Water surrounding the lipophile is apparently structured and restricted in its movement [5]. Thus its removal from water results in an increase in entropy and provides the interesting paradox in which an aggregated apparently ordered system has a greater entropy than one in which the amphiphilic compounds are distributed randomly throughout the aqueous phase.

B. Micellar Structure and Shape

Solutions containing micelles are thermodynamically stable, but because each individual aggregate consists of a relative small number of amphiphilic molecules, a particular aggregate has a finite lifetime. In the words of Winsor [6], "micelles are of a statistical character, and it is important to guard against a general picture of micelles as persistent entities having well-defined geometrical shapes". This view of the dynamical character of micelles has been substantiated by the finding that micellar solutions are characterized by two relaxation times [7,8]. As depicted by Fig. 2.2, τ_1 is believed to be characteristic of the lifetime of an individual surfactant molecule in a particular aggregate and τ_2 is thought to be a measure of the lifetime of an aggregate. These relaxation times both can differ by as much as three orders of magnitude, depending on the surfactant concentration as well as the molecular structure of the surfactant, electrolyte concentration, and other variables, but τ_1 is usually in the range 10^{-8} to 10^{-3} s and τ_2 in the range 10^{-3} to 1 s.

In addition to the dynamic nature of micelles, it is also true that even in an equilibrium solution the number of amphiphiles associated to form a micelle is not the same for all micelles. There exists a distribution of micellar aggregation numbers. Figure 2.3 illustrates a distribution of aggregate sizes at a surfactant concentration in excess of the CMC. Essentially, as shown, two dominant "species" exist in solution: unaggregated surfactant, which may be called monomer, and micelles with an aggregation number n. Micelles having small aggregation numbers do not persist [10,11]. If $\bar{n}$ is defined as the average aggregation number and σ as standard deviation of the micellar aggregate distribution, the values shown in Table 2.2 apply.

It is interesting that, except for the amphiphiles with very short lipophilic chains, the distribution is sharply centered about the mean aggregation number. This means that in some cases it is possible to

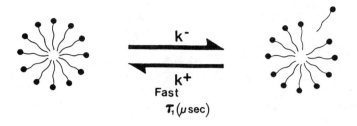

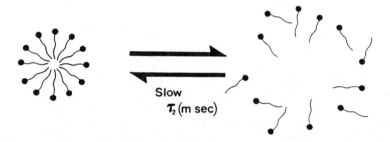

FIGURE 2.2 Two relaxation mechanisms proposed for dilute micellar solutions of pure surfactants. The slow process is thought to take place through a sequence of increasingly smaller aggregates and not suddenly as shown. (See also Ref. 9.)

think in terms of a micellar solution in which all of the micelles have a single aggregation number $\bar{n}$ and are in equilibrium with surfactant monomer. Furthermore, if $\bar{n}$ is large, the micelle can to a good approximation be treated as a separate thermodynamic phase in equilibrium with monomer. Thermodynamic models in which the micelle is treated as a separate phase will be termed "phase separation models" and the micelle will be deemed a "pseudophase" to emphasize that it is not truly a thermodynamic phase with an infinite lifetime, and in this sense the models are approximations.

Although the question of micellar shapes is still under study, a considerable body of data is available on micellar aggregation number. These have been determined primarily by classical light-scattering techniques [12] but there are now a number of other approaches: intrinsic viscosity, tracer diffusion, quasi-elastic light scattering [1,5,13-15], and others which can all be applied. The range of

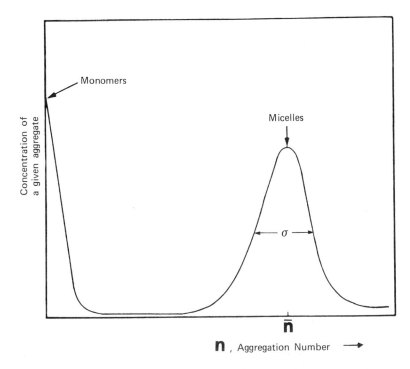

FIGURE 2.3 Form of distribution curve for aggregates which is responsible for existence of two relaxation times: τ_1 and τ_2.

TABLE 2.2 Micellar Size Distributions at 25°C

Surfactant	Aggregation number, $\bar{n}$	CMC (M)	σ
$C_6H_{13}SO_4Na$	17	0.42	6
$C_7H_{15}SO_4Na$	22	0.22	10
$C_{12}H_{25}SO_4Na$	64	8.2×10^{-3}	13
$C_{14}H_{29}SO_4Na$	80	2.05×10^{-3}	16.5
$C_{16}H_{33}SO_4Na$	100	4×10^{-4}	11

Source: These values are taken from Ref. 7.

TABLE 2.3   Average Aggregation Numbers in Water

Amphiphilic compound	Temperature (°C)	Aggregation number, $\overline{n}$	Reference
$C_{12}H_{25}O(C_2H_4O)_6H$	15	140	17
$C_{12}H_{25}O(C_2H_4O)_6H$	25	400	17
$C_{12}H_{25}O(C_2H_4O)_6H$	35	1400	17
$C_{12}H_{25}O(C_2H_4O)_6H$	45	4000	17
$C_{14}H_{29}O(C_2H_4O)_6H$	35	7500	18

reported aggregation numbers is large.   Selected values are given
in Table 2.3.   Rosen [16] has compiled an extensive list.

The aggregation numbers shown in Tables 2.2 and 2.3 indicate
the diversity of micellar sizes.   Based on concepts which relate primarily to simple constraints on the packing geometry, it can be argued that those micelles with modest aggregation numbers (ca. 50 to
60) are likely to be roughly spherical in form, but as the aggregation number increases, the shapes will tend to become disks, rods,
or perhaps even vesicles (see Fig. 2.4).

The aggregation number of the amphiphilic ions listed in Table
2.2 is seen to decrease with decreasing lipophilic chain lengths.
Considering the standard deviations together with the average aggregation numbers, it is clear that those amphiphiles with short lipophiles
form a distribution of aggregates having a large fraction of small aggregation numbers (trimers, tetramers, etc).   One is tempted not to
include small aggregates exhibiting such a wide degree of dispersity
and such short lifetimes within the definition of a micelle and perhaps
to exclude those molecules having short lipophilic chains from the
class of molecular structures which are called amphiphiles.   However,
the observed trends with decreasing lipophile size are regular and
continuous.   It is not possible to distinguish sharply the properties
of molecules exhibiting small aggregation numbers from other systems
having large aggregation numbers without introducing an arbitrary
criterion which will become complex to apply since the aggregation
number of a particular surfactant molecule is not an intrinsic function
of the molecular structure.   It also depends on the type and amount
of added electrolyte, the temperature, the presence of organic additives, and so on.   Therefore, the word "micelle" will be applied here
to describe all associated amphiphilic compounds in solution regardless
of their size, dispersity, or lifetime, although in most examples cited
here, systems with larger aggregation numbers are to be visualized.

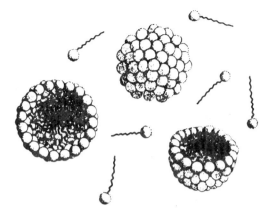

HARTLEY SPHERICAL MICELLES

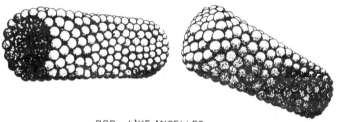

ROD - LIKE MICELLES

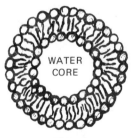

VESICLE

FIGURE 2.4 S_1 Forms

If the aggregation number is small, the phase separation model
describing the formation of micelles is poor at best and perhaps to-
tally misleading at worst.   Property/amphiphile concentration curves
do not exhibit a sharp inflection point, but instead show a more
gradual change in slope over a wider range of concentrations.   It
becomes more difficult, or perhaps even impossible, to define a CMC.

Different experiments will yield large differences in the measured
CMC, and perhaps if the degree of cooperative association is too
small, even theoretical definitions of the CMC will not apply [19,20].

Tanford [5] and Israelachvili et al. [21] have both considered
the shape of micelles using simple geometric arguments. They con-
clude that large aggregations of amphiphilic molecules consisting of
a single lipophilic chain tend to be disks or rods, whereas smaller
aggregates tend to be spherical. Those lipophiles consisting of two
separate chains (as, e.g., sodium di-2-ethylhexylsulfosuccinate,
Table 2.1.) tend to form vesicles or bilayers. These forms are depicted in Fig. 2.4. A thermodynamic analysis to help determine
whether rods or spheres form is presented in Section 2.VI.C.

Detailed studies of micellar size and shape using both classical
[22] and quasi-linear light scattering [14] have been presented for
sodium dodecyl sulfate (SDS) in solutions of relative large electrolyte
concentration (0.15 to 0.6 M NaCl) to reduce intermicellar interactions.
The aggregation numbers calculated based on estimated micellar dimensions were found to range from 60 at low salt concentrations to more
than 1500 at higher electrolyte and SDS concentrations. It was thought
that the results for larger aggregation numbers corresponded to a
shape closely approximating rodlike micelles.

Young et al. [23] measured the angular dissymmetry of the scattered light to obtain the radius of gyration of micelles. Their results
for SDS in 0.6 M NaCl are shown in Fig. 2.5. The radius of gyration
was found to vary with the hydrodynamic radius in a manner consistent with rod-like structure rather than ellipsoids or spheres.

Thus micelles can take on a variety of forms, which may change
with temperature, added electrolyte, or surfactant concentration.
Geometrical calculations to determine micellar size assume that the interior of a micelle is practically "liquidlike." In the next section some
of the experimental evidence available to support this view is examined.

C. Internal Fluidity of Micelles

Hartley [24] was one of the first to argue for a liquidlike micellar interior based primarily on their ability to solubilize large quantities of
nonpolar substances. His contention was that this is only possible if
the interior is fluidlike. Other thermodynamic measurements have also
demonstrated the close similarity between the micellar core and liquid
hydrocarbons. For example, the partial molar volumes and compressibilities of aqueous amphiphilic solutions with and without solubilized
alkanes are very close to those of the liquid alkanes [25,26].

Studies of the liquid-crystalline state also tend to confirm the dis-
ordered nature of the lipophilic chains. From x-ray [27] and NMR
[28,29] measurements it is clear that there is a high degree of alkyl
chain mobility.

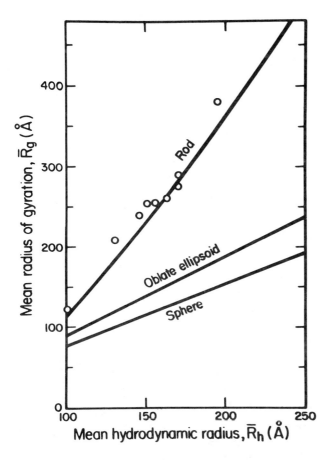

FIGURE 2.5 Mean radius of gyration plotted versus the mean hydrodynamic radius for different micelle shapes according to theoretical
predictions (solid lines). Circles give experimental quasi-elastic
light scattering results for SDS micelles at different temperatures in
0.6 M NaCl. (After Ref. 23.)

Although it is probably not reasonable to view the interior of
a micelle composed of alkyl chains as having precisely the physical
and chemical properties of the corresponding liquid alkanes, it seems
that the evidence points to more liquidlike than solidlike behavior
[30]. Lindman and Wennerstrom [31] have concluded that a liquidlike interior with some of the alkyl chains extended to a greater degree than in the corresponding pure alkane liquid is the most probable state.

Another hypothesis which will prove useful is that for many systems of interest, there is very little penetration of water into the core of an S_1 micelle. This is a difficult hypothesis to establish, but NMR studies of some nonionic polyethylene oxide amphiphilic compounds indicate that there is little or no contact between the alkyl chains and water [32]. Such studies are difficult to interpret and perhaps lack the sensitivity to discern a small degree of penetration when it does occur [33]. In the thermodynamic analyses presented in subsequent sections (see Section 2.VI.C), the contact between water and the lipophilic chains will be assumed to be confined to the interfacial region. Since the lifetime of an amphiphile in a micelle is finite, it is clear that there must be many contacts between the methylene groups of the amphiphile and water as amphiphiles move in and out of aggregates. These are, however, to be counted as surface interactions.

The micellar interior can therefore be considered as an essentially water-free, liquidlike region. This picture of a micelle will be shown to be entirely consistent with the observed free energy change attending the transfer of a lipophile from water into the interior of a micelle. This change is remarkably similar to the free energy change associated with transfering a hydrocarbon molecule from water into a bulk liquid phase (see Section 2.V).

III. FACTORS INFLUENCING THE CRITICAL MICELLE CONCENTRATION

A. Amphiphile Structure

For many micellar systems, the aggregation number is relatively large (>50) and the lifetime of a micelle becomes large. These systems exhibit sharp property/concentration inflection points (Fig. 2.1) and the critical micelle concentration is practically independent of the type of experiment used for its determination. To an excellent approximation, a single CMC may be assigned. In doing so, micelles are then considered to be a separate phase (phase separation model) in equilibrium with a solution of monomeric amphiphiles.

Mukerjee and Mysels [34] have compiled an extensive list of CMCs and have also provided valuable critical discussion regarding the experimental methods used and their probable accuracy. Table 2.4 lists selected CMC values.

The CMC depends on a variety of factors, including the number of carbon atoms and the structural arrangement of the lipophile. As a rule the CMC is halved by the addition of one methylene group to a linear chain (n-alkane) group attached to a single terminal ionic hydrophilic group. For nonionic surfactants an increase in two methylene groups reduces the CMC to about 1/10 of its previous value [16]. The accuracy of these approximate rules of thumb may be ascertained by comparing various of the entries given in Table 2.4.

TABLE 2.4 Selected Critical Micelle Concentrations

Structure	Temperature (°C)	CMC (mM)	Reference
	Anionics		
$C_8H_{17}SO_3^- Na^+$	40	160	36
$C_8H_{17}SO_4^- Na^+$	40	140	37
$C_{10}H_{21}SO_3^- Na^+$	40	41	36
$C_{10}H_{21}SO_4^- Na^+$	40	33	37
$C_{12}H_{25}SO_3^- Na^+$	40	9.7	38
$C_{14}H_{29}SO_3^- Na^+$	40	2.5	36
$C_{12}H_{25}SO_4^- Na^+$	25	8.2	39
$C_{12}H_{25}SO_4^- Li^+$	25	8.9	40
$C_{12}H_{25}SO_4^- K^+$	40	7.8	41
$(C_{12}H_{25}SO_4^-)_2 Ca^{2+}$	70	3.4	42
$C_{12}H_{25}SO_4^- N(CH_3)_4^+$	25	5.5	40
$C_{12}H_{25}SO_4^- N(C_2H_5)_4^+$	30	4.5	43
$C_{16}H_{33}SO_4^- Na^+$	40	5.8×10^{-1}	37
$C_{12}H_{25}CH(SO_4^- Na^+)C_3H_7$	40	1.7	37
$C_{10}H_{21}CH(SO_4^- Na^+)C_5H_{11}$	40	2.4	37
$C_{13}H_{27}CH(CH_3)CH_2SO_4^- Na^+$	40	8×10^{-1}	44
$C_{12}H_{25}CH(C_2H_5)CH_2SO_4^- Na^+$	40	9×10^{-1}	44
$C_{11}H_{23}CH(C_3H_7)CH_2SO_4^- Na^+$	40	1.1	44

TABLE 2.4 (Continued)

Structure	Temperature (°C)	CMC (mM)	Reference
Fluorinated anionics			
$C_7F_{15}COO^-K^+$	25	29	45
$C_9F_{19}COO^-NH_4^+$	≈25	38	46
Cationics			
$C_8H_{17}N(CH_3)_3^+Br^-$	25	140	36
$C_{10}H_{21}N(CH_3)_3^+Br^-$	25	68	36
$C_{10}H_{21}N(CH_3)_3^+Cl^-$	25	61	47
$C_{12}H_{25}N(CH_3)_3^+Br^-$	25	16	36
$C_{12}H_{25}N(CH_3)_3^+Cl^-$	25	20	48
Zwitterionics			
$C_8H_{17}N^+(CH_3)_2CH_2COO^-$	27	250	49
$C_8H_{17}CH(COO^-)N^+(CH_3)_3$	60	86	50
$C_{10}H_{21}CH(COO^-)N^+(CH_3)_3$	27	13	50
$C_{12}H_{25}CH(COO^-)N^+(CH_3)_3$	27	13	50
Nonionics			
$\underline{n}\text{-}C_4H_9(OC_2H_4)_6OH$	20	800	51
$\underline{n}\text{-}C_4H_9(OC_2H_4)_6OH$	40	710	51
$(CH_3)_2CHCH_2(OC_2H_4)_6OH$	20	910	51
$\underline{n}\text{-}C_6H_{13}(OC_2H_4)_6OH$	20	74	51
$(C_2H_5)_2CHCH_2(OC_2H_4)_6OH$	20	100	51
$(C_4H_9)_2CHCH_2(OC_2H_4)_6OH$	20	3.1	51

TABLE 2.4 (Continued)

Structure	Temperature (°C)	CMC (mM)	Reference
$(C_4H_9)_2CHCH_2(OC_2H_4)_9OH$	20	3.2	51
$\underline{n}\text{-}C_{12}H_{25}(OC_2H_4)_4OH$	25	4×10^{-2}	52
$\underline{n}\text{-}C_{12}H_{25}(OC_2H_4)_{14}OH$	25	5.5×10^{-2}	52
$\underline{n}\text{-}C_{12}H_{25}(OC_2H_4)_{23}OH$	25	6.0×10^{-2}	52
$C_{16}H_{33}O(C_2H_4O)_9H$	25	2.1×10^{-3}	53

When the hydrocarbon tail is branched, the carbon atoms on the branches appear to have, to a first approximation, about one-half the effect as a carbon atom added to a linear alkane chain. The addition of substituents such as a polar group, as for example, -O- or -OH into the lipophilic chain, generally results in an increased value of the CMC. Carbon-carbon double bonds also result in increased values of the CMC.

In aqueous media, ionic surfactants have higher CMCs than nonionic surfactants composed of equivalent lipophilic groups. The values presented in Table 2.4 reflect this difference. For example, the CMC of ionic surfactants having a 12-carbon straight-chain lipophilic moiety is approximately 10 mM, whereas nonionics with similar lipophiles have CMCs of approximately 0.1 mM. As the hydrophilic group is moved from a terminal position to a more central one, the CMC increases, as seen by comparing the values given for the hexadecyl sulfates listed in Table 2.4

There are marked differences in the CMCs of the various alkane salts. The order of decreasing CMC is ammonium salts > carboxylates > sulfonates > sulfates. The order is predictable because the difference in the hydrophilic solvent affinity of these ionic groups is generally attributed to the proximity of the charge on the ion to the α-carbon of the lipophilic chain [35]. Higher charge densities associated with charges localized near the α-carbon tend to increase the electrostatic contribution to the free energy of micellization (see Section 2.IV.D), thereby increasing the free energy and hence the CMC. It should be remembered, however, that the localization of the charge on a single amphiphile is not the only factor that determines the electrostatic contribution to the free energy of micellization. The density of charge per unit area of micellar surface which depends on the micellar geometry is also a critical factor, as are the electrolyte concentration and the temperature.

To represent the CMC as a function of the molecular structure, it has been found useful to tabulate the constants A and B which appear in the following empirical equation:

$$\log CMC = A - BN \qquad\qquad\qquad\qquad (2.1)$$

where N is the number of methylene groups in the lipophilic chain. Values of A and B for a selected series of surfactants are given in Table 2.5.

The constants in Table 2.5 reflect two important features of the micellization process. B is practically the same for all ionic amphiphiles, reflecting the fact, noted above, that adding two methylene groups to a linear alkyl lipophile reduces the CMC by about a factor of 2 irrespective either the length of the lipophile or the structure of the ionic head group. It will be seen that this constancy has its origins in the way water molecules are structured in the presence of dissolved polar compounds (the hydrophobic effect).

For the amphiphile $\underline{n}C_nH_{2n+1}(OC_2H_4)_6OH$ there is a considerable difference in the influence of an additional methylene group. This is a reflection of the fact that for nonionic amphiphiles, water is apparently more effectively excluded from the core of the micelle as compared to ionic ones. The reason for this may be the better packing that is possible when the hydrophile is not charged.

The quantity A shown in Table 2.5 measures the difference between different hydrophiles. Generally speaking, ionic amphiphiles having the smaller charge density, as for example the sulfates and sulfonates, also exhibit the smallest A [35].

For nonionic polyethylene oxide surfactants, increasing the length of the ethyleneoxide chain (EO number) increases the CMC, but not markedly. Typical values are given in Table 2.4.

B. Type of Counterion

The counterions are those ions of charge opposite to that of the amphiphile. Their main influence on the CMC of ionic surfactants is a function of their valence, although other factors, such as the degree of hydration, may have an observable influence. The CMCs of several dodecyl sulfates with divalent counterions (Ca^{2+}, Mg^{2+}, Pb^{2+}, etc.) is around 2 mM (see Table 2.4), while the alkali dodecyl sulfates have CMCs which are roughly four times larger. Differences in the CMC among the different alkali counterions are small and difficult to measure. Mukerjee et al. [54] noted that the CMC of alkali

TABLE 2.5   Constants for the Relationship: log CMC = A − BN

Surfactant series	Temperature (°C)	A	B
Na carboxylates (soaps)	20	1.8_5	0.30
K carboxylates (soaps)	25	1.9_2	0.29
Na (K) n-alkyl 1-sulfates or -sulfonates	25	1.5_1	0.30
Na n-alkane-1-sulfonates	40	1.5_9	0.29
Na n-alkane-1-sulfonates	55	1.1_5	0.26
Na n-alkane-1-sulfonates	60	1.4_2	0.28
Na n-alkyl-1-sulfates	45	1.4_2	0.30
Na n-alkyl-1-sulfates	60	1.3_5	0.28
Na n-alkyl-2-sulfates	55	1.2_8	0.27
Na p-n-alkylbenzenesulfonates	55	1.6_8	0.29
Na p-n-alkylbenzenesulfonates	70	1.3_3	0.27
n-Alkylammonium chlorides	25	1.2_5	0.27
n-Alkylammonium chlorides	45	1.7_9	0.30
n-Alkyltrimethylammonium bromides	25	1.7_2	0.30
n-Alkyltrimethylammonium chlorides (in 0.1 M NaCl)	25	1.2_3	0.33
n-Alkyltrimethylammonium bromides	60	1.7_7	0.29
n-Alkylpyridinium bromides	30	1.7_2	0.31
n-$C_nH_{2n+1}(OC_2H_4)_6OH$	25	1.8_2	0.49

Source: Ref. 16.

dodecyl sulfates increases slightly with decreasing counterion atomic number.

For cationic amphiphiles the influence of the counterion is slightly more pronounced than that of the anionics. For n-dodecyltrimethylammonium salts, the CMC follows the sequence $NO_3^- < Br^- < Cl^-$ [55]. For alkylpyridinium halides, the CMC increases appreciably with increasing counterion size [34].

Organic counterions can greatly reduce the CMC. Thus $C_{12}H_{25}SO_4^-$ has a CMC of 8.2 mM when the counterion is Na^+ and a CMC of 4.5 when the counterion is $N(C_2H_5)_4^+$ (see Table 2.4). The organic portion of the counterion is thought to help stabilize the counterion charge within the palisade layer of the micelle, thereby greatly reducing the repulsive forces between charged surfactant hydrophiles and consequently decreasing the free energy of micellization. These counterions are thought to be better bound to the micelle (see Section 2.IV.E). The CMC of hexadecyltrimethylammonium salts is lowered by about one order of magnitude on substitution of salicylate or certain other substituted benzoates for bromide [56].

C. Added Electrolyte

The presence of added electrolyte depresses the CMC [1,57-60]. Shinoda [1], using a phase separation model, developed a semiempirical equation to describe this behavior for ionic surfactants. This relationship was later modified by Moroi [57] in the following form:

$$\ln CMC = \frac{Nw_{HC}}{kT} - k_g \ln(CMC + C_s) + K_1 \qquad (2.2)$$

where K_1 and k_g may be regarded as empirical constants, although Moroi developed K_1 further to display an explicit dependence on the surface charge density and the temperature. NW_{HC} is the free energy associated with the transfer of a lipophile of length N from an aqueous phase to a micelle g and C_s is the concentration of added electrolyte.

k_g is the counterion binding coefficient and is thought to represent the degree to which counterions are bound to the micelle. Because of its importance, counterion binding is a subject treated more fully in a subsequent section. It will be seen that there is no stoichiometric binding as is sometimes suggested. In fact, the concentration of counterions in excess of the coions is a decreasing function of distance measured from the center of a micelle. The number of ions bound to the micelle will have no strict definition and will vary depending on the type of experiment used for its determination.

Equation (2.2) reduces to Eq. (2.1) in the absence of added electrolyte (C_s = O), where

$$B = -\frac{w_{HC}}{(1 + k_g)kT}$$

$$A = \frac{K_1}{1 + k_g}$$

(2.3)

The meaning of these terms will be explored further when the thermodynamics of micelle formation is formulated (Section 2.VI).

Equation (2.2) predicts that the ln CMC should be a linear function of ln C_S when C_S is much larger than the CMC. Figures 2.6 and 2.7 show that both anionic and cationic surfactant ions satisfy the predicted behavior over a wide range of salt concentrations. The counterion binding coefficient is obtained as the slope of the lines drawn in Figs. 2.6 and 2.7. Note that the slope of the line depends on the molecular structure of the surfactant. For potassium laurate shown in Fig. 2.6, $k_g \approx 0.570$ is obtained by fitting the data. A smaller number, $k_g \approx 0.458$, is obtained from the variation of dodecane-1 sodium sulfonate CMC with electrolyte concentration. These two binding coefficients are practically independent of whether the cation is sodium or potassium.

The counterion binding coefficients would tend to indicate that counterions are more strongly held to potassium laurate micelles than to dodecyl sulfate micelles. This is consistent with the observation noted in the preceding section that the charge density associated with carboxylates exceeds (is more localized) that of sulfates. One might anticipate that stronger counterion binding would appear in laurate micelles than in sulfates, irrespective of the experiment, since the charge of laurate is more localized than that of sulfonate. Values of k_g for other hydrophiles are given in Table 2.6.

The depression of the CMC due to added electrolyte is mainly the result of a decrease in the thickness of the ionic atmosphere surrounding the ionic headgroups and the consequent decreased electrical repulsion between these charged groups within the micellar structure (see Section 2.IV.D). The CMC of nonionic surfactants is also depressed slightly by the addition of electrolyte. This may be attributed primarily to a salting-out effect rather than to a specific interaction between the added electrolyte and the hydrophile, although Becher [61] has suggested that the polyethylene oxide chain possibly possesses a small positive charge. The basis for his argument is shown by Fig. 2.8, which is a graph of the constant A given in Eq. (2.1) as a function of the electrolyte concentration. It is seen that a rise in electrolyte concentration decreases A, with the divalent anions giving the larger decrease.

At equal concentrations divalent anions have a greater effect than do monovalent ones. This supports the view that the polyethylene oxide chain group in water exhibits a small positive charge.

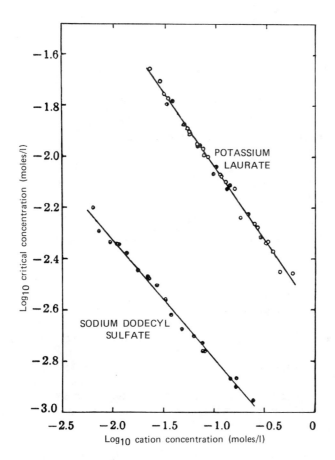

FIGURE 2.6   Salt effect with anionic detergents ○:KCl; ●:NaCl;
◐:K_2SO_4; ◓ :Na_2SO_4; ◔ :$Na_4P_2O_7$.   (After Ref. 62.)

D. Organic Additives

Small amounts of organic materials may produce marked changes in
the CMC.   Sometimes these are present as impurities or as by-prod-
ucts in the manufacture of the surfactant.   At least two mechanisms
whereby organic materials influence the surfactant CMC can be identified.   Some organic materials markedly reduce the CMC by being
incorporated into the palisade layer of the micelle.   These are generally polar compounds such as alcohols.   When this tendency is
marked, the polar compounds must be considered to be amphiphiles.
Other materials, such as urea and short-chain (methanol and ethanol)
alcohols, may alter the solvent-surfactant interaction.   Long-chain alcohols, amines, and similar compounds containing a polar group produce

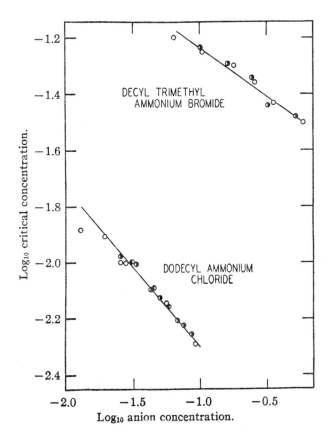

FIGURE 2.7   Log-log plot of the salt effect with cationic detergents
o: NaCl; ●:BaCl$_2$; ●:LaCl$_3$.   (After Ref. 62.)

TABLE 2.6   Some Values of the Counterion
Binding Coefficient

Hydrophile	k_g
Alkali carboxylates	0.58
Alkylammonium halides	0.56
Alkali alkyl sulfates	0.46
Alkyltriethylammonium halides	0.37

Source: Ref. 63.

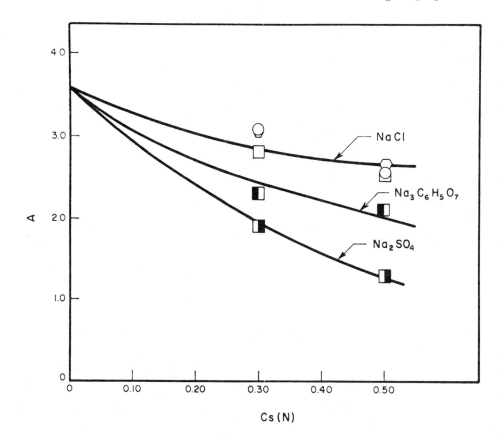

FIGURE 2.8   Relation between the constant A defined in Eqn. (2.1)
and concentration of electrolyte (C_S).   □ : Lauryl alcohol derivatives;
○ :   tridecyl alcohol derivatives; and o: nonyl phenol derivatives.
After Ref. 61.)

a considerable lowering of the CMC. The effect is increased by increasing the concentration of the polar additive or by increasing its
chain length.   These trends are shown by Figs. 2.9 and 2.10, where
the CMC of potassium myristate (tetradecanoate) is plotted against
the concentration of alcohols of different chain length.   Shorter-chain-
length alcohols are thought to be adsorbed mainly in the outer portion of the micelle close to the micelle/water "interface." The longer-
chain members are thought to be incorporated into the micelle mainly
in the outer portion of the core between the surfactant molecules.
Adsorption of the additives in this fashion decreases the electrical
work to form a micelle of ionic surfactants by decreasing the mutual
repulsion of the ionic heads in the micelle.

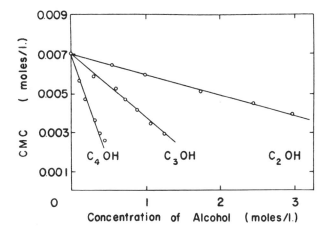

FIGURE 2.9   Effect of ethanol, propanol-1 and butanol-1 on the cmc of potassium myristate at 18°C. (After Ref. 1.)

Depression of the CMC appears to be greater for straight-chain compounds than for branched ones and increases with chain lengths to a maximum when the length of the alcohol approaches that of the surfactants hydrocarbon lipophile. Also, a straight-chain alcohol is more effective than a branched one in reducing the CMC.

Additives that reduce hydrogen bonding or the dielectric constant of water will both tend to increase the CMC. Examples are urea and

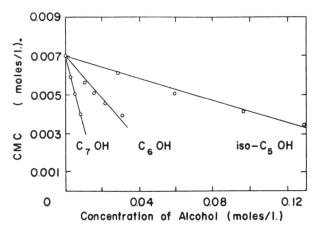

FIGURE 2.10   Effect of heptanol, hexanol, and isopentanol of the CMC of potassium myristate at 18°C. (After Ref. 1.)

formaldehyde and both are believed to increase the CMC of surfactants (especially nonionics) by disrupting the structure of water.

E.   Temperature

The effect of temperature on CMC is complex.   The CMC of sodium
dodecyl sulfate initially decreases and then increases with rising temperature as shown in Fig. 2.11.   The existence of an extremum is
the consequence of competing forces one dominating at lower temperatures, whereas the other tends to prevail at higher ones.   The two
main contributions to the free energy of micellization of ionic amphiphiles are the hydrophobic effect, which tends to promote micelle formation, and the electrostatic repulsion, which opposes it.   The latter
contribution tends to increase somewhat with increasing temperatures
because the effective range of the electrostatic repulsions increases
[the Debye length increases; see Eq. (2.7)] and because counterion
binding decreases with temperature [64].

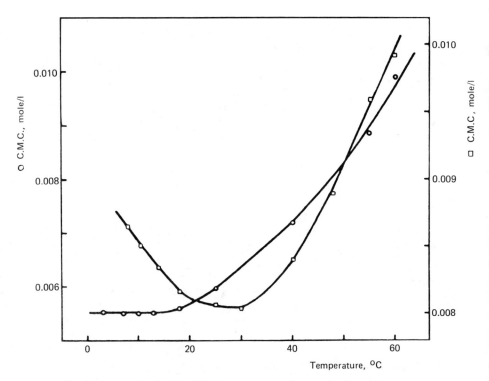

FIGURE 2.11   Effect of temperature on the cmc of sodium dodecyl sulfate in water (□) and in 9.27% ethanol (O).   (After Ref. 65.)

The hydrophobic effect promoting the formation of micelles appears to become more pronounced, reach a maximum at temperatures near 300 K, and then weaken with further increases in temperature [5]. Thus at temperatures in excess of 300 K, the effect of temperature on both the electrostatic and the hydrophobic contributions to the free energy of micelle formation is to increase the CMC. Therefore, for most ionic amphiphiles, as shown for dodecyl sulfate by Fig. 2.11, the CMC increases steadily with increasing temperature [66]. This trend will result in a reduced solvency of all amphiphiles at elevated temperatures, a fact that is stressed in some detail in Chap. 7.

A CMC minimum may also occur for nonionic amphiphiles, but the minimum value, if it occurs, is usually found at higher temperatures than the minimum for ionic amphiphiles [67]. An increase in the temperature tens to decrease the degree of hydrogen bonding between water and nonionic hydrophiles. This trend is evidenced by the precipitation of nonionic surfactants as an entirely separate surfactant-rich phase at elevated temperatures. Nonionic amphiphiles experience an inverse solubility and exhibit an upper consolute temperature known as the cloud point (see Section 2.VIII.A). For this reason, increasing the temperature tends to decrease the CMC because the hydrophilic interaction opposing micelle formation is weakened.

On the other hand, as noted, the hydrophobic effect tending to promote micelle formation is also weakened by increased temperatures. Thus, for nonionic amphiphiles, there are competing tendencies at all temperatures, and the CMC may or may not exhibit a minimum. The CMC of many of the polyethylene oxide glycols shown in Fig. 2.12 does exhibit a minimum at about 45°C. As expected, the temperature at which minima occur in the CMC of nonionic amphiphiles is generally larger than that associated with ionic compounds because for temperatures near room temperature both the electrostatic and the hydrophobic effects change in a way so as to decrease the forces causing ionic micelles to form.

F. Pressure

Partial molar (or molal) volumes of surfactants in aqueous solutions, measured at 1 atm by a variety of techniques, range in magnitude from 1.2 cm^3/mol for dimethylnonylamine oxide to 16 cm^3/mol for sodium tetradecyl sulfate [68]. The volume change upon micellization is positive and generally increases with the length of the alkyl chain of amphiphile [69]. As a consequence of the greater volume of amphiphiles in the micellar state compared to the unassociated state, compression of these solutions through externally applied pressure inhibits micelle formation (Le Châtelier's principle). Thus the CMC is expected to increase with rising pressure. All reported results show the CMC to increase up to a pressure of 1 to 2 kbar, but then in

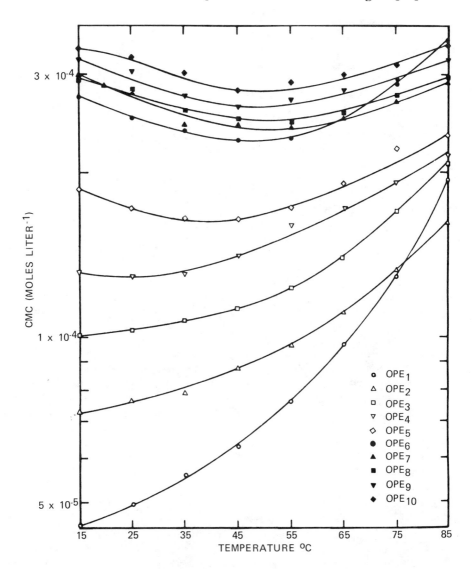

FIGURE 2.12 Effect of temperature on the CMC on single species octylphenoxyethoxyethanols (OPE_m) where m refers to the ethylene-oxide chain length. (After Ref. 67.)

some cases a maximum is attained and the CMC then decreases with increasing pressures [70,71]. In other cases, no maximum is seen [72]. This difference may be due to the technique used for the determination of the CMC [72].

While the existence of a maximum at higher pressures may be a subject of some debate, the trends at lower pressure seem well established. Figure 2.13 shows the trend. It is seen that the increase in CMC is modest over a wide pressure range.

The increase in volume as micelles form provides evidence that water is structured in the presence of the lipophilic portion of the amphiphile. Thus like the melting of ice, a volume increase occurs when the lipophiles are removed from water. For a further discussion, refer to Section 2.V, which deals with the hydrophobic effect.

IV. COUNTERION BINDING AND ELECTRO-
STATIC CONSIDERATIONS

A micelle composed of ionic amphiphiles is a highly charged entity and ions of opposite charge are drawn into the vicinity of the micellar surface by the electrical field that exists because of the surface charge. These oppositely charged ions, called counterions, are distributed in a diffuse layer around the micelle and in close association with the ionic amphiphiles in an arrangement resembling a Stern layer [73,74]. Their concentration in excess of the coion concentration is a function of the distance from the center of the micelle. This separation of charge between amphiphile and counterion endows the micelle with a number of interesting and important properties and in addition represents an important factor contributing to the thermodynamics of

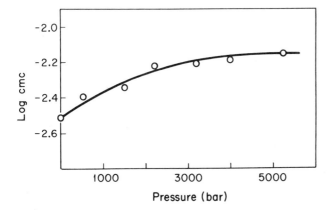

FIGURE 2.13 Pressure dependence of the critical micellar concentration of sodium dodecyl sulfate. (After Ref. 72.)

micellization. In this regard there are a number of issues to be considered in this section. The first concerns the distribution of counterions about the micelle. This entails calculation of the concentration of counterion as a function of radial position. Both spherical and cylindrical micelles will be considered. Also, the relationship between the electrical potential at the interface relative to the surface charge is developed. Given this relationship, the electrical free energy of a micelle can be calculated. This contribution to the total free energy is positive and will play a primary role in considerations of the thermodynamics of micellization. Finally, the issue of counterion binding will be addressed. Counterion binding is a term often used to describe those counterions that reside near the micellar surface containing the ionic headgroups and are therefore strongly attracted to the micelle. Of course, there are no real bonds that link the counterions to the micelle, and the number of counterions "bound" to the micelle will therefore depend to some extent on the experiment used to determine the degree of binding.

One such experiment has been discussed. It has been noted that the dependence of the CMC on the electrolyte concentration contains a factor k_g [see Eq. (2.2)], which is called the binding coefficient. It is thought to represent the fraction of the ionic amphiphiles which are neutralized by bound charge. Thus measurement of CMC as a function of the electrolyte concentration provides one method for determining the binding. Electrical conductivity represents another. The latter method will, among others, be described in this section.

A. The Poisson–Boltzmann Equation

The equation that describes the variation of the electrical potential with distance from an interface, which is a classical electrostatic problem, is the Poisson equation [75]:

$$\underline{\nabla} \cdot \underline{\nabla} \psi = - \frac{\rho_{ex}}{\varepsilon} \tag{2.4}$$

In this expression ρ_{ex} is the net charge density, a quantity that is a function of position, and ε is the permittivity of the medium. To specify the boundary conditions, the potential is arbitrarily assigned a zero value at an infinite distance from the interface. In the discussion that follows, an isolated surface implies that only individual aggregates are considerd and thus the discussion relates to dilute solutions. More concentrated solutions in which micelle-micelle interactions become important are complex and are not considered in this section. Some discussion of these interactions is presented in the treatment of microemulsion thermodynamics (Chap. 8).

The primary difficulty is to express the charge density as a function of the potential so that Eq. (2.4) can be solved. The usual procedure is to relate the ion concentrations in terms of the potential by means of a Boltzmann factor in which the work required to bring an ion to a position with the potential ψ is given by $z_i e \psi$, where z_i is the ion valence and e is the unit charge. The probability of finding an ion at this position is given by

$$n_i = n_i^{(\infty)} \exp \left(\frac{-z_i e \psi}{kT} \right) \tag{2.5}$$

In this expression n_i is the concentration of ion i of a point near the surface where the potential is ψ and $n_i^{(\infty)}$ is the concentration far from the surface, that is, the bulk concentration. It is not difficult to show that this expression is equivalent to assuming the electrochemical potential of an ideal solution to be a constant.

The charge density is therefore obtained by summing the local ion concentrations multiplied by their respective valences. The resultant summation can be greatly simplified when the electrostatic energy is small as compared to the thermal energy, kT (Debye-Hückel approximation) and because in the electrical neutral solution $\Sigma \ z_i n_i^{(\infty)} = 0$. This yields the linear Poisson-Boltzmann equation [76]

$$\underline{\nabla} \cdot \underline{\nabla} \psi = \frac{\psi}{\lambda^2} \tag{2.6}$$

where λ is the Debye length, defined by

$$\lambda^0 = \left(\frac{\varepsilon kT}{e^2 \sum n_i^{(\infty)} z_i^2} \right)^{1/2} \tag{2.7}$$

B. The Spherical Micelle

The simplest model considers the ionic charge of the amphiphilic molecules to be confined to a spherical surface of fixed radius. The surface charge density is assumed to be small enough that the Debye-Hückel approximation is valid, the counterions considered to be point charges, and the dielectric constant for water is taken to be constant. Although this is a highly idealized model, the essential features of the diffuse portion of the double layer are included and many of the observed trends are qualitatively predicted based on the results of this analysis. Other more comprehensive, and correspondingly more complex models have been proposed [74,77] but these are not required for the purposes of this discussion.

For the simplest case, the solution of Eq. (2.6) is

$$\psi = \frac{a\psi_0}{r} e^{-(r-a)/\lambda}$$

(2.8)

where the potential approaches a fixed value ψ_0 as $r \rightarrow a$, the micellar radius. The potential is therefore seen to decrease rapidly as r, the distance measured from the center of the micelle, increases. Within a distance from the micellar surface that is three or four times the Debye length the potential declines essentially to zero. The charge density also decreases rapidly. To the approximation used here,

$$\rho_{ex} = -\frac{\varepsilon\psi}{\lambda^2} = \frac{a\varepsilon\psi_0}{\lambda^2 r} e^{-(r-a)/\lambda}$$

(2.9)

Note that the charge in the diffuse layer has a sign opposite to that of ψ_0 and that much of the counterion charge is located near the micellar surface.

A relationship between n, the aggregation number, and the potential can easily be developed since the counterion charge must be equal but opposite in sign to the micellar charge, σ_s. Thus

$$\sigma_s = ez_{ion}n = -\int_a^\infty \rho_{ex} 4\pi r^2 \, dr$$

(2.10)

where z_{ion} is the valence of the amphiphilic compound. This equation can be integrated to yield

$$\sigma_s = 4\pi\varepsilon a\psi_0\left(1 + \frac{a}{\lambda}\right)$$

(2.11)

The relationship between σ_s and ψ_0 is seen to be a linear one, although this is not generally the case. It is true because the Debye-Hückel approximation has been invoked. Other, more accurate solutions to the Poisson-Boltzmann equation do not yield a linear relationship.

C. Cylindrical Micelles

For micelles composed of a large number of amphiphiles, the micelles can no longer be considered to be spherical, and for very large aggregation numbers, rod-like shapes are generally envisioned. The distribution of counterions in the diffuse layer about a rod-like micelle can readily be ascertained if essentially the same idealizations as applied to spherical micelles are valid. The solution of the Poisson-Boltzmann equation for a long cylinder is

$$\psi = \psi_0 \frac{K_0(r/\lambda)}{K_0(a/\lambda)} \tag{2.12}$$

where a is the radius of the rod and K_0 is the modified Bessel function of the second kind and zero order. For large values of r/λ [78],

$$K_0\left(\frac{r}{\lambda}\right) \simeq \sqrt{\frac{\pi\lambda}{2r}} \exp\left(\frac{-r}{\lambda}\right)\left(1 - \frac{8\lambda}{r} + \cdots\right) \tag{2.13}$$

so that again the potential rapidly approaches zero within a few Debye lengths of the surface of the cylinder.

Calculation of the surface charge per unit of length yields

$$\sigma_s = \frac{2\pi\varepsilon a\psi_0}{\lambda} \frac{K_1(a/\lambda)}{K_0(a/\lambda)} \tag{2.14}$$

where $K_1(a/\lambda)$ is the modified Bessel function of order 1. K_0 and K_1 are tabulated functions [78]. Note that in this equation σ_s is the charge per unit length of rod, whereas in the spherical case σ_s represents the total charge. Thus σ_s will denote surface charge, but the dimensions will differ depending on the geometry of the micelle.

D. Electrostatic Free Energy

The isothermal work required to create or charge the double layer is a quantity required in the formulation of the free energy of micellization. This free energy may be calculated using different approaches [79] and for complex relationships between the surface charge and the surface potential different results may be obtained. However, when the Debye-Hückel approximation is applied, the relationship between ψ_0 and σ_s is linear and the answer is unique independent of the approach. Thus an approach used by Stigter [80] will be applied here.

The micelle is imagined to be formed initially under neutral conditions. Charge, both positive and negative, is brought from infinity in small steps, $\sigma_s\, d\xi$, where ξ is a charging parameter varying between 0 and 1 during the charging process. When an amphiphile is charged, a counterion in the diffuse region of the double layer is charged simultaneously so that electrical neutrality is maintained at every point in the buildup of the charges. Expressed mathematically, one writes

$$0 = \sigma_s \xi + \int \rho'_{ex}\, dV \tag{2.15}$$

The electrical free energy can be found by incorporating the values of the potential at the surface, ψ_0', and in the double layer, ψ', where the primes indicate an arbitrary stage in the charging process. Using these in Eq. (2.15) gives an integral expression for the electrical free energy [80],

$$\Delta f_{el} = \sigma_s \int \psi_0' \, d\xi + \int \frac{d\xi}{\xi} \int \psi' \rho_{ex}' \, dV \tag{2.16}$$

which after application of the divergence theorem can be written as

$$\Delta f_{el} = \varepsilon \int \frac{d\xi}{\xi_-} \int \underline{\nabla}\psi' \cdot \underline{\nabla}\psi' \, dV \tag{2.17}$$

The first term on the right of Eq. (2.16) represents the work necessary to charge the amphiphiles, and the second is the work gained by positioning the counterions in the diffuse layer. The first term is positive and always larger than the second term, which is negative, and therefore the electrical free energy is a positive quantity.

Neither Eq. (2.16) nor (2.17) requires the Debye-Hückel approximation, but as remarked earlier, unless this condition is imposed, the integration is greatly complicated and different approaches for calculating the free energy will yield different results with no sure way of ascertaining which is better. Applying Eq. (2.17) to the spherical micelle, one finds that

$$\Delta f_{el} = \frac{\sigma_s^2}{2\varepsilon a(1 + a/\lambda)} \tag{2.18}$$

Doing the same for the cylindrical case leads to the following expression:

$$\Delta f_{el} = \frac{\sigma_s^2 \lambda}{\varepsilon a} \frac{K_0(a/\lambda)}{K_1(a/\lambda)} \tag{2.19}$$

These free energy expressions show that the electrostatic contribution of ionic micelles is positive, opposing their formation.

E. Counterion Binding

The density of counterions surrounding a micelle is large near the micellar surface but rapidly (exponentially) decreases away from the

surface as shown by Eq. (2.8). The degree to which the counterions are associated with a micelle depends on a number of factors,
including the distance measured from the surface. At low surfactant
concentrations less than the CMC, the equivalent conductance of a
n-alkane salt lies close to the Onsager slope for a univalent strong
electrolyte. This is typically and rather abruptly at the CMC followed
by a steeply descending portion of the curve. Figure 2.14 shows
this typical trend. It was early pointed out by McBain [81] that if
the formation of micelles involves aggregation of a number of ions of
like charge by the process

$$nA^- \rightleftharpoons (A_n)^{n-} \qquad\qquad (2.20)$$

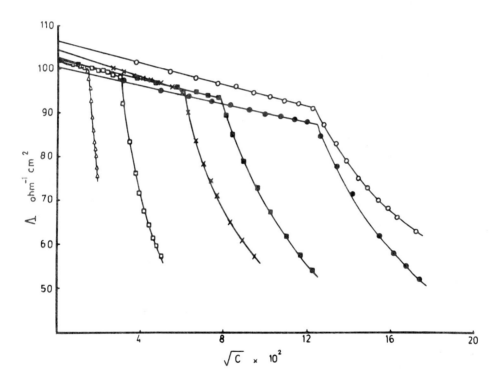

FIGURE 2.14 Equivalent conductance, Λ, of anionic ethoxylated sur-
factants at 25°C. $\circ$:$C_{10}E_2SO_4Na$; $\blacktriangle$:$C_{12}E_0SO_4Na$; $\bullet$:$C_{12}E_1SO_4Na$;
$\triangle$:$C_{12}E_2SO_4Na$; $\square$:$C_{14}E_2SO_4Na$; $\blacksquare$:$C_{16}E_2SO_4Na$. Ordinate: Λ,
$ohm^{-1}cm^{-2}$. Abscissa: $\sqrt{C} \times 10^2$. (After Ref. 83.)

it should lead to an increase rather than a decrease in electrical conductivity. His argument was straightforward. Micelle formation would
give rise to a single entity having a charge and volume n times that
of a single amphiphilic ion. From Stokes' law the viscous resistance
to its movement in an electric field is inversely proportional to the
radius of the micelle and will therefore be $1/n^{1/3}$ of that exerted on
a single ion, whereas the force is proportional to the net charge.
The micelle should therefore move more rapidly than the single ions
in an electric field and lead to a conductivity of $n^{2/3}$ times that due
to n individual ions.

The fall in equivalent conductivity accompanying micelle formation
must necessarily be due to the inclusion within the micelle of a considerable number of the counterions. These counterions effectively
neutralize an equivalent charge on the micelle and are themselves
withdrawn from the conduction process. Both effects lead to a reduction in the equivalent conductance. The degree to which counterions
are "bound" to the micellar surface will depend on a number of thermodynamic factors, including the temperature, the electrolyte concentration, and the surface charge density, as well as the magnitude of
the applied electric field.

If the surface charge density is small the fraction of the counterions bound to the micelle is correspondingly small and an increase in
equivalent conductance when micelles form rather than the decrease
shown in Fig. 2.14 may occur. For some cationic surfactants an increase in equivalent conductance is sometimes, but not always [82],
observed. Examples are shown in Fig. 2.15. The maximum in the
equivalent conductance can be attributed either to increased binding
at higher surfactant concentrations or to a change in the micellar
shape at concentrations above the CMC.

The model that considers micelle formation as being equivalent to
a phase transition suggests that the activity of the amphiphile remains
constant at total concentrations in excess of the CMC. Surfactant
added to solution at concentrations above the CMC is incorporated
into the micelle without increase in monomer activity. Figure 2.16
shows measurement of the activity sodium ions and dodecyl sulfate
ions in solution using ion-specific electrodes. The sodium dodecyl
sulfate activity is given by [85]

$$a_{\pm} = a_{+}a_{-} \qquad\qquad (2.21)$$

where a_+ and a_- are the individual ion activities of the sodium and
dodecyl sulfate ions, respectively. Because not all sodium ions are
bound to the micelle, the sodium activity rises as sodium dodecyl
sulfate molecules are added to a solution at concentrations above the
CMC. This rise in sodium ion activity evidenced by the increase in

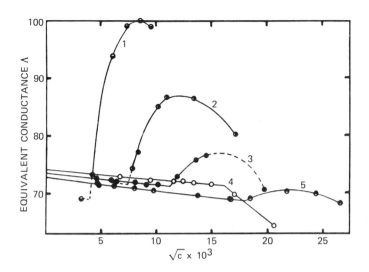

FIGURE 2.15 Conductance of long chain quaternary ammonium bromates: (1) octadecyltriamylammonium, (2) octadecyltributylammonium, (3) octadecyltripropylammonium, (4) octadecyltriethylammonium, (5) hexadecyltributylammonium bromate. (After Ref. 84.)

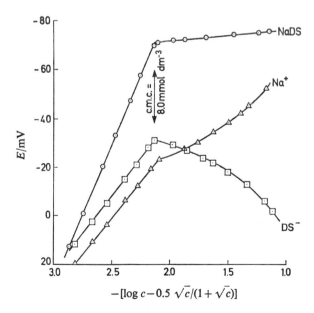

FIGURE 2.16 Results from measurements of the activity of sodium dodecylsulfate, of sodium ions and of dodecylsulfate ions in solutions of sodium dodecylsulfate. (After Ref. 89.)

electrode potential shown in Fig. 2.16 causes the dodecyl sulfate
ion activity to be decreased as the proportion of the amphiphile that
remains in solution decreases with increased Na^+ activity. This trend
can be understood in terms of Eq. (2.2). Increased electrolyte concentration tends to decrease the CMC, which is equivalent to decreasing the monomer concentration. The resultant amphiphile activity,
a_+, does, however, remain essentially constant, as shown in Fig. 2.16.

The counterion binding coefficient can be inferred from measurements of CMC as a function of added electrolyte (see Section 2.III.C),
electrical conductivities, NMR relaxation times [86], tracer diffusion
studies [87,88], and other experiments. It is generally found that
k_g varies within the range 0.4 to 0.8. It also varies with micellar
geometry and tends to increase with any change that decreases the
Debye length. For example, increasing the electrolyte concentration
or decreasing the temperature both tend to decrease k_g.

An important variable is also the degree to which the charge of
the amphiphile is localized. For example, since for sulfates the charge
can reside on any one of three oxygens with equal probability, one
would expect the density of the charge to be smaller than for quaternary ammonium, for which the charge may be associated with any one
of three hydrogens since hydrogen is a much smaller atom than oxygen
[35]. Table 2.6 shows k_g for ammonium halides to be larger than
those for sulfates.

The important point is that there does exist a binding between
the counterions and the micelle, but this binding does not obey a
well-defined stoichiometry. It varies depending on the system as
well as the experiment used to determine its value.

V. HYDROPHOBIC EFFECT

As shown by Eq. (2.18), the electrostatic forces contribute positively
to the free energy of micelle formation. Furthermore, the standard
enthalpy change associated with the formation of ionic micelles, which
includes, of course, an electrostatic contribution, is often small and
sometimes positive. What, then, is the force promoting micelle formation? Tanford [5] has summarized the available scientific evidence
relevant to this question, and the surprising conclusion is that systems in which surfactant aggregates form are higher entropy states
than those in which the surfactant molecules are distributed randomly
throughout the system. This concept was apparently first proposed
by Frank and Evans [90] and considers that the presence of the
lipophilic portion of the amphiphile in water requires that the water
molecules orient themselves in a much less random fashion to accommodate this intrusion than they otherwise would. This is the so-called
"iceberg" effect [90,91]. Micellization may increase the local ordering
of amphiphilic molecules and thereby decrease the entropy, but the

gain in the freedom of the previously restricted water molecules more than compensates for any increase in order due to the aggregation process. Restriction of the movement of water molecules in the presence of a hydrocarbon or any apolar compound has been termed the hydrophobic effect or the hydrophobic interaction. It is responsible for the small solubilities of apolar compounds in water.

This effect is difficult to characterize based on theoretical calculations since the collective motion of a large number of water molecules must necessarily be considered. Thus it will be necessary to rely on an experimental evaluation of the free energy changes associated with the transfer of the lipophile from the interior of a micelle into water. It is first interesting and instructive to consider, as a model, the standard free energy changes associated with the transfer of a hydrocarbon from its liquid state into water in a state of infinite dilution. This standard free-energy change is

$$\Delta\mu^0_{HF} = \mu^0_{HC} - \mu^0_{AQ} \tag{2.22}$$

where $\Delta\mu^0_{HF}$ represents the difference between the standard free energies at infinite dilution (μ^0_{AQ}) and in the hydrocarbon phase (μ^0_{HC}). Based on solubility measurements in water it is found that for a linear alkane there is a contribution to $\Delta\mu^0_{HF}$ of -2100 cal/mol of CH_3 groups and -850 cal/mol of CH_2 groups [5], all at 25°C. A more general relationship applicable to cyclic and branched chains as well as linear alkanes relates the free energy of transfer to the surface area of contact between water and the hydrocarbon molecules, which is 25 ± 5 cal/Å^2 of surface area measured at the distance of closest approach of water molecules to the hydrocarbon [92,93]. This result, showing that the free energy of transfer is intimately coupled to the surface area of contact, provides further confirmation of the iceberg model of the hydrophobic effect.

To separate that portion of $\Delta\mu^0_{HF}$ which is entropic from that which is enthalpic, the solubility of the hydrocarbon must be measured as a function of temperature or $\Delta\bar{H}^0_{HF}$ measured directly by calorimetric methods. Thus the relationship between $\mu^0_{HC} - \mu^0_{AQ}$ and the corresponding partial molar enthalpies and entropies is

$$\mu^0_{HC} - \mu^0_{AQ} = \bar{H}^0_{HC} - \bar{H}^0_{AQ} - T(\bar{S}^0_{HC} - \bar{S}^0_{AQ}) \tag{2.23}$$

The separate contributions can be determined using the thermodynamic result

$$-\frac{d}{dT}\left(\frac{\mu^0_{HC} - \mu^0_{AQ}}{T}\right) = T^2(\bar{H}^0_{HC} - \bar{H}^0_{AQ}) \tag{2.24}$$

The effect of temperature on the aqueous solubility of hydrocarbons always leads to curved plots, as indicated by Fig. 2.17. At lower temperatures near 25°C, the solubility decreases as the temperature rises. This shows that $\bar{H}^0_{HC} - \bar{H}^0_{AQ}$ is positive and solubility in water is energetically favored. In this range the large negative values of $\Delta\mu^0_{HF}$ therefore result entirely from entropic contributions. $\bar{S}^0_{HC} - \bar{S}^0_{AQ}$ is a large positive number indicating the large entropy increase referred to previously as the hydrophobic effect.

The rapid increase in solubility with increase in temperature shown in Fig. 2.17 is consistent with the concept that the primary contribution to $\Delta\mu^0_{HF}$ is related to a structuring of water surrounding the hydrocarbon because one would expect this structuring to become less pronounced with increasing temperature because of thermal disruptions. Thus the main force promoting the formation of micelles is weakened at elevated temperatures. This will have a number of consequences. As shown by Figs. 2.11 and 2.12, CMC of both ionic and nonionic amphiphiles increases with increasing temperature at sufficiently high temperatures. Also, the solvency of

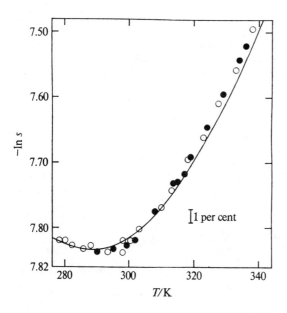

FIGURE 2.17 The effect of temperature on the solubility of benzene in water. The solid line is drawn assuming that the heat capacity change is independent of temperature. (After Ref. 94.)

amphiphilic compounds may be expected to decrease at elevated temperatures. This trend is amply documented in Chap. 7.

The results cited for linear alkanes do not apply directly to the transfer of the lipophilic moiety of an amphiphile from an aqueous phase into the interior of a micelle. First, the thermal motion of the lipophile is more restricted in the interior of a micelle than the corresponding hydrocarbon in a liquid state. It is also possible that some water may penetrate into the interior of the micelle. Thus the observed values at $25°C$ of -2000 cal/mol of CH_3 groups and -700 cal/mol of CH_2 groups associated with the transfer of the lipophile into a micelle, although differing slightly from the corresponding values for pure hydrocarbons, are reasonable. These values will apply when formulating the thermodynamics of micelle formation.

It is perhaps appropriate to note at this point that while the formulation of micellar thermodynamics presented in the following section is both interesting and instructive, the predictive values of the results are strictly qualitative. Because knowledge of the free energy related to the hydrophobic effect is strictly empirical, not theoretical, and because its temperature dependence is complex, the equations developed are not in any sense fundamental and extrapolation of them to different temperatures must be carried out with caution.

Furthermore, the electrostatic free energy as expressed by Eqs. (2.18) and (2.19) represents a simplified result and does not truly reflect the distribution of counterions near the micellar surface. This further weakens the quantitative value of the thermodynamic results.

VI. THERMODYNAMICS OF MICELLE FORMATION

A. Spherical or Almost-Spherical Micelles

The aggregation of amphiphiles to form micelles may be considered equivalent to a chemical reaction leading to the formation of a complex. It is this approach that is outlined here. More details have been given by Tanford [5] and Israelachvili et al. [21]. In this treatment a micelle composed of n amphiphilic compounds is considered to be a distinct species characterized by a standard free energy of formation $n\mu_n^0$. If x_n is taken to be the mole fraction of amphiphile incorporated into those micelles having an aggregation number n, then x_n/n is the mole fraction of the micelles and their chemical potential is given by

$$\mu_n = n\mu_n^0 + kT \ln \frac{x_n}{n} \qquad (2.25)$$

Ideality is not necessarily assumed in writing this expression. If μ_n^0 depends on the environmental variables as well as n, Eq. (2.25) will also include nonidealities. For example, the degree of counterion binding to ionic micelles (see Section 2.III.C) is reflected as an effect of counterion concentration on μ_n^0. In this section dilute solutions are considered and μ_n^0 will, in these cases, depend only on n, not on the concentration of amphiphile.

If the formation of micelles is represented as the reaction

$$nA_1 \rightleftharpoons A_n \tag{2.26}$$

then the condition for chemical equilibrium is

$$\mu_n = n\mu_1 \tag{2.27}$$

where μ_1 is the chemical potential of the monomer. Equation (2.27) can also be written as

$$\mu_n^0 + \frac{kT}{n} \ln \frac{x_n}{n} = \mu_1^0 + kT \ln x_1 \tag{2.28}$$

where μ_1^0 is a standard state chemical potential of amphiphile monomer at infinite dilution. This can be rearranged to yield

$$x_n = nx_1^n \exp\left[\frac{n(\mu_1^0 - \mu_n^0)}{kT}\right] \tag{2.29}$$

Once $n(\mu_1^0 - \mu_n^0)$, the standard free energy associated with the transfer of n unassociated amphiphiles to an aggregate, is established, x_n can be calculated, subject, however, to the restriction that the mole fraction of amphiphile S, which is specified, be given by

$$S = \sum x_n \tag{2.30}$$

The contributions to the standard free energy of transfer are the hydrophobic interaction, the interfacial energy of the hydrocarbon/water contact, and the electrostatic repulsion. Because of their profound importance with respect to the formation of micelles, the electrostatic and hydrophobic contributions have been discussed in separate paragraphs. The remaining contribution arises from the assumption that water is excluded from the micellar core and recognition that the area of the amphiphilic headgroups is insufficient to cover the entire surface of a spherical micelle [92]. Thus there is contact between the oil-like interior of a micelle and the aqueous phase. The free energy per amphiphile of this contact can be represented as γA, where A is the area per amphiphile on the spherical surface and γ is an energy per area [≈ 50 ergs/cm^2 [21]].

Equation (2.18) represents the electrostatic free energy to form a spherical micelle containing n charges, and therefore $\Delta f_{el}/n$ is expressed per unit charge. Noting that $A = 4\pi a^2/n$ and $\sigma_s = en$, then

$$\frac{\Delta f_{el}}{n} = \frac{2\pi ae^2}{\varepsilon A(1 + a/\lambda)} \qquad (2.31)$$

Summing the three contributions, the standard free energy of transfer can be written as

$$\mu_1^0 - \mu_n^0 = -\Delta\mu_{HF}^{0(n)} - \gamma A - \frac{2\pi ae^2}{\varepsilon A(1 + a/\lambda)} \qquad (2.32)$$

Note the possibility that the hydrophobic effect may depend on n by designating $\Delta\mu_{HF}^0$ as a function of n. This would imply that the packing of molecules may not be the same in micelles with different aggregation numbers.

To relate A to n, note that the entire volume of the micelle interior must be filled; there can be no void spaces. Thus

$$\frac{4}{3}\pi a^3 = nv \qquad (2.33)$$

where v is the volume of the lipophile. Also, by definition,

$$A = \frac{4\pi a^2}{n} = (3v)^{2/3}\frac{(4\pi)^{1/3}}{n^{1/3}} = \frac{3v}{a} \qquad (2.34)$$

Then

$$\mu_1^0 - \mu_n^0 = -\Delta\mu_{HF}^{0(n)} - \gamma\left[A + \frac{A_o}{A}\right] \qquad (2.35$$

where

$$A_o = \sqrt{\frac{2\pi ae^2}{\varepsilon\gamma(1 + a/\lambda)}}$$

Here A_o, the optimum area per amphiphile, defines that area for which the free energy per amphiphile is minimum.

The essential concepts required to develop a thermodynamic model are embodied in Eq. (2.35). Using Eq. (2.29), it is possible to consider in more detail the size distribution of spherical micelles. Suppose that m represents the aggregation number for which $A = A_o$. Then from Eq. (2.29),

$$x_n = n\left\{\frac{x_m}{m} \exp\left[-\frac{m}{kT}(\mu_n^0 - \mu_m^0)\right]\right\}^{n/m} \tag{2.36}$$

Since m is the optimum aggregation number, $\mu_1^0 - \mu_m^0 = -\Delta\mu_{HF}^{0(m)} - 2A_0$ and $m = 4\pi(3v)^2/A_0^3$. Assuming that $\Delta\mu_{HF}^{0(m)} = \Delta\mu_{HF}^{0(n)}$, it can be shown [21] that

$$x_n = n\left(\frac{x_m}{m} \exp\left\{-\frac{\gamma A_o m}{kT}\left[\left(\frac{m}{n}\right)^{1/6} - \left(\frac{n}{m}\right)^{1/6}\right]^2\right\}\right) \tag{2.37}$$

The independent parameters appearing in this expression are v, x_m, m, and γ. Figure 2.18 shows a plot of x_n for m = 60, v = 350 Å^3, γ = 50 ergs/cm^2, and $x_m = 10^{-4}$.

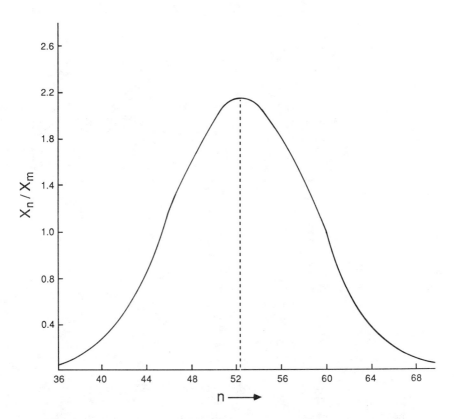

FIGURE 2.18 Concentration of amphiphiles in spherical micelles of aggregation number n. The parameters are m = 60, v = 350Å^3, γ = 50 ergs/cm^2, and $x_m = 10^{-4}$.

The position of the maximum of the distribution is not sensitive to small variations in x_m and therefore to the total amphiphile concentration. The curve is essentially Gaussian distributed, so that the mean aggregation number $\bar{n}$ corresponds to the position of the maximum. For the result shown, $\bar{n}$ and m differ only by a few percent, and hence the area per amphiphile A is only slightly different from A_0. These trends are consistent with the experimental results presented in Table 2.2. All analyses that take m to be large and include opposing forces yield curves similar to those shown by Fig. 2.18.

Since the distribution is sharply distributed about the mean aggregation number $\bar{n}$, it can be assumed with little error that all surfactant molecules exist in aggregates n or in monomer form. Thus the mole fraction of amphiphile is, from Eq. (2.30),

$$S = x_1 + x_n \tag{2.38}$$

Based on this result and a definition of the CMC as being that mole fraction of amphiphile for which the proportion of amphiphile present in monomer form equals that present as aggregates, Eq. (2.29) gives as an approximation for the CMC,

$$\ln \text{ CMC } = -\frac{\ln \bar{n}}{\bar{n} - 1} - \ln 2 - \frac{\bar{n}}{\bar{n} - 1} \frac{\mu_1^0 - \mu_n^0}{kT} \tag{2.39}$$

Equation (2.39) can be written as

$$\ln \text{ CMC } = \frac{K_1}{\bar{n}^{1/3}} + K_2 \tag{2.40}$$

when $\bar{n}$ is related to A_0 through Eq. (2.34). This result is instructive. It shows that the smaller values of the CMC are in general related to the larger aggregation members. This trend is seen in Table 2.3. Since both K_1 and K_2 depend on the structure of the lipophilic and hydrophilic portions of the amphiphiles ions, it is difficult to make quantitative comparisons; however, the qualitative trends on changing molecular structure and electrolyte concentration are correctly predicted by Eq. (2.40). To assess correctly the temperature variation of the CMC, $\Delta\mu_{HF}^0$ must be known as a function of the temperature. As noted in Section 2.V, this relationship is complex and is difficult to quantify.

B. Micelle Dimensions

The dominant factor in the consideration of micellar shape is the fact that one dimension of the core cannot exceed the length of two fully

extended alkyl chains. Simple shapes that a micelle might assume
are those of oblate or prolate spheroids of revolution, vesicles, bilayers, or toroids [95]. Tanford [92] notes that for ellipsoids of
revolution, the minor axis is limited to a length less than ℓ_{max}, the
length of the lipophilic chain, while the major axis can grow indefinitely. With the approximations

$$\ell_{max} = 1.5 + 1.265N \quad (\text{Å}) \tag{2.41}$$

where N is the number of carbon atoms and

$$v = 27.4 + 26.9N \quad (\text{Å}^3) \tag{2.42}$$

given an aggregation number n, then both the major and the minor
axis (= ℓ_{max}) are determined. It can also be shown that for a given
value of n, oblate spheroids of revolution have a smaller area than
do prolate spheroids.

Since micelles will take a form to minimize the free energy, it is
reasonable to expect that the oblate spheroids will be preferred to
prolate spheroids. However, depending on the molecular structure
of the amphiphile, other shapes may yield even smaller values of the
free energy [21].

C. Cylindrical Micelles

The shape of a micelle is not easy to determine based on purely thermodynamic considerations. In this section the issues related to the
shape are brought into focus by considering the tendency of rod-like
micelles to form. A simplified micelle shown in Fig. 2.19 is considered.
In this analysis micelles having an aggregation number less than n_0
do not exist, since as shown by Fig. 2.19 there are $n_0/2$ amphiphiles
in each hemispherical end cap. For micelles containing n amphiphiles,
$n - n_0$ reside within the cylindrical region. For $n > n_0$,

$$\mu_1^0 - \mu_n^0 = \mu_1^0 - \frac{n_0\mu_{n_0}^0 + \mu^0(n - n_0)}{n} \tag{2.43}$$

The chemical potential of a molecule in the cylindrical region, μ^0,
is taken to be independent of the number of such molecules already
present in that region. Once $n - n_0$ is sufficiently large, this is
clearly an excellent approximation. The $\mu_{n_0}^0$ is the standard chemical
potential per amphiphile defined by Eq. (2.32). To simplify, define

$$\Delta = (\mu_{n_0}^0 - \mu_1^0)n_0 \tag{2.44}$$

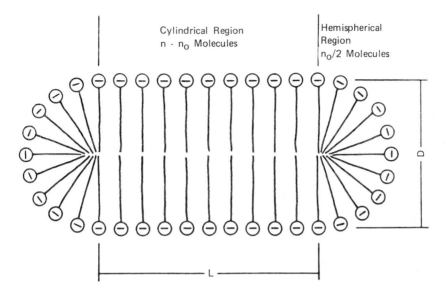

FIGURE 2.19 Idealized structure of a rod-like micelle consisting of
two hemispherical end caps and a cylindrical region.

and

$$\delta = \mu^0 - \mu_1^0 \qquad\qquad (2.45)$$

The quantity Δ is the "gap spacing," that is, the free energy difference between the smallest micelle possible and the monomer chemical
potential of n_0 molecules. The second parameter, δ, is the ladder
spacing; that is, it represents the increment of free energy associated with the addition of a surfactant monomer to the cylindrical region [96].

Define two "characteristic mole fractions" x_A and x_B whose magnitudes are related to the energy gap spacing per monomer $\Delta/(n_0 - 1)$
and the ladder spacing as follows:

$$x_A = \exp\left[\frac{\Delta}{(n_0 - 1)kT}\right] \qquad\qquad (2.46)$$

and

$$x_B = \exp\left(\frac{\delta}{kT}\right) \qquad\qquad (2.47)$$

It is also useful to define

$$K = \frac{1}{x_A} \left(\frac{x_A}{x_B}\right)^{n_0} = \exp\left[\frac{\Delta - n_0 \delta}{kT}\right] = \exp\left[\frac{n(\mu_{n_0}^0 - \mu^0)}{kT}\right] \quad (2.48)$$

Starting with Eq. (2.29), which applies to any shape of micelle, x_n can now be expressed as

$$\frac{x_n}{n} = x_A \left(\frac{x_1}{x_A}\right)^{n_0} \left(\frac{x_1}{x_B}\right)^{n-n_0} \quad (2.49)$$

Equation (2.49) shows that x_n is a decreasing function of n since x_B is less than unity. Note that to determine the distribution of micellar sizes, one must also satisfy the material balance equation (2.30), which for this model can be written as

$$S = x_1 + \sum_{n_0}^{\infty} x_n \quad (2.50)$$

where S is the mole fraction of amphiphile in the solution. Substituting Eq. (2.49) into Eq. (2.50) gives

$$S = x_1 + \sum n x_A \left(\frac{x_1}{x_A}\right)^{n_0} \left(\frac{x_1}{x_B}\right)^{n-n_0} \quad (2.51)$$

Summing yields

$$S = x_1 + \frac{1}{K}\left(\frac{x_1}{x_B}\right)^{n_0+1} \left[\frac{n_0}{(1 - x_1/x_B)} + \frac{1}{(1 - x_1/x_B)^2}\right] \quad (2.52)$$

From this equation as S becomes large, $x_1 \to x_B$. Then

$$K(S - x_B) \approx \left(\frac{x_1}{x_B}\right)^{n_0+1} \left[\frac{n_0}{(1 - x_1/x_B)} + \frac{1}{(1 - x_1/x_B)^2}\right] \quad (2.53)$$

Therefore, it follows that the quantity $K(S - x_B)$ determines x_1/x_B and thereby establishes the decrease of x_n with increasing n [see Eq. (2.49)].

If $K(S - x_B)$ is large, that is, $K(S - x_B) \gg n_0^2$, it is easy to demonstrate that

$$\frac{x_1}{x_B} = 1 - \frac{1}{[K(S - x_B)]^{1/2}} \tag{2.54}$$

and to a further approximation,

$$x_n \simeq \frac{n}{K} \exp \left\{ - \frac{n}{[K(S - x_B)]^{1/2}} \right\} \tag{2.55}$$

Thus the micellar size distribution is a monotonically decreasing exponential function of n. The rate of decrease is in direct proportion to $K(S - x_B)^{1/2}$.

Several conclusions can be drawn from this result. First, to increase $K(S - x_B)$, one can increase the surfactant concentration S. Thus higher surfactant concentrations favor a broader distribution of micellar sizes, that is, a greater tendency toward rod-like micelles. One can also reduce x_B by making μ^0 a more negative number. This can be accomplished by more effective shielding of the micellar surface charge, as, for example, higher electrolyte concentrations or by using a more stable counterion (one with a hydrophobic attachment) or by adding a polar compound such as alcohol.

The value $K(S - x_B)$ can also be increased by increasing x_A. Since Δ is a negative number ($\mu^0_{n_0}$ is a larger negative number than is μ^0_1), x_A is increased by increasing Δ or, equivalently, reducing the driving force for micelle formation.

An average aggregation number can be estimated. By definition for micelle formation,

$$\bar{n} = \frac{\displaystyle\sum_{n_0}^{\infty} x_n}{\displaystyle\sum_{n_0}^{\infty} (x_n/n)} \tag{2.56}$$

or

$$\bar{n} = \frac{S - x_1}{\displaystyle\sum_{n_0}^{\infty} (x_n/n)} \tag{2.57}$$

which simply represents the molecules of amphiphile associated into aggregates of any size divided by the total number of aggregates. This can be shown to be

$$\bar{n} = n_0 + [K(S - x_B)]^{1/2} \tag{2.58}$$

which again shows that the mean size of the cylindrical micelles is larger for larger $K(S - x_B)$.

The quantity K is evidently a most important factor contributing to polydispersity. The physical meaning is simple. Taking the logarithm of Eq. (2.48) yields

$$kT \ln K = (\mu^0_{n_0} - \mu^0)n_0 \tag{2.59}$$

This is the difference in chemical potential associated with n_0 surfactant molecules in the cylindrical region of the micelle as compared with n_0 in the hemispherical regions of the micelle. Under conditions favorable for cylindrical micelles, this term will be positive, that is, the free energy per amphiphile molecule must be less in the cylindrical region than in the hemispherical caps. The hydrophobic interaction should be approximately the same whether the amphiphile is incorporated into a spherical or a cylindrical region. Therefore, the free energy difference per amphiphile is given by

$$\mu^0_{n_0} - \mu_0 = (\Delta\hat{f}^{sphere}_{el} - \Delta\hat{f}^{cylinder}_{el})$$

$$+ \gamma(A_{sphere} - A_{cylinder}) \tag{2.60}$$

where A_{sphere} and $A_{cylinder}$ are the areas per amphiphile in the spherical and cylindrical regions, respectively, and $\Delta\hat{f}_{el}$ is the electrostatic free energy per amphiphile.

Based on the electrostatic contributions to the electrostatic free energies given by Eqs. (2.18) and (2.19), it can be shown that for $\lambda/a \ll 1$ and $a = \ell_{max}$,

$$\frac{\Delta\hat{f}^{sphere}_{el}}{\Delta\hat{f}^{cylinder}_{el}} = \frac{L_c}{2\ell_{max}} \tag{2.61}$$

where L_c is that length of cylinder which is long enough to contain n_0 amphiphiles. Since the density in the interior of both the cylindrical and the spherical micelles must be the same, then

$$\frac{4}{3} \pi \ell_{max}^3 = \pi \ell_{max}^2 L_c \tag{2.62}$$

Since both contain n_0 amphiphiles, it is concluded that

$$\frac{\Delta \hat{f}_{el}^{sphere}}{\Delta \hat{f}_{el}^{cylinder}} = \frac{2}{3} \tag{2.63}$$

the electrical contribution to the chemical potential is greater at a cylindrical surface than at a spherical one. Thus electrostatics favor spherical micelles. (There is also a small entropic advantage in having a number of spherical micelles as compared to one cylindrical one.) The driving force for cylindrical micelles must therefore stem from the surface area of hydrocarbon in contact with water. Note that

$$A_{cylinder} = \frac{8}{3} \pi \ell_{max}^2 \tag{2.64}$$

and

$$A_{sphere} = 4 \pi \ell_{max}^2 \tag{2.65}$$

Thus

$$A_{cylinder} < A_{sphere} \tag{2.66}$$

and cylinders are favored.

It is interesting to note that γ is not sensitive to the temperature. Thus increased temperatures which increase λ tend to favor spheres. On the other hand, increasing ℓ_{max} makes $A_{sphere} - A_{cylinder}$ a larger number. Thus longer chain surfactants favor cylinders. Finally, increasing the salt concentration at first rapidly decreases $\mu_{n_0}^0 - \mu_0$ (remember that this is a positive number), thereby favoring cylinders.

Thus to create cylindrical micelles, one would use lower temperatures, longer surfactant tails, and higher salinities. The result, according to this theory, is a broad distribution of micellar sizes.

D. The Energetics of Micelle Formation

Equation (2.39) shows the dependence of the CMC on the standard free-energy difference $\mu_1^0 - \mu_n^0$. Because this intimate relationship exists, models such as the one given by Eq. (2.32) have been constructed to help clarify the origin of the various mechanisms that contribute to the process of micelle formation. In addition to modeling, the factors contributing to $\mu_1^0 - \mu_n^0$ can be divided into enthalpic and entropic terms using classical thermodynamics. The enthalpy of micelle formation is given by

$$\Delta \bar{H}_{mic}^0 = \bar{H}_n^0 - \bar{H}_1^0 = - T^2 \frac{d}{dT} \left(\frac{\mu_n^0 - \mu_1}{T} \right) \tag{2.67}$$

For large n, Eq. (2.39) can be substituted in Eq. (2.67) to yield

$$\Delta \bar{H}_{mic}^0 = - T^2 \frac{d}{dT} (k \ln CMC) \tag{2.68}$$

This equation is accurate for nonionic amphiphiles. One comparison of $\Delta \bar{H}_{mic}^0$ measured directly using calorimetric methods with a corresponding value obtained using Eq. (2.68) is shown in Table 2.7. The two are in fair agreement. It should be stressed, however, that very accurate and closely spaced CMC data are required to evaluate accurately the temperature derivative required by Eq. (2.68) since the CMC temperature plot has a large curvature (see Fig. 2.13).

The standard states in Eq. (2.67) refer to states at infinite dilution. The chemical potentials are given by Eqs. (2.25) and (2.28). An examination of these equations reveals that for ionic amphiphiles, the effect of added electrolyte or even the increasing concentration of those counterions not bound to the micelle (see Fig. 2.12) is not explicitly accounted for and hence is included together with the standard free energy of transfer. In the model expressed by Eq. (2.32)

TABLE 2.7   Thermodynamic Properties of n-Alkylmethylsulfoxides at 296.7 K

Compound	$\Delta \bar{H}_{mic}^0$ (kcal/mol)	ΔG_{mic}^0 (kcal/mol)	$T\Delta S_{mic}^0$ (kcal/mol)
C_8MSO	1.8[a] 1.9[b]	-4.50	6.3
C_9MSO	1.7[b]	-5.26	7.0
$C_{10}MSO$	1.3[b]	-6.09	7.4
$C_{11}MSO$	0.72[b]	-6.86	7.6

[a]Calorimetric determination.
[b]Application of Eq. (2.68), differentiating CMC as a function of temperature.
Source: Ref. 98.

the counterion concentration is included within the Debye length and therefore contributes to the standard free energy.

When increasing the concentration of amphiphile, $k_g n$ of the counterions surrounding an aggregate containing n amphiphiles are strongly associated or bound to the aggregate, whereas $(1 - k_g)n$ (see Section 2.III.C) of the counterions reside in the diffuse layer and may be regarded as unbound. As shown by Table 2.6, k_g is approximately one-half.

The dependence of the standard free energies on the counterion concentration requires that the derivatives in Eq. (2.67) be partial derivatives holding the effective counterion concentration constant. This is difficult if not impossible to achieve since k_g is most certainly a function of the temperature [64]. Equation (2.69) should therefore not be applied to ionic amphiphiles. Another approach, one that effectively removes the influence of the counterions from the standard free energy, is required.

Since a fraction k_g of the counterions are associated with the micelle, it is tempting to include this fact in writing the chemical equation expressing the association of the amphiphilic compounds. As noted in Section 2.III.C, counterion binding is not really a stoichiometric quantity since its value depends to some extent on the experiment. Nevertheless, it will be considered here to be a stoichiometric quantity with the advantage gained that the standard-state free energy will be roughly independent of the counterion concentration.

Thus a modified description of the chemical association mechanism leading to micelle formation of an anionic amphiphile is proposed as follows [97]:

$$nA_1^- + nk_g C^+ \rightleftharpoons \left(A_n C_{nk_g} \right)^{n(1-k_g)-} \qquad (2.69)$$

where C^+ are the counterions.

This expression differs from Eq. (2.26) in that the number of counterions associated with a micelle is smaller than the number of charged amphiphiles. Based on Eq. (2.69), the condition for chemical equilibrium is

$$n\mu_n = n\mu_1 + nk_g\mu_c \qquad (2.70)$$

Using Eq. (2.28) and taking $\mu_c = \mu_c^0 + kT \ln \bar{\gamma}_c x_c$, Eq. (2.70) can be written in the form

$$\Delta G_{mic}^0 = \mu_n^0 - \mu_1^0 - k_g\mu_c^0 = kT \ln x_1$$
$$+ k_g kT \ln(x_c + S)\bar{\gamma}_c - \frac{kT}{n} \ln \frac{x_n}{n} \qquad (2.71)$$

where $\bar{\gamma}_c$ is the activity coefficient of the counterion and x_c is the mole fraction of the added electrolyte.

The standard free energy change represented in Eq. (2.71) differs from that of Eq. (2.26) by virtue of the appearance of terms that relate specifically to the counterion concentration. It is hoped that Eq. (2.71) correctly accounts for the counterion concentration, so that ΔG^0_{mic} is <u>independent</u> of x_c. For this to be the case, k_g must itself be independent of the counterion concentration. If this is not the case, there is no advantage gained by representing micelle formation by Eq. (2.69) as compared to Eq. (2.26). As will be seen, Eq. (2.69) does lead to the observed CMC variation with increased electrolyte and does, therefore, offer advantages in the study of the energetics of micellization.

It is convenient at this point to consider the energetics of micellization in the limit of large $\bar{n}$ (phase separation model) and thereby negate the need to be concerned with $\bar{n}$ and its variation with temperature. It should, however, be evident that unless $\bar{n}$ is large (>50), this approximation will not apply and the results will be subject to considerable error.

As $\bar{n} \to \infty$, then at the CMC where there are few micelles,

$$\Delta G^0_{mic} = \mu^0_n - \mu_1 - k_g \mu^0_c = kT \ln[CMC(\bar{\gamma}_c x_c + \bar{\gamma}_c CMC)k_g] \quad (2.72)$$

since CMC = x_1. Rewriting Eq. (2.72) gives

$$\ln CMC = \frac{\Delta G^0_{mic}}{kT} - k_g \ln(\bar{\gamma}_c x_c + \bar{\gamma}_c CMC) \quad (2.73)$$

This equation has precisely the same form as Eq. (2.2) provided that the counterion activity is taken to be ideal. This form accurately represents the CMC (see Fig. 2.6) as a function of added electrolyte. It is this agreement that justifies the initial chemical formulation expressed by Eq. (2.69) and suggests that k_g is a constant independent of the electrolyte composition. This justification is purely empirical and in certain cases may therefore fail. Thus some care should be exercized in using Eq. (2.73).

If k_g depends only on the temperature, Eq. (2.67) applies and

$$\Delta \bar{H}^0_{mic} = -kT^2 \frac{d}{dT}[\ln CMC + k_g \ln(CMC + x_c)\bar{\gamma}_c] \quad (2.74)$$

For the case in which x_c is zero, $\bar{\gamma}_c$ is 1, and k_g independent of temperature, the often used expression

$$\Delta \bar{H}^0_{mic} = -kT^2(1 + k_g) \frac{d}{dT}(\ln CMC) \quad (2.75)$$

is obtained. This expression has been applied to determine the enthalpy of both ionic and nonionic (k_g = 0) micelle formation. This equation should be used with some care since many of the assumptions imposed in its derivation have not been entirely justified. It has, however, been applied to ionic surfactants with some degree of success.

Table 2.8 shows various reported enthalpies for sodium n-dodecyl sulfate. The most striking feature is the relatively small value of the enthalpies found either by differentiating the CMC with respect to temperature or by direct calorimetric measurement. For lower temperatures, $\Delta \bar{H}^0_{mic}$ is a positive quantity indicating the CMC to be a decreasing function of temperature. At a temperature between 25 and 30°C the sign of $\Delta \bar{H}^0_{mic}$ for ionics changes, indicating the existence of a minimum. This is consistent with the trends shown in Fig. 2.11.

Regardless of its sign, $\Delta \bar{H}^0_{mic}$ is small, thereby indicating the main contribution to the free energy to be the entropic or hydrophobic effect. Indeed, it was the discovery of the small values of $\Delta \bar{H}^0_{mic}$ that originally prompted the development of the hydrophobic concept. A comparison of ΔG^0_{mic}, ΔH^0_{mic}, and $T \Delta S^0_{mic}$ is shown in Table 2.7. These data apply to a series of nonionic surfactants (k_g = 0) so that

TABLE 2.8 Enthalpy of Micellization of Sodium n-Dodecyl Sulfate

Additive	Temperature (°C)	$\Delta \bar{H}^0_{mic}$ (cal/mol)	Method	Reference
None	23	349	Cal.[a]	100
	25	87	Cal.	100
	30	-609	Cal.	100
	10	1200	CMC[b]	101
	20	400	CMC	101
	30	-300	CMC	101
	40	-1100	CMC	101
NaCl 0.023 mol/dm^3 23°C		-15	Cal.	100
NaCl 0.023 mol/dm^3 25°C		-153	Cal.	100
NaCl 0.023 mol/dm^3 30°C		-678	Cal.	100

[a]Calorimetric determination.
[b]Application of Eq. (2.75), differentiating CMC as a function of temperature.

the primary requirement for Eq. (2.75) to be applicable is that the micelles have a large average aggregation number. This condition may not apply to those amphiphiles in Table 2.7 having short lipo-philic groups; however, the excellent agreement between the ΔH^0_{mic} measured using a calorimeter with that obtained by application of Eq. (2.75) would indicate that the results are reliable.

The results presented in Table 2.7 confirm two assertions made previously. The main driving force for the formation of micelles is entropy. The enthalpies given in Table 2.7 are positive and micell-ization is energetically opposed. The entropy contribution is large and positive. It is the main contribution to ΔG^0_{mic}.

By comparing the ΔG^0_{mic} for the various lipophilic chain lengths the results given in Table 2.8 show that adding one methylene group does decrease ΔG^0_{mic} by approximately -750 cal/mol, as claimed previ-ously.

The change in CMC with temperature provides an indication of the enthalpy of micellization. The results reported in this section show that the enthalpies so obtained are small. They may be posi-tive or negative, but they are nevertheless small. The results are consistent with previous concepts presented, especially the hydro-phobic effect, which remains as the main driving force for micelliza-tion. For polyoxyethylenated nonionics, presumably because both the amount of water structured by the lipophile and the amount of water bound by the hydrophilic polyoxyethylene group in the mono-meric species decrease with an increase in temperature, there is a decrease in both ΔH_{mic} and ΔS_{mic} [99].

VII. MIXED MICELLES

A. General

When two or more different types of surfactant molecules are dissolved in water, the micelles that form are mixed; that is, an aggregate con-tains representatives of each type of amphiphile present in the aque-ous solution. Furthermore, the proportions of each type of amphiphile in a given micelle may be very different from its overall proportions present in the solution. Indeed, one might anticipate that in a binary mixture of surfactants, with one having a longer lipophilic chain than the other, the amphiphile with the longer chain will partition preferen-tially into the micelle. Thus in such a mixture the micelle will contain a larger proportion of longer-chain molecules than is present in solu-tion. For mixtures of surfactant it can be shown that the composition of the monomer changes continuously upon increase of the surfactant concentration, that one of the surfactant components reaches a sharp maximum in concentration at the CMC, and that the total monomer con-centration continues to increase as the surfactant concentration is

increased. All of these trends are in sharp contrast with the behavior of aqueous solutions containing a single amphiphile. For a single
amphiphile in solution we have seen that the chemical potential of the
surfactant changes little with increased concentration beyond the CMC
(Fig. 2.16), and indeed in accordance with the phase separation model,
it does not vary at all. For mixed surfactant systems even using the
phase separation approximation, the chemical potential of the surfactants will vary substantially with increasing overall concentration, even
at concentrations in excess of the mixture CMC. The variation in the
surfactant chemical potential at concentrations in excess of the mixture CMC is an important issue. It has been shown to give rise to
sharp surface tension and interfacial tension minima near the point at
which micelles first form [102,103], to be associated with adsorption
maxima [104], and to contribute to unexpected behavior when attempt-
ing to separate surfactants at concentrations above the CMC using
chromatographic techniques [105]. All of these seemingly unrelated
phenomena have their origins in the variation of chemical potential
with concentration. This is one aspect of mixed micelle behavior that
is pursued in this section.

Certain surfactants form ideal mixed micelles; that is, the mixture
properties can be predicted assuming that the micellar pseudophase is
an ideal mixture of surfactants. Beginning with this assumption, the
chemical potentials of surfactant in the micellar phase can be determined as a function of the pure component CMCs [106,107]. This
yields a mixture CMC as well as the relationship between the composition of both the monomer and the micellar pseudophase. The mixture CMC is often used as a test of ideality of the mixed micelles.
If the predicted and experimental CMC values agree, the micelles are
regarded as ideal.

There are surfactant mixtures that do not obey this definition of
ideality. For example, mixtures of anionic and nonionic surfactants
are nonideal [107,108]. The mixture CMC deviates markedly from the
ideal predictions, and therefore calculations to obtain an adequate prediction of the surfactant chemical potential must in some way take into
account the nonideality in micelle formation. An example of one method
of treating nonideal mixed micelles is presented in this section.

The thermodynamics of mixed micelles requires special consideration. The heats of mixing that are accessible to calorimetric measurement are not directly the standard enthalpy of micellization when mixed
micelles are involved. Furthermore, to obtain thermodynamic properties from mixture CMC variation with temperature requires special pre-
cautions. These are developed here. The basis for development to
follow is the phase separation model. It is clearly possible to include
a finite micellar aggregation number and a distribution of aggregation
numbers, but very little benefit accrues from doing so. The essential
concepts and results follow neatly from the phase separation model.

B. Phase Separation Model: Equilibrium Between Ionic Amphiphile
    Monomer and Micelles

In the phase separation model the micellar pseudophase is considered
to be composed entirely of surfactant. For a mixed surfactant system let y_i be the mole fraction of surfactant of type i contained in
the micellar phase. This mole fraction refers, therefore, to the moles
of component i per mole of amphiphile within the micellar pseudophase.
It will also be convenient to define a mole fraction related to those
molecules that exist as monomer. If $C_{i,mon}$ is the monomer concentration of component i, then the mole fraction is related to the mixture CMC by

$$C_{i,mon} = x_i CMC_M \qquad (2.76)$$

where CMC_M is the total monomer concentration which is considered
to be a unique function of x_i. This mole fraction, x_i, is the fraction
of the total monomer concentration that is component i. Finally, if
z_i is the mole fraction of component i in the original surfactant system, an overall material balance yields

$$z_i C_T = (C_T - CMC_M)y_i + x_i CMC_M \qquad (2.77)$$

where C_T is the total concentration of surfactant added to the system. Equation (2.77) shows that when micelles first form $z_i = x_i$ and
CMC_M can also be regarded as the CMC of the mixture. CMC_m can,
therefore, be measured by determining the mixture CMC for various
mixture compositions.

The analysis will require the equality of chemical potential between the monomer and the micellar pseudophases. Thus

$$\mu^0_{i,mic} + kT \ln \gamma_i y_i = \mu^0_{i,mon} + kT \ln(x_i CMC_M) \qquad (2.78)$$

where γ_i is the activity coefficient within the micellar pseudophase.
To evaluate $\mu^0_{i,mic} - \mu^0_{i,mon}$, CMC_i is defined to be the critical micelle concentration of component i in a solution containing counterions at a concentration $CMC_M + C_s$, the sum of the added electrolyte
plus the total of all the surfactant counterions, but in the absence
of all other surfactants. Thus in the absence of all other surfactants, $y_i = 1$, $\gamma_i = 1$, and

$$\mu_{i,mic} = \mu^0_{i,mon} + kT \ln CMC_i \qquad (2.79)$$

where CMC_i must be evaluated at the counterion concentration of the
solution. Based on Eq. (2.79), Eq. (2.78) takes the form

$$\gamma_i y_i CMC_i = x_i CMC_M \tag{2.80}$$

This equation has been written by Mysels and Otter [106]. Since $\Sigma \ x_i = 1$, the mixture CMC can be represented as

$$CMC_M = \sum \gamma_i y_i CMC_i \tag{2.81}$$

or because $\Sigma \ y_i = 1$, one also has

$$CMC_M = \left(\sum \frac{x_i}{\gamma_i CMC_i} \right)^{-1} \tag{2.82}$$

so that the CMC_i play an important role in determining the mixture CMC. For ionic amphiphiles the counterions must be considered and Eq. (2.2) or (2.73) (they are, in fact, identical) can be used to determine CMC_i. Let CMC_i^0 be the CMC in the absence of both added electrolyte and other surfactants. Then Eq. (2.2) can be written as

$$\ln \frac{CMC_i}{CMC_i^0} = -k_{gi} \ln \frac{CMC_M + C_s}{CMC_i^0} \tag{2.83}$$

Note that $CMC_M + C_S$ represents total counterion concentration at which CMC_i is evaluated. The CMC_i^0 and k_{gi} are treated in this analysis as known parameters.

The variables of interest are x_i, y_i, and CMC_M. To solve for these variables given the total amount of surfactant and the composition of the surfactant system, it is necessary to have an additional relationship giving γ_i as a function of composition.

It will be seen that mixed surfactant systems containing mixtures of anionic and nonionic surfactants form highly nonideal micellar pseudophases and γ_i different from unity are required to fit the experimental observations. However, it is first interesting to consider simpler ideal systems so that the general features of the process can be understood.

C. Ideal Mixtures of Nonionic Surfactants

For an ideal micellar pseudophase, $\gamma_i = 1$, and for nonionic surfactants, $k_{gi} = 0$. For these conditions Eq. (2.80) reduces to

$$y_i CMC_i^0 = x_i CMC_M \tag{2.84}$$

and Eqs. (2.81) and (2.82) become

$$CMC_M = \sum_i y_i CMC_i^0 \tag{2.85}$$

and

$$CMC_M = \frac{1}{\sum_i (x_i/CMC_i^0)} \tag{2.86}$$

Equations (2.85) and (2.86) express the mixture CMC in terms of either the monomer mole fraction or the mole fraction in the micellar pseudophase. It is most important to understand clearly that CMC_M is a function of the total surfactant concentration and the proportions of each type of amphiphile present in the original surfactant mixture. Thus the CMC_M is not an intrinsic property of the molecular structure of the amphiphile but is instead a variable that depends on the state of the system. The changes in CMC_M, y_i, and x_i with C_T for a given z_i have significant consequences which will be described in subsequent chapters.

To understand those changes in CMC_M that take place as C_T is increased, consider two extreme limits. At lower limit suppose that only a few aggregates have formed and practically the entire surfactant inventory is present as monomer. For this case $C_T = CMC_M$ and $x_i = z_i$. In this limit Eq. (2.86) yields

$$CMC_M^L = \frac{1}{\sum_i (z_i/CMC_i^0)} \tag{2.87}$$

At the other extreme $C_T \gg CMC_M$ and $y_i = z_i$; Eq. (2.85) then becomes

$$CMC_M^U = \sum_i z_i CMC_i^0 \tag{2.88}$$

Thus $CMC_M(C_T)$ is bound between the limits

$$CMC_M^L \leq CMC_M(C_T) \leq CMC_M^U \tag{2.89}$$

Thus the monomer concentration increases as C_T increases.

These trends are best illustrated with the help of a simple example as shown by Fig. 2.20 and Table 2.9, in which a binary system is considered. Component 1 has a $CMC_1^0 = 350$ μM and for component 2, $CMC_2^0 = 100$ μM. Thus, for example, component 1 is a molecule with more ethylene oxide units than component 2. Based on the results displayed in Fig. 2.20, it is seen that a surfactant blend composed of equal proportions of component 1 and component 2 has a mixture CMC

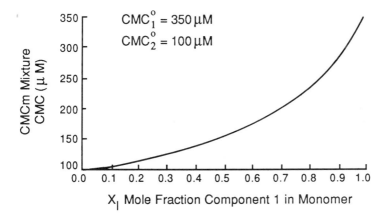

FIGURE 2.20 Nonionic binary surfactant system mixture critical micelle concentration.

of 155 μM. By changing the proportions of components in the blend, the mixture CMC is correspondingly changed. Even for this ideal case, the surfactant having the greatest tendency to form micelles (smallest CMC) strongly dictates the value of the mixture CMC. This is evidenced by the smaller slope at $x_1 = 0$ compared to that at $x_1 = 1$. The slope increases monotonically with increasing x_1.

TABLE 2.9 Micellar Solutions of Mixed Surfactants at Concentrations in Excess of the CMC[a]

x_1	CMC_M (μM)	C_T (μM)	$C_{1,mon}$ (μM)	$C_{2,mon}$ (μM)	y_1
0.50	155	155	77.5	77.50	0.221
0.55	164.7	198.8	90.6	74.1	0.259
0.58	170.7	210.0	99.0	71.7	0.283
0.60	175.0	262.5	105.0	70.0	0.300
0.65	186.7	369.4	121.4	63.3	0.347
0.70	200.0	600.0	140.0	60.0	0.400
0.75	215.4	1615	161.5	53.9	0.462
0.777	224.7	∞	174.6	50.1	0.500

[a]$z_1 = z_2 = 0.5$; $CMC_1^o = 350$ μM; $CMC_2^o = 100$ μM.

In accordance with the phase separation model, Fig. 2.20 provides the relationship between CMC_M and x_1 irrespective of the quantity of surfactant present in solution. This curve can therefore be measured once and for all (but not without difficulty, see below) and presumably holds even when the total surfactant concentration is in excess of the mixture CMC. As has been noted, the most significant feature of mixed micelle behavior is that as the concentration of a surfactant blend is increased, the composition and concentration of both the monomer and micellar pseudophases changes so that the proportions of the component having the largest CMC_i^0 increases in both the monomer phase and the micellar phase. This is shown in Table 2.9. To calculate the results shown in this table, one must use the material balance, Eq. (2.77). Note that as C_T is increased, a point is reached at which micelles form. In this case $C_T = C_M$ at 155 μM. Since very few micelles are present, the monomer concentration is the same as that in the original surfactant blend. Thus $x_1 = z_1 = 0.5$. The small amount of micellar pseudophase present contains only 22% of component 1. It is composed primarily of component 2. A further increase in C_T results in a decrease in the monomer concentration of component 2 since most of component 2 is incorporated into the micelles. The total monomer concentration increases, as does the monomer concentration of component 1, the amphiphile having the larger CMC of the two components. The decrease in monomer concentration of the more surface active of the two amphiphiles is a rather surprising result and is now understood to be the origin of the surface tension or interfacial tension minimum often observed for mixed surfactant systems [102,103].

Figure 2.21 shows the surface tension plotted as a function of the logarithm of solution concentration for two pure surfactants. It is seen that the break is distinct and without any evidence of a minimum. The surface tension of mixtures of the two pure components shows a distinct minimum, which is to be anticipated in accordance with the theory presented here. Actually, the solid lines represent theoretical predictions by Clint [102], who has extended the mixed micelle equilibrium theory to include surface tensions.

The theory developed here for ideal mixed micelles and extended by Clint [102] to include a prediction of surface tension should apply to the class of ethoxylated nonylphenols, which are studied extensively in this book. These compounds are mixtures. In fact, the manufacturing process produces molecules having a distribution of polyethylene oxide units which closely approximates a Poisson distribution [109]. This is given by

$$n_i = \frac{m^{i-1} \exp(-m)}{(i-1)!} n_T \qquad (2.90)$$

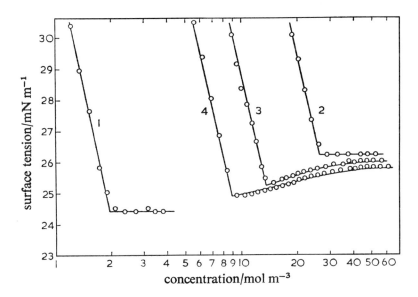

FIGURE 2.21   The surface tension for pure n-decyl methyl sulphoxide (curve 1) and pure n-octyl methyl sulphoxide (curve 2) and for
two mixtures all shown as function of concentration.   Curve 3, mole
fraction of n-decyl component = 0.075; curve 4, mole fraction of n-
decyl component = 0.156.   (After Ref. 102).

where m is the average number of ethylene oxide units, i represents
the number of ethylene oxide units in compound i, and n_T is the
total moles of surfactant.

Table 2.9 shows that the mole fraction of monomer is, at small
surfactant concentrations, equal to that of the system mole fraction.
The micelles at this point are quite rich in that component having
the smallest CMC.   Clearly, as C_T is increased, the fraction of the
total surfactant inventory found in the micellar phase increases, and
as $C_T \to \infty$, it is the micellar phase that has a composition equal to
that of the monomer (in Table 2.9, y_1 = 0.5 at $C_T \to \infty$).   In this
limit the monomer phase is composed primarily of component 1.   Thus
the proportion of component 1 in both the micellar and the monomer
phase increases with increasing total concentration.

This behavior of surfactant mixtures differs markedly from that
exhibited by a solution containing a single surfactant molecule.   This
difference will play an important role in the behavior of micellar solutions, as noted subsequently.

D. Mixtures of Anionic Surfactants

If a mixture of ionic surfactants forms an ideal pseudophase, $\gamma_i = 1$, then based on Eqs. (2.80), (2.81), and (2.82), it can be shown that

$$CMC_M = \frac{\prod\limits_{i=1}^{N} (CMC_i^0)^{k_{gi}+1}}{\sum\limits_{i=1} x_i \left[\prod\limits_{j=1,j \neq i}^{N} (CMC_j^0)^{k_{gj}+1} (CMC_M + C_s)^{k_{gj}} \right]}$$

$$(2.91)$$

This equation provides an implicit relationship giving the mixture CMC as a function of the monomer composition. This equation has been verified experimentally by ultrafiltration, where the monomer composition is measured directly [110]. Its derivation does assume that the micellar pseudophase is ideal.

E. Mixed Micelles of Anionic and Nonionic Surfactants

The mixture CMC for mixed anionic-nonionic micelles deviates markedly from ideal behavior. This dictates the need to introduce an activity coefficient [107,108,111,112]. Rubingh [108] and Scamehorn et al. [111] have shown that the mixture CMC data can be correlated by assuming the mixed micellar pseudophase to be a regular solution. This model proposed by Rubingh for binary mixtures can be extended to multicomponent systems [113].

Regular solutions are those mixtures whose excess entropy of mixing vanishes. The excess free energy of mixing is therefore equal to the enthalpy of mixing and may be represented in terms of a single parameter, W. Thus the nonideal behavior of a binary mixture surfactant molecules is expressed as follows:

$$\Delta G_E = \Delta H_E = y_1 y_2 W \qquad\qquad (2.92)$$

The activity coefficients are given by

$$kT \ln \gamma_1 = y_2 \frac{d \Delta G_E}{dy_1} + \Delta G_E = y_2^2 W \qquad\qquad (2.93)$$

and

$$kT \ln \gamma_2 = -y_1 \frac{d \Delta G_E}{dy_1} + \Delta G_E = y_1^2 W \qquad\qquad (2.94)$$

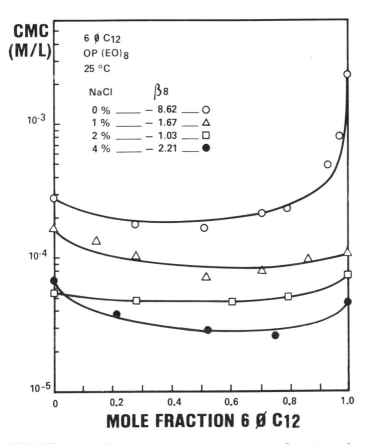

FIGURE 2.22 The mixture CMC shown as a function of monomer mole fraction of sodium dodecyl benzene sulfonate [$6\phi C_{12}$] mixed with octaoxyethylene octyl phenol [OP(EO)$_8$]. β_8 denotes W/kT. (After Ref. 117.)

Figure 2.22 shows the CMC$_M$ of a mixed anionic/nonionic surfactant system at varying levels of added electrolyte. The solid lines are the predictions based on the regular solution model using the interaction parameter shown. This interaction parameter used with Eqs. (2.93) and (2.94) gives the activity coefficients needed to calculate the mixture CMC. The comparison between prediction and experiment is acceptable; however, other thermodynamic properties, such as the ΔH_E as measured calorimetrically [114] or the actual composition of the micellar phase [115], are not well predicted by the regular solution model. Other more precise models have been proposed [115], but these will not be considered here. The use of the regular solution model gives reasonable results provided that the temperature is not varied.

The mixed anionic/nonionic systems generally appear to be non-ideal and require activity coefficient corrections. The interaction parameter does become larger (less negative) as electrolyte is added (see Fig. 2.22). The system tends to behave more ideally in higher concentrations of electrolyte. This trend indicates that the nonideality has its origins in the electrostatic contribution to the free energy [77,116].

F. Thermodynamics of Mixed Micelle Formation

A brief accounting of the thermodynamics of mixed micelle formation is given with emphasis on those features which differ from those presented previously for a single surfactant molecule. A mixed surfactant system consisting of anionic and nonionic surfactants is, to be definite, considered here. The changes that would be required to treat other possible combinations will be obvious.

The process of micellization involves the aggregation of n_1, molecules of anionic surfactant, and n_2, molecules of nonionic surfactant; k_{gM} is the degree of counterion binding to the mixed micelle. Thus considering the formation of a mixed micelle as a chemical reaction requires that

$$\mu_{mic} = \frac{n_1}{n_1 + n_2} \mu_1 + \frac{n_2}{n_1 + n_2} \mu_2 + k_{gM} \frac{n_1}{n_1 + n_2} \mu_c \qquad (2.95)$$

where μ_1, μ_2, and μ_c are the monomer chemical potentials of components 1, 2, and c, respectively. Here c denotes the counterion. Equation (2.95) can be written for ideal chemical potentials as follows:

$$\Delta G^0_{mic} = (\mu^0_{mic} - y_1 \mu^0_1 - y_2 \mu^0_2 - k_{gM} y_1 \mu^0_c)$$
$$= kT(y_1 \ln C_1 + y_2 \ln C_2 + k_{gM} y_1 \ln C_e) \qquad (2.96)$$

or

$$kT \ln \left(C_1^{y_1} C_2^{y_2} C_c^{k_{gM} y_1} \right) = \Delta G^0_{mic} \qquad (2.97)$$

This form of the free energy is complex and is justifiable only if k_{gM} depends on temperature and the micellar composition y_1 and y_2 but is independent of the solution counterion composition and the solution monomer concentration. It is known that k_{gM} depends on the composition of the micelles [77].

Equation (2.97) can be put into slightly more convenient form by noting that

$$C_1 = x_1 CMC_M \tag{2.98}$$

and

$$C_2 = x_2 CMC_M \tag{2.99}$$

Also, the counterion concentration can be expressed as the sum of three terms: the added electrolyte concentration, the counterions associated with the anionic monomer, and the counterions not incorporated into the mixed micelle. Thus

$$C_c = C_s + x_1 CMC_M + (1 - k_{gM})y_1(C_T - CMC_M) \tag{2.100}$$

Equations (2.98) and (2.99) substituted into Eq. (2.97) yields

$$\Delta G_{mic} = \ln CMC_M + \ln(x_1^{y_1} x_2^{y_2}) + y_1 k_{gM} \ln C_c \tag{2.101}$$

The heat of micellization is given by

$$\Delta \bar{H}^0_{mic} = -T^2 k \frac{\partial}{\partial T} \left[\ln CMC_M + \ln(x_1^{y_1} x_2^{y_2}) + y_1 k_{gM} \ln C_c \right]_{y_1} \tag{2.102}$$

We note that experiments designed to make use of this expression require measurement of CMC_M, x_1, and k_{gM} all as a function of T. Furthermore, ΔG^0_{mic} is a function of y_1 as well as temperature. Thus the partial derivative with respect to temperature appearing in Eq. (2.102) requires that the micellar compositions be held constant. Measurement of the micellar composition is, in general, difficult. In some cases extensive surface tension measurements can be used to infer the micellar composition [112]. Ultrafiltration can also be applied to obtain micellar compositions for a limited number of systems [115]. Because of these profound difficulties there are no comprehensive reported studies of the thermodynamics of nonionic/anionic surfactant systems.

VIII. SOLUBILITY

A. Crystal Solubility

1. Krafft Temperature

Ionic amphiphiles that have a definite crystalline structure, such as the 1-alkyl or 2-alkyl sulfates, which exhibit a CMC also often show a break in the solubility/temperature curve. The temperature at which this sharp break occurs is called the Krafft temperature [118].

The solubility/temperature curve evidently denotes a solution concentration at which the chemical potential of the amphiphile equals
that of the crystalline solid.   In the absence of changes in the crystalline form, the chemical potential of a crystal increases regularly
with temperature.   If the temperature is less than the Krafft temperature, the solution concentration of the amphiphile in equilibrium with
the crystal is less than the critical micelle concentration.   Thus the
solution contains few, if any, micelles and the solubility is small.   A
further increase in temperature will result in a corresponding increase
in the solubility, but until micelles form, the increase in concentration
will be correspondingly small.   At the Krafft temperature, the solubility increases rapidly because large numbers of molecules are required
to increase the chemical potential of the monomer slightly.   In the
phase separation model the rate of increase in solubility as a function
of temperature is predicted to be essentially infinite.   In fact, it is
not infinite, but as shown by Fig. 2.23, the increase in solubility at
the Krafft temperature is striking.

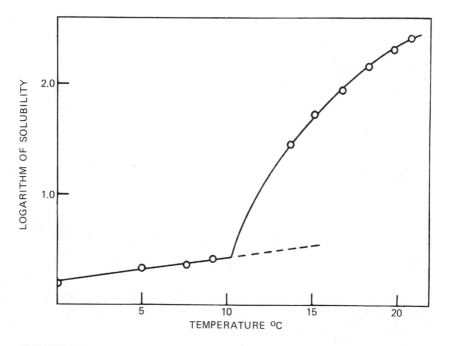

FIGURE 2.23   Logarithm of the solubility of sodium dodecyl sulfate in
water as a function of temperature.   The dashed line represents the
expected behaviour in the absence of micelle formation.   (After Ref.
119.)

It should be emphasized that the solubility of a substance is determined by the equality of chemical potential between the solid and liquid phases. Conclusions concerning the intermolecular relationships in liquid phases can be drawn from comparison of solubilities only if the solubilities refer to comparable crystalline phases. In this way the lower water solubility of tetradecane-1 and -2 sodium sulfates as compared with the tetradecane-3 to -7 sulfates does not imply a reduction in the hydrophobe/hydrophilic ratio as the branches of the chain become more nearly equal. Actually, the reverse is true (see Chap. 5). The solubilities of the tetradecane-3 to -7 sulfates are greater than those of the tetradecane-1 to -2 sulfates because, for steric reasons, the former compounds produce much less stable crystals. They therefore show a greater solubility not only in water but also in organic solvents. The tendency to crystallize of amphiphilic salts in which the ionic group does not occupy a near-terminal position in the molecule is often exceedingly small. Thus sodium sulfo-di-(2-ethylhexyl) succinate (Aerosol OT in Table 2.1) is a colorless gel even in rather pure form. The sodium tetradecane-3 to -7 sulfates are obtained by evaporation of their solutions in acetone as gummy materials that do not crystallize or are difficult to crystallize. These are but a few of the considerable body of facts that provide an indication of the difficulty that one has in crystallizing these midchain-substituted compounds.

The amphiphile with the hydrophile group attached midchain have a greater tendency to form conjugate liquid phases than do those amphiphiles with endpoint attachments. This increased tendency is evident by observation of the surfactant concentration-phase relationship at a fixed temperature. In some cases a second liquid-crystalline phase forms more readily for those anionic amphiphiles with midchain attachments than for those having more nearly terminal attachments. This will be described in subsequent discussion relating to the mesophases that form in higher-concentration surfactant solutions.

When two surfactants of like charge are blended, the Krafft temperature of the mixture is intermediate between the pure components or shows a very shallow minimum [121]. For mixtures of nonionic and anionic surfactants, the Krafft temperature has been found to decrease slightly upon addition of small quantities of the nonionic surfactant and then to reach a plateau [122]. The lowering of the Krafft temperature is enhanced by increasing the number of polyoxyethylenes in the hydrophile.

2. Cloud Point Temperature

The sudden onset of turbidity of a nonionic surfactant solution on raising the temperature is called the cloud point. At a somewhat higher temperature the solution tends to separate into two phases. One of the phases is surfactant-rich, whereas in the other the surfactant concentration is normally quite small. The nature of the micellar shape as this point is approached is somewhat controversal at the present time. It is sometimes stated that the micellar size of nonionic surfactants increases with temperature [17]. This conclusion is obtained mainly from light-scattering experiments. Recent studies [123–125] indicate that secondary aggregation of micelles may occur and that light-scattering data have been misinterpreted. According to the thermodynamic theory [see Eq. (2.40)], the increase in critical micelle concentration with temperature heralds a decrease rather than an increase in the aggregation number. Thus the turbidity that is observed is then related to a lower consolute critical point [124] which is characterized by large concentration fluctuations. The increased light scattering would then simply be a manifestation of the same critical opalescence observed for simple fluids near their critical points (see also Zulauf and Rosenbusch [125]).

The temperature at which clouding occurs depends on the structure of the polyoxyethylenated nonionic surfactant and the composition of the aqueous solution. It is not a strong function of the surfactant concentration. For a particular hydrophobic group, the larger the percentage of ethylene oxide in the surfactant molecule, the higher the cloud point temperature, although the relationship between the percentage and the cloud point is not linear. Table 2.10 lists the cloud points of some ethoxylated nonylphenols. It is seen that with increasing proportions of ethylene oxide, the cloud point at first increases rapidly and then increases more slowly [126]. A study of the effect of structural changes in the surfactant molecule on the cloud point indicates that at constant ethylene oxide content the cloud point is lowered by the following: increased molecular weight of the

TABLE 2.10   Cloud Point of Polyethyleneoxide Nonylphenols in Aqueous Solutions

	Cloud point (°C)			
	Mole ratio	ethylene oxide	to	nonylphenol
Solvent	9	10.5	15	20
Water	55	72	98	>100
3%NaCl	45	61	84.5	95
3%Na$_2$CO$_3$	32	48	70	78
3%NaOH	31	45.5	67	73
3%HCl	60.5	78	>100	>100

Source: Ref. 128.

surfactant [51], broader distribution of ethylene oxide chain lengths
such as exists in commercial mixtures, branching of the hydrophobic
group, more central position of the ethylene oxide hydrophilic group,
replacement of the terminal hydroxyl of the hydrophilic group by a
methoxyl, and replacement by an ester linkage of the ether linkage
between the hydrophilic and hydrophobic groups.

The effect of inorganic additives on solutions of nonionic surfactants has been reported by Schott and co-workers [127–129]. Schott
has identified two categories of additive that cause considerable increase in cloud points: (1) urea and salts with anions known to disrupt the structure of water, such as iodides, thiocyanates, and per-
chlorates, and (2) salts with cations capable of forming complexes
with model ethers, such as dioxane. In the latter case the resulting
complexation increases the solubility of the surfactant molecules in
water, thereby increasing the cloud point temperature. This phenomenon is referred to as salting in. Electrolytes in the second cate-
gory include strong acids and salts of lithium and polyvalent cations
such as lead, cadmium, magnesium, aluminum, and calcium. Lowering
of the cloud point (salting out) was noted for relatively few electrolytes. These are restricted to those with noncomplexing cations (sodium, potassium, ammonium, and cesium) in association with anions
such as chloride and nitrates. Table 2.10 shows that addition of sodium chloride does decrease the cloud point temperature.

B. Conjugate Liquid Phases

Solutions containing micelles which are essentially noninteracting, the
class of systems that has been the focus of attention in this chapter,
are limited to a narrow concentration range centered in regions of
small surfactant concentration. The phenomena that occur in this
narrow region are characteristic of a wide range of molecular structures and are of general interest. However, at intermediate to high
concentrations, conjugate liquid phases called mesophases (from the
Greek "mesos", meaning middle or intermediate) form which are intermediate between the dilute, isotropic micellar phases and the crystalline phases of pure amphiphiles. These mesophases are often called
lyotropic phases to denote the fact that the solvent molecules are not
uniformly distributed throughout the structure but are more numerous
in some regions than others. In aqueous solutions, which are of primary concern here, water is concentrated in those parts containing
the hydrophile.

Even for binary mixtures of water and amphiphile, a wide variety
of structures and phases form [130,131]. Often the various structures follow each other in a definite sequence which is related to the
proportion of water and amphiphile. Starting with an isotropic micellar solution which is labeled S_1, lyotropic liquid-crystalline phases
may appear. As shown by Fig. 1.2, often the first to appear is the

"middle phase" or M_1. This phase consists of infinitely long rods in hexagonal array, as depicted in Fig. 2.24. The sodium laurate/water system shown in Fig. 2.25 exhibits the M_1 phase at concentrations of 30 to 40 wt % sodium laurate. At small concentrations the S_1 phase is seen.

At higher concentrations the "neat phase" with lamellar structure may appear. This structure designated as G is depicted in Fig. 2.26. Between M_1 and G other mesophases sometimes occur—a rectangular phase in which indefinitely long, mutually parallel rods in orthorhombic array and with a rectangular cross section may appear. This phase is, as are M_1 and G, anisotropic. Isotropic mesophases which are quite viscous may also appear intermediate between M_1 and G. The phase denoted V_1 in Fig. 2.27 is such a phase. This phase is believed to consist of short rod-like elements having an axial ratio nearly equal to 1 arrayed to form a viscous three-dimensional network.

The mesophases noted above for ionic amphiphiles compounds are also observed in the presence of added electrolyte, but the phase boundaries are often shifted by a significant extent. Furthermore, the chemical nature of the salt often makes a great difference. For

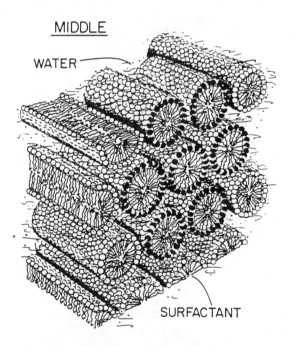

FIGURE 2.24 M_1 Mesophase. Rod-like micelles of indefinite length and arranged in hexagonal array.

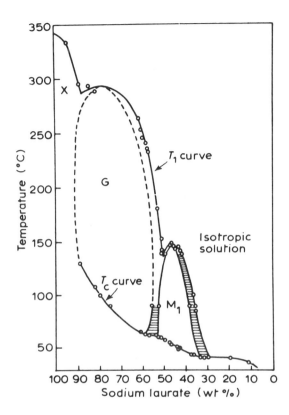

FIGURE 2.25 Phase diagram for sodium laurate/water system. (After Refs. 132,133.)

example, the cubic V_1 phase shown in Fig. 2.27 does not appear if the chlorides are replaced by the corresponding bromides. This is in contrast with the small influence of various monovalent alkali metal ions on the CMC of anionic amphiphiles (see Section 2.III). The structure of the mesophases that form in the case of ionic surfactants has been reviewed by Winsor [6] and by Skoulios [134].

Figure 2.28 shows that nonionic amphiphiles exhibit the same sequence of mesophases as displayed by ionics. Both the middle and the neat phases are seen in Fig. 2.28. This depends strongly on the molecule structure of the amphiphile. Husson et al. [136] studied the various derivatives of nonylphenol polyoxyethylenes with the general formula $C_9H_{19}C_6H_4O(CH_2CH_2O)_nH$ and found that with short-chain compounds (n = 6 and 8) only the neat phase appears, whereas both neat and middle phases are found when n = 9. Higher-chain-length compounds, n > 10, show a middle phase only.

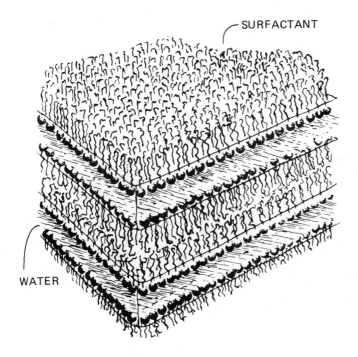

FIGURE 2.26   G Phase lamellar structures; sometimes called the neat phase.

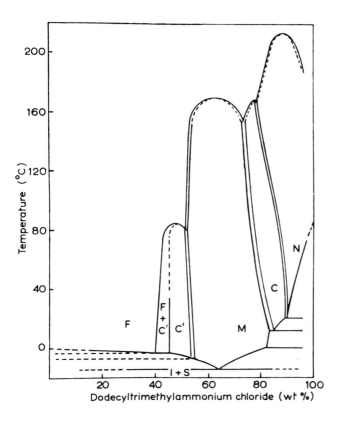

FIGURE 2.27 Phase diagram for the dodecyltrimethyl ammonium chloride ($NC_{12}Me_3Cl$)/water system. [F ≡ S_1 (isotropic); C′ and C ≡ V (isotropic cubic); and N ≡ G (neat).] (After Ref. 27.)

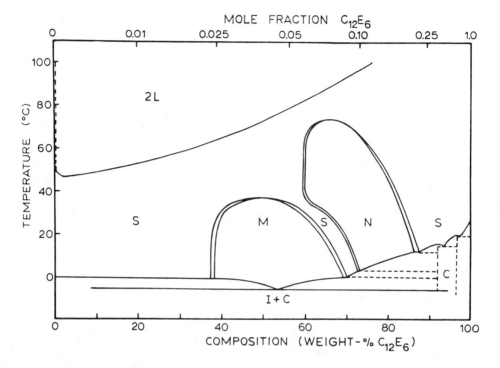

FIGURE 2.28   PHase diagram for ethoxylated dodecyl alcohol.   [The
S isotropic phase separating M from N is a cubic mesophase.]   (From
Ref. 135.)

REFERENCES

1.   K. Shinoda, in <u>Colloidal Surfactants</u> (K. Shinoda, B. Tamamushi,
     T. Nakagawa, and T. Isemura, eds.), Academic Press, New
     York (1963).

2.   J. J. Christensen, R. W. Hanks, and R. Izatt, <u>Handbook of
     Heats of Mixing</u>, Wiley, New York, p. 1410 (1982).

3.   G. S. Hartley, <u>Paraffin-Chain Salts</u>, Hermann & Cie, Paris
     (1936).

4.   W. C. Preston, <u>J. Phys. Colloid Chem.</u>, <u>52</u>:84 (1948).

5.   C. Tanford, <u>The Hydrophobic Effect: Formation of Micelles
     and Biological Membrane</u>, Wiley, New York (1980).

6.   P. A. Winsor, <u>Chem. Rev.</u>, <u>68</u>:1 (1968).

7. E. A. G. Aniansson, S. N. Wall, M. Almgren, H. Hoffmann,
    I. Kielmann, W. Ulbricht, R. Zana, J. Lang, and C. Tondre,
    J. Phys. Chem., 80:905 (1976).

8. J. Lang, C. Tondre, R. Zana, R. Bauer, H. Hoffmann, and
    W. Ulbricht, J. Phys. Chem., 79:276 (1975).

9. A. Mallioris, J. Lang, and R. Zana, J. Chem. Phys., 90:655
    (1986).

10. E. A. G. Aniansson and S. N. Wall, J. Phys. Chem., 78:1024
    (1974).

11. E. A. G. Aniansson and S. N. Wall, J. Phys. Chem., 79:857
    (1975).

12. P. Debye and E. W. Anacker, J. Phys. Chem., 55:644 (1951).

13. E. W. Anacker, in Cationic Surfactants (E. Jungermann, ed.),
    Marcel Dekker, New York, p. 205 (1970).

14. N. A. Mazer, G. B. Benedek, and M. C. Carey, J. Phys.
    Chem., 80:1075 (1976).

15. P. Missel, N. A. Mazer, G. B. Benedek, and M. C. Carey,
    J. Phys. Chem., 87:1264 (1983).

16. M. J. Rosen, Surfactants and Interfacial Phenomena, Wiley,
    New York, p. 88 (1978).

17. R. R. Balmbra, J. S. Clunie, J. M. Corkill, and J. F. Good-
    man, Trans. Faraday Soc., 58:1661 (1962).

18. R. R. Balmbra, J. S. Clunie, J. M. Corkill, and J. F. Good-
    man, Trans. Faraday Soc., 60:979 (1964).

19. E. Ruckenstein and R. Nagarajan, J. Phys. Chem., 79:2622
    (1975).

20. E. Ruckenstein and R. Nagarajan, J. Phys. Chem., 85:3010
    (1981).

21. J. N. Israelachvili, J. D. Mitchell, and B. W. Ninham, J.
    Chem. Soc. Faraday Trans. 2, 72:1525 (1976).

22. R. J. Williams, J. N. Phillips, and K. J. Mysels, Trans. Fara-
    day Soc., 51:728 (1955).

23. C. Y. Young, P. J. Missel, N. A. Mazer, G. B. Benedek, and
    M. C. Carey, J. Phys. Chem., 82:1375 (1978).

24. G. S. Hartley, J. Chem. Soc., 1938 (1968).

25. T. S. Brun, H. Hoiland, and E. Vikingstad, J. Colloid Inter-
    face Sci., 63:89 (1978).

26. E. Vikingstad and H. Hoiland, J. Colloid Interface Sci., 64:510
    (1978).

27. R. R. Balmbra, J. S. Clunie, and J. F. Goodman, Nature, 222:1159 (1969).

28. H. Wennerstrom, Chem. Phys. Lett., 18:41 (1973).

29. H. H. Mautsch, H. Saito, and I. C. P. Smith, Prog. Nucl. Magn. Reson. Spectrosc., 11:211 (1977).

30. D. W. R. Gruen, J. Phys. Chem., 89:153 (1985).

31. B. Lindman and H. Wennerstrom, Topics in Current Chemistry: Micelles, Springer-Verlag, Berlin, p. 1 (1980).

32. P. Stenius and P. Ekwall, Acta. Chem. Scand., 21:1767 (1967).

33. P. Mukerjee, J. R. Cardinal, and N. R. Desai, in Micellization, Solubilization, and Microemulsions, Vol. 1 (K. L. Mittal, ed.), Plenum Press, New York, p. 241 (1977).

34. P. Mukerjee and K. J. Mysels, Critical Micelle Concentrations of Aqueous Surfactant Systems, NSRDS-NBS 36, U.S. Government Printing Office, Washington, D. C. (1971).

35. D. Stigter, J. Phys. Chem., 78:2480 (1974).

36. H. B. Klevens, J. Phys. Chem., 52:130 (1948).

37. H. C. Evans, J. Chem. Soc., 579 (1956).

38. J. E. Bujake and E. D. Goddard, Trans. Faraday Soc., 61:190 (1965).

39. P. H. Elworthy and K. J. Mysels, J. Colloid Sci., 21:331 (1966).

40. K. J. Mysels and L. H. Princen, J. Phys. Chem., 63:1696 (1959).

41. K. Meguro and T. Kondo, J. Chem. Soc. Jpn. Pure Chem. Sec., 77:1236 (1956).

42. J. M. Corkill and J. F. Goodman, Trans. Faraday Soc., 58:206 (1962).

43. K. Meguro and T. Kondo, Nippon Kagaku Zasshi, 80:818 (1959).

44. E. Goette and M. J. Schwuger, Tenside, 6:131 (1969).

45. K. Shinoda and K. Katsura, J. Phys. Chem., 68:1568 (1964).

46. C. H. Arrington, Jr. and G. D. Patterson, J. Phys. Chem., 57:247 (1953).

47. H. W. Hoyer and A. Marmo, J. Phys. Chem., 65:1807 (1961).

48. J. Osugi, M. Sato, and N. Ifuku, Rev. Phys. Chem. Jpn., 35:32 (1965).

49. K. Tori and T. Nakagawa, Kolloid Z. Z. Polym., 188:47 (1963).

50. K. Tori and T. Nakagawa, Kolloid Z. Z. Polym., 189:50 (1963).

51. P. H. Elworthy and C. McDonald, Kolloid Z., 195:16 (1964).

52. M. J. Schick and A. H. Gilbert, J. Colloid Sci., 20:464 (1965).

53. P. H. Elworthy and C. B. Macfarlane, J. Pharm. Pharmacol. Suppl., 14:100 (1962).

54. P. Mukerjee, K. J. Mysels, and P. Kapauan, J. Phys. Chem., 71:4166 (1967).

55. M. F. Emerson and A. Holtzer, J. Phys. Chem., 71:1898 (1967).

56. M. L. Corrin and W. D. Harkins, J. Am. Chem. Soc., 69:679 (1947).

57. Y. Moroi, N. Nishikido, M. Saito, and R. Matuura, J. Colloid Interface Sci., 52:356 (1975).

58. E. J. R. Sudholter and J. B. F. N. Engberts, J. Phys. Chem., 83:1854 (1979).

59. M. E. Hobbs, J. Phys. Chem., 55:675 (1951).

60. E. K. Mysels and K. J. Mysels, J. Colloid Sci., 20:315 (1965).

61. P. Becher, J. Colloid Sci., 17:325 (1962).

62. M. L. Corrin and W. D. Harkins, J. Am. Chem. Soc., 69:683 (1947).

63. P. Molyneux, C. T. Rhodes, and J. Swarbrick, Trans. Faraday Soc., 509:1043 (1965).

64. Y. Moroi, N. Nishikido, H. Uehara, and R. Matuura, J. Colloid Interface Sci., 50:254 (1975).

65. B. D. Flockhart, J. Colloid Sci., 12:557 (1957).

66. E. D. Goddard and G. C. Benson, Can. J. Chem., 35:986 (1957).

67. E. H. Crook, G. F. Trebbi, and D. B. Fordyce, J. Phys. Chem., 68:3592 (1964).

68. K. Shinoda and T. Soda, J. Phys. Chem., 67:2072 (1963).

69. J. M. Corkill, J. F. Goodman, and T. Walker, Trans. Faraday Soc., 63:768 (1967).

70. S. D. Hamann, J. Phys. Chem., 66:1359 (1962).

71. R. F. Tuddenham and A. E. Alexander, J. Phys. Chem., 66:1839 (1962).

72. S. Rodriguez and H. Offen, J. Phys. Chem., 81:47 (1977).

73. D. Stigter, J. Phys. Chem., 68:3603 (1964).

74. D. Stigter, J. Phys. Chem., 79:1008 (1975).

75. J. D. Jackson, Classical Electrodynamics, Wiley, New York, p. 33 (1975).

76. P. C. Hiemenz, Principles of Colloid and Surface Chemistry, Marcel Dekker, New York, p. 360 (1977).

77. J. F. Rathman and J. F. Scamehorn, J. Phys. Chem., 88:5807 (1984).

78. M. Abramowitz and I. A. Stegun, eds., Handbook of Mathematical Functions, p. 374, National Bureau of Standards Applied Mathematics Series, No. 55, U.S. Department of Commerce, Washington, D.C. (1964).

79. E. J. W. Verwey and J. T. G. Overbeek, Theory of the Stability of Lyophobic Colloids, Elsevier, New York (1948).

80. D. Stigter, Rec. Trav. Chim. Pays Bas, 73:593 (1954).

81. J. W. McBain, Trans. Faraday Soc., 9:99 (1913).

82. A. W. Ralston and C. W. Hoerr, J. Am. Chem. Soc., 64:772 (1942).

83. B. W. Barry and R. Wilson, Colloid Polym. Sci., 256:251 (1978).

84. M. J. McDowell and C. A. Kraus, J. Am. Chem. Soc., 73:2173 (1951).

85. K. G. Denbigh, Principles of Chemical Equilibrium, Cambridge University Press, London, p. 303 (1963).

86. T. Yoshida, K. Taga, H. Okahayashi, K. Matsushita, H. Kamaya, and I. Ueda, J. Colloid Interface Sci., 109:336 (1986).

87. B. Lindman and B. Brun, J. Colloid Interface Sci., 42:388 (1973).

88. N. Kamenka, M. Chorro, H. Fabre, B. Lindman, J. Rouviere, and C. Cabos, Colloid Polym. Sci., 257:757 (1979).

89. S. G. Cutler, P. Meares, and D. G. Hall, J. Chem. Soc. Faraday Trans. 1, 74:1758 (1978).

90. H. S. Frank and M. W. Evans, J. Chem. Phys., 13:507 (1945).

91. G. Nemethy and H. A. Scheraga, J. Chem. Phys., 36:3382 (1962).

92. C. Tanford, J. Phys. Chem., 78:2469 (1974).

93. Y. Murata, K. Motomura, and R. Matuura, Mem. Fac. Sci. Kyushu Univ., C11:29 (1978).

94. S. J. Gill, N. F. Nichols, and I. Wadso, J. Chem. Thermodyn., 8:445 (1976).

95. R. B. Hermann, J. Phys. Chem., 76:2754 (1972).

96. P. J. Missel, N. A. Mazer, G. B. Benedek, C. Y. Young, and M. C. Carey, J. Phys. Chem., 84:1044 (1980).

97. J. N. Phillips, Trans. Faraday Soc., 51:561 (1955).

98. J. H. Clint and T. Walker, J. Chem. Soc. Faraday Trans. 1, 71:946 (1975).

99. R. A. Hudson and B. A. Pethica, in Chemistry, Physics and Application of Surface Active Substances, Vol. 4, Proceedings of the Fourth International Congress (J. T. G. Overbeck, ed.), Gordon and Breach, New York, p. 631 (1964).

100. G. Pilcher, M. N. Jones, L. Espada, and H. A. Skinner, J. Chem. Thermodyn., 1:381 (1969).

101. B. D. Flockhart, J. Colloid Sci., 16:484 (1961).

102. J. H. Clint, J. Chem. Soc. Faraday Trans. 1, 71:1327 (1975).

103. E. I. Franses, M. S. Bidner, and L. E. Scriven, in Micellization, Solubilization, and Microemulsions, Vol. 2 (K. L. Mittal, ed.), Plenum Press, New York, p. 855 (1977).

104. F. J. Trogus and R. S. Schechter, J. Colloid Interface Sci., 70:293 (1979).

105. J. H. Harwell, F. G. Helfferich, and R. S. Schechter, AIChE J., 28:448 (1982).

106. K. J. Mysels and R. J. Otter, J. Colloid Sci., 16:474 (1961).

107. H. Lange and K. Beck, Kolloid Z. Z. Polym., 251:424 (1973).

108. D. N. Rubingh, in Solution Chemistry of Surfactants, Vol. 1 (K. L. Mittal, ed.), Plenum Press, New York, p. 337 (1979).

109. M. J. Schick, Nonionic Surfactants, Marcel Dekker, New York, p. 45 (1966).

110. I. W. Osborne-Lee, W. H. Wade, and R. S. Schechter, J. Colloid Interface Sci., 94:179 (1983).

111. J. F. Scamehorn, R. S. Schechter, and W. H. Wade, J. Disp. Sci., 3:261 (1982).

112. N. Funasaki and S. Hada, J. Phys. Chem., 83:2471 (1979).

113. P. M. Holland and D. N. Rubingh, J. Phys. Chem., 87:1984 (1983).

114. P. M. Holland, in Structure/Performance Relations in Surfactants, ACS Symposium Series No. 253 (M. J. Rosen, ed.), American Chemical Society, Washington, D.C., p. 141 (1984).

115. I. W. Osborne-Lee and R. S. Schechter, J. Colloid Interface Sci., 108:60 (1985).

116. C. P. Kurzendorfer, M. J. Schwuger, and H. Lange, Ber. Bunsenges. Phys. Chem., 82:962 (1978).

117. A. Graciaa, J. Lachaise, M. Bourrel, I. W. Osborne-Lee, R. S. Schechter, and W. H. Wade, SPE Reservoir Eng., 2:305 (1987).

118. P. A. Winsor, Solvent Properties of Amphiphilic Compounds, Butterworth, London, p. 31 (1954).

119. M. E. L. McBain and E. Hutchinson, Solubilization, Academic Press, New York (1955).

120. M. Hato and K. Shinoda, J. Phys. Chem., 77:378 (1973).

121. N. Nishikido, H. Akisada, and R. Matuura, Mem. Fac. Sci. Kyushu Univ., C10:91 (1977).

122. E. J. Staples and G. J. T. Tiddy, J. Chem. Soc. Faraday Trans. 1, 74:2530 (1978).

123. C. Tanford, Y. Nozaki, and M. F. Rohde, J. Phys. Chem., 81:1555 (1977).

124. G. Karlstrom, J. Phys. Chem., 89:4962 (1985).

125. M. Zulauf and J. P. Rosenbusch, J. Phys. Chem., 87:856 (1983).

126. L. I. Osipow, Surface Chemistry: Theory and Industrial Applications, R. E. Krieger, Melbourne, Fla., p. 223 (1977).

127. H. Schott, J. Colloid Interface Sci., 43:150 (1973).

128. H. Schott and S. K. Han, J. Pharm. Sci., 64:658 (1975).

129. H. Schott and S. K. Han, J. Pharm. Sci., 65:975 (1976).

130. J. M. Corkill and J. F. Goodman, Adv. Colloid Interface Sci., 2:279 (1969).

131. R. G. Laughlin, in Surfactants (T. F. Tadros, ed.), Academic Press, Orlando, Fla., p. 53 (1984).

132. J. W. McBain and W. W. Lee, Oil Soap Chicago, 20:17 (1943).

133. P. Ekwall, in Advances in Liquid Crystals (G. H. Brown, ed.), Academic Press, New York, p. 1 (1975).

134. A. E. Skoulios, Adv. Colloid Interface Sci., 1:79 (1967).

135. J. S. Clunie, J. F. Goodman, and P. C. Symons, Trans. Faraday Soc., 63:2839 (1967).

136. F. Husson, H. Mustacchi, and V. Luzzati, Acta Crystallogr., 13:668 (1960).

3

Nonpolar Solutions Containing Amphiphiles

I. REVERSE MICELLES

The primary mechanism responsible for micelle formation in aqueous solutions is the hydrophobic effect (see Section 2.V). For typical amphiphiles, the micelles that persist are those composed of a relatively large number of amphiphiles. As shown by Fig. 2.3, small aggregates (trimers, tetramers, etc.) do not exist in perceptible concentrations. The distribution of aggregates represented by Fig. 2.3 arises because a minimum number of amphiphilic compounds must act collectively to shield the lipophiles effectively from water. Smaller aggregates such as tetramers cannot be arranged so as to achieve a minimum area of contact between the lipophile and water. Thus larger aggregation numbers are energetically favored as compared to micelles containing few molecules. This explains the distribution shown by Fig. 2.3.

Amphiphilic molecules in nonpolar solvents do not experience a hydrophobic effect. Singleterry [1] has noted that there is a substantial decrease in the free energy when polar hydrophiles (nonionic or ionic molecules that are ion paired) aggregate and has listed a number of surfactants that form micelles in nonpolar solvents. The lipophiles remain dispersed among the hydrocarbon molecules, while the hydrophiles interact collectively.

Thus in nonpolar solvents the environment of the lipophile prior to and after aggregation is not changed in any substantial way. The force responsible for the association of amphiphiles is directly attributable to the interaction between hydrophiles, and since it is generally sterically difficult to arrange a large number of amphiphilic compounds so that their hydrophiles are in close proximity one with

the other, one characteristic feature of the micelles formed in nonpolar solvents is the small aggregation number. Micelle formation is
a balance of the free energy of hydrophile association against the
tendency for the amphiphiles to be molecularly dispersed throughout
the solvent (entropy of mixing). These mechanisms clearly do not
support large aggregates such as those often found in aqueous solutions (see Table 2.3).

Aggregation numbers for representative amphiphiles in nonpolar
solvents are shown in Table 3.1 They seldom range above 20 and
are often much less. Thus it can be imagined that because of the
small aggregation numbers relative to aqueous micelles, the associated
dimers, trimers, and so on, exist in measurable quantities and begin
to form at quite small solute concentrations. Because of these gradual trends, sharp changes in a property/concentration curve may not
be observed and a critical micelle concentration difficult to assign.
Figure 3.1 shows the absorbance of dodecylammonium benzoate in
cyclohexane at three different wavelengths [5]. A slight break is
seen and a CMC of approximately 0.01 M can be assigned; however,
because the variation in the property/concentration curve is not
marked, this value lacks real precision. Difficulties in assigning a
CMC have been discussed [6,7]. Solubilization methods for CMC
determination have been widely used particularly in early work, but
these are open to some criticism [7]. One method that appears to be
promising is that of positron annihilation. This method is based on
the observation that the reactions, as well as the formation of the
positronium atom (the bound state of an electron and a positron) are
greatly dependent on the environment in which these reactions occur.
Using this technique, abrupt CMCs were noted for Aerosol OT and
dodecylammonium propionate in benzene and cyclohexane [8,9] at concentrations that agreed well with "CMC" values determined by light
scattering [10], dye adsorption techniques [11], dielectric increment
measurements [12], and H_{NMR} [13,14].

Ionic amphiphiles are expected to exist in nonpolar solvents as
ion pairs. The counterions are strongly associated with the amphiphile, essentially creating a neutral but highly polar molecule. The
ion pairs are believed to exist molecularly dispersed in equilibrium
with associated species. There appears to be two separate and distinct modes of aggregation [15]. In one class populated primarily
by cationic surfactants, the aggregation numbers are quite small
($n \approx$ 3 to 6), there is no real identifiable CMC, and n increases
continuously with increasing solute concentration.

The second class, consisting primarily of anionic surfactants,
exhibits larger values of $\bar{n}$ (25 to 30), a CMC is more or less apparent, and $\bar{n}$ approaches a plateau value as the solute concentration
is increased. The difference between the two classes according to

TABLE 3.1 Aggregation Numbers in Hydrocarbon Solvents

Amphiphile	Solvent	Temperature (°C)	Solute concentration (M)	Aggregation number, $\bar{n}$	Reference
Sodium dinonylnaphthalene-sulfonate	Benzene	25	10^{-5} to 10^{-3}	12	2
Barium dinonynaphthalene-sulfonate	Benzene	25	10^{-5} to 10^{-3}	7	2
Sodium di-2-ethylhexyl-sulfosuccinate (Aerosol OT)	Benzene	28	2.7×10^{-3}	24	3
	Benzene	25	4×10^{-4}[a]	13	4
	Cyclohexane	28	1.3×10^{-3}	56	3

[a]Concentration equals the CMC.

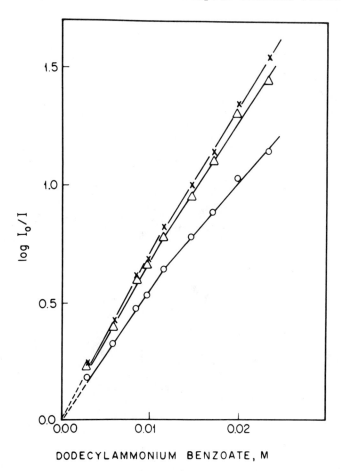

FIGURE 3.1 Absorbance of dodecylammonium benzoate solutions in cyclohexane as a function of concentration at 20°C 270 nm (X), 277 nm (O), and 363 nm (△). (After Ref. 5.)

Muller [15] relates to the ionic radius of the hydrophile and to the dielectric constant of the solvent. His model uses an approach similar to that applied to the formation of lattices by gaseous alkali halides and predicts a stepwise formation of open-chain oligomers with approximately equal equilibrium constants for the binding of additional monomers if the sum of the radii of the ionic headgroups of one surfactant is large or if the dielectric constant of the solvent is large. If, on the other hand, the dielectric constant and the sum of the radii are small, compact clusters (small aggregation numbers) are preferred and the association is more analogous to micellization in aqueous solutions. Although this categorization correctly predicts the association patterns of dodecylammonium propionate and Aerosol OT in benzene, it was not found to apply to solutions of alkylammonium carboxylates in cyclohexane, all of which exhibit an apparent CMC as detected by positron annihilation [16].

The aggregation process is known to be sensitive to the presence of small quantities of water or other impurities which are difficult to remove and then exclude from the system [17,18]. Thus the determination of a "true" CMC in a binary system excluding water is difficult. Indeed, IR and NMR studies of lithium and cesium salts of dinonylnaphthalene sulfonic acids in heptane [19] and of alkali and alkylammonium polystyrene sulfonates [20] strongly suggest hydrogen bonding to be the predominant, if not decisive mechanism causing micellization in nonpolar media. In this case a part of the difference between the two classes of behavior in nonpolar media can be attributed to the presence of water as an impurity.

Irrespective of the primary mechanism responsible for the association of amphiphiles in nonpolar solvents, it is clear that the structure of the aggregates is such that the lipophiles are mixed with the solvent while the hydrophiles are in some way associated together. This structure is designated as a reverse or inverted micelle. These comprise the S_2 system represented in Fig. 1.2.

II. FACTORS DETERMINING AGGREGATION

A. Thermodynamic Considerations

The association of n monomeric amphiphiles to form an aggregate can be represented as follows:

$$nA_1 = A_n \qquad\qquad\qquad (3.1)$$

This is the same equation as that written for the formation of micelles in aqueous solutions (see Section 2.VI.A) as there is no difference in principle. In this case, however, the standard free energy of aggregate formation will not include the same contributions— hydrophobic, surface, and electrostatic—as given by Eq. (2.38). Here the primary contributions are less precisely known, but they have been considered by Ruckenstein and Nagarajan [21], who included head interactions, the loss of translational and rotational degrees of freedom, and hydrogen or metal coordination bonding. We will not give these individual contributions here. The standard free energy of formation, ΔG_{mic}, will simply be represented in terms of an equilibrium constant. Thus the chemical reaction represented by Eq. (3.1) yields the equilibrium condition

$$\beta_n = \exp\left(-\frac{\Delta G_{mic}}{kT}\right) = \frac{C_n}{C_1^n} \qquad (3.2)$$

where β_n is an equilibrium constant and C_1 and C_n are the concentrations of the unassociated amphiphile and aggregates consisting of n amphiphiles, respectively. Again the solutions are presumed to be ideal. There exists an equilibrium constant β_n to be associated with each reaction forming aggregates with n amphiphiles.

It is also noted that the total concentration of amphiphile is given by

$$C_T = \sum n\beta_n C_1^n \qquad (3.3)$$

and the average aggregation number defined as

$$\bar{n} = \frac{\Sigma\, n\beta_n C_1^n}{\Sigma\, \beta_n C_1^n} = \frac{C_T}{\Sigma\, \beta_n C_1^n} \qquad (3.4)$$

The determination of the β_n is difficult, and most often two or three predominant oligomers are identified and the stability constants associated with the formation of these oligomers determined from experimental data assuming that the others are small and do not contribute [7]. For a particular system in which only three oligomers containing $p < q < r$ monomers are formed, Eq. (3.3) becomes

$$C_T = C_1 + p\beta_p C_1^p + q\beta_q C_1^q + r\beta_r C_1^r \qquad (3.5)$$

For example, tetra-n-heptylammonium chloride in benzene at 50°C is reported to exhibit two primary oligomers, characterized by the equilibrium constants $\beta_3 = 9.12 \times 10^4$ and $\beta_{18} = 1.023 \times 10^{41}$ [7]. Because one of the oligomers is an association complex consisting of 18 monomers, this system tends to exhibit a plateau value for $\bar{n}$ and a rather sharp change in the slopes of some property/concentration curves. This system falls into the second class as defined by Muller [15] and as described above.

Figure 3.2 shows the fraction of the total surfactant inventory present as monomers, trimers, and octadecomers. The concentration of octadecomers rises rapidly and over a narrow concentration range becomes the dominant species. The trends shown in Fig. 3.2 should be compared with those shown in Fig. 3.3. For this system it is

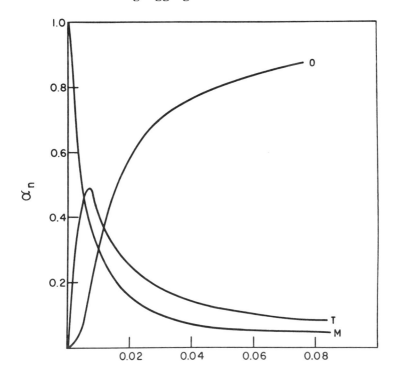

TETRAHEPTYLAMMONIUM CHLORIDE, M

FIGURE 3.2   The degree of formation of monomers (M), trimers (T) and octadecomers (O) of tetra-n-heptylammonium chloride as a function of the total amphiphile concentration in benzene at 50°C. (After Ref. 7.)

seen that dimers and hexamers are dominant associated species. Initially dimers form as the surfactant concentration increases and then hexamers begin to form slowly as the dimer concentration is gradually decreased.   For this system $\bar{n}$ will increase over a wide range of concentrations and will not appear to achieve a plateau value.   Fig. 3.3 therefore illustrates the type of behavior expected for the first class of systems.

The equilibrium constants β_n become difficult to determine as the number of predominant oligomers increases.   The errors in even very precise vapor pressure or other data are amplified, making the uncertainties in the values of β_n large [7].   Fendler [18] has noted that in some cases a better approach takes all of the β_n to be equal. This leaves but one adjustable parameter.   Although it seems quite unlikely that the free energies are independent of the aggregation

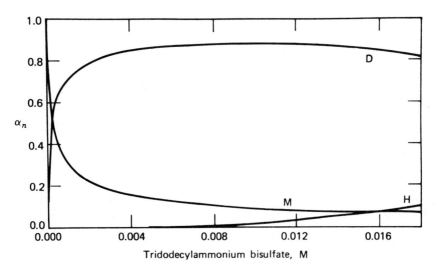

FIGURE 3.3 The degree of formation of monomers (M), dimers (D) and hexamers (H) of tri-n-dodecylammonium bisulfate as a function of the total amphiphile concentration in benzene at 25°C. (After Ref. 7.)

number especially for small n, Fendler has, using this approach, obtained a better fit of the mean aggregation numbers as a function of surfactant concentration for some systems than is possible based on Eq. (3.5).

B. Factors Influencing the Standard Free Energy Formation

1. The Effect of the Amphiphile

The factors that contribute to the β_n have not been quantified, but some qualitative trends are known. The primary factor contributing to the aggregation process is apparently the polarity of the amphiphile in the nonpolar solvent. It seems reasonable to suppose that as the dipole moment of the surfactant increases, the tendency to aggregate should increase. Extensive listings of the dipole moments have been reported [22,23]. In comparing the dipole moments of copper, nickel, and lead oleates, the values increase with increasing cation size, but whether the aggregation number increases or not is not certain based on existing data. Part of the difficulty stems from the fact that molecules with large dipole moment may also tend to polarize the solvent [see Debye intermolecular energies, Eq. (1.3)] and give rise to solvation interactions that do not favor aggregation.

The amphiphilic compounds, which have been studied extensively, are often assymmetric, and steric factors also play an important role in determining the degree of aggregation. Thus it is difficult to define the full impact of increasing the dipole moment. Figure 3.4 shows the average aggregation number of di-2-ethylhexylsulfosuccinate (Aerosol OT) in isooctane [17]. The interesting point is that the aggregation number does increase with the size of the cation, thereby indicating the expected trend. Furthermore, by replacing NA^+ with NH_4^+ it is seen that a plateau value for $\bar{n}$ is no longer found, indicating the complexity of theoretically assigning free-energy values to β_n.

A general trend seems to be that increasing the chain length of the lipophile tends to decrease the aggregation number [24–27], all other factors being fixed. This trend is expected to the extent that

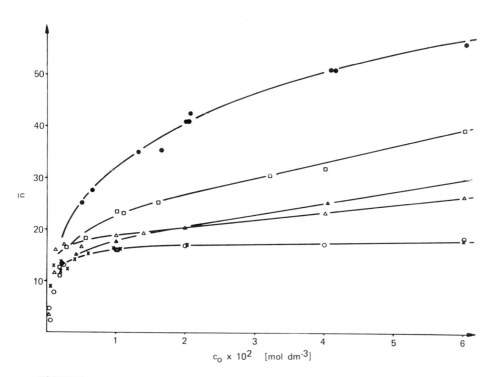

FIGURE 3.4   Average aggregation number ($\bar{n}$) of Li(x); Na(o); K(△); NH_4(▲); Rb(□); Cs(●) di-2-ethylhexylsulfosuccinate (AOT) in isooctane at 25°C based on vapor pressure osmometry. (After Ref. 17.)

increasing the lipophile increases the tendency of the lipophile to dissolve into the solvent. However, for very long lipophiles, the cohesive energy between the lipophiles, $A_{\ell\ell}$, may oppose the mixing of the lipophile with the solvent and for these long chains, further increasing lipophile lengths should tend to increase, rather than decrease, the aggregation number. There do not appear to be any reported data showing this trend, although in the series of sodium dialkylnaphthalenesulfonates, the influence of the lipophilic chain length is minimal [28].

2. The Effect of the Solvent

One important factor governing the formation of inverse micelles is the structure of the solvent. Perri [29] has found the aggregation number of Aerosol OT micelles to decrease with increasing molar volume of the solvent. He further concluded that the micelles were not spherical. Effectively, the influence of the molar volume of the solvent can be anticipated. As noted above, any change that increases the tendency of the lipophilic chains to "dissolve" reduces the aggregation number.

The theory developed by Ruckenstein and Nagarajan [21] supports the trend established by Perri. They note, however, that molar volume is not the only property of the solvent that determines the degree of aggregation. Since the driving force promoting amphiphile association is often a dipole-dipole interaction, it is clear that the dielectric constant of the solvent is also a factor. The tendency to form reverse micelles generally decreases with increasing solvent polarity; however, it is not possible to simply correlate the degree of aggregation with the solvent dielectric constant. For example, although cyclohexane, carbon tetrachloride, and benzene all have about the same dielectric constant, an amphiphile will generally not exhibit similar degrees of aggregation in the three solvents. It appears that a part of the difference in solvency can be attributed to the strong solvation of the amphiphile ion pair by the π electrons of the benzene [30, 31]. A study reported by Little and Singleterry [32] concluded that the aggregation number of alkali dinonylnaphthalenesulfonates in different solvents correlated with the solubility parameter of the solvent. Thus, as shown by Fig. 3.5, disparate results for different solvents having the same dielectric constants can be understood in terms of their solubility parameter. As the solubility parameter of the solvent is increased, the micelles tend to assume a smaller size. The interpretation of this result is simply that the larger solubility parameters tend to favor the solubility of the lipophile in the solvent since the solubility parameter of the lipophile generally has a very large effective value. Therefore, an increase in the solubility parameter of the solvent will better match that of the amphiphile and favor the solubility of the amphiphile in the solvent, thereby reducing the aggregation number.

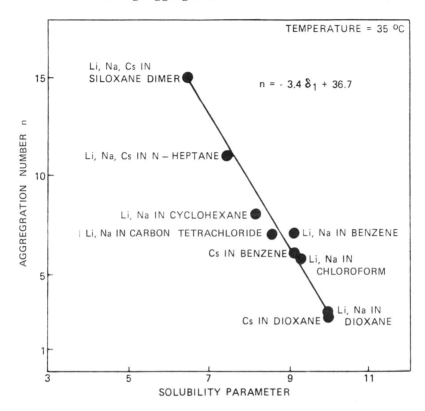

FIGURE 3.5   Aggregation number of alkali-dinonylnaphthalene sulfonates shown as a function of the solubility parameter of the solvent.
(After Ref. 32.)

Highly polarizable solvents such as carbon tetrachloride can solvate the amphiphilic compounds to some extent and lead to reduced
aggregation numbers [33].   This emphasizes the need to consider
mechanisms leading to specific interactions between the solvent and
the amphiphile.   The issue of solvent effects on the aggregation
number is a complex one and is not fully resolved.   For example,
even though the correlation between the solvent solubility parameter
and the aggregation number shown by Fig. 3.5 appears to be a reasonable one, little is known about the way this curve will shift for
different amphiphiles.

3.   Temperature Dependence

It seems clear from a consideration of the mechanism promoting the
aggregation of amphiphiles in nonpolar solvents that an increase in

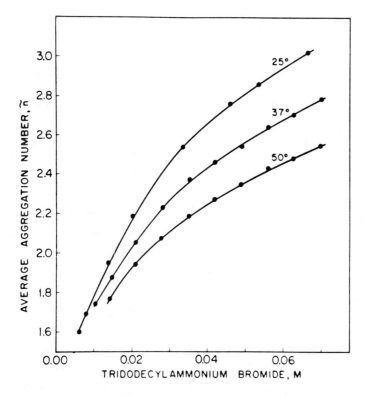

FIGURE 3.6   Average aggregation number of tri-n-dodecylammonium bromide in benzene as a function of concentration for various temperatures.   (After Ref. 7.)

temperature should decrease the aggregation number.   This is the trend generally observed, but not in all cases [34].   Figure 3.6 shows one example, but many others have been represented [35,36]. It has been suggested that cationic surfactants show a more pronounced temperature-dependent aggregation number [7] than do the anionic amphiphiles.   This is perhaps related to the tendency for cationics to form smaller aggregates, but the origin of the difference has not been definitely established.

    Nonionic surfactants exhibit behavior similar to that of cationic surfactants but an even more pronounced temperature-dependent average aggregation number.   They also exhibit a wide dispersity of aggregate sizes [17].

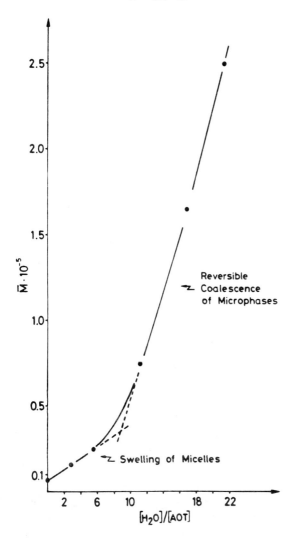

FIGURE 3.7 Mean weight average of sodium bis-(2-ethylhexyl) sulfosuccinate (AOT) as a function of the ratio of water to AOT. The temperature is 25°C. (After Ref. 38.)

C. Solubility Behavior

The rough generalization that inorganic and organic salts freely soluble in water are only sparingly soluble in organic solvents, and vice versa, also applies to amphiphilic compounds. Thus to enhance their solubility in apolar media, many of the amphiphiles discussed in this chapter have substantially greater lipophilic molecular weights than those discussed in Chap. 2 in connection with aqueous solutions. Fendler and Fendler [37] noted that if both ions are large molecules with high solubilities in apolar media, micelle formation is not expected. Conversely, aggregation is expected if one of the ions is relatively insoluble in organic solvents.

The addition of water has a dramatic effect on the solubility of amphiphiles and their aggregation in apolar solvents [38]. The initial added water serves to hydrate the ions, and this water is considered bound. Water added subsequently can be visualized as forming a pool within a spherical inverse micelle [18]. Figure 3.7 shows the weight-average molecular weight of Aerosol OT determined by light-scattering and ultracentrifugal measurements plotted against small aliquots of water at constant temperature. The two regions shown represent the two types of water.

The water added initially serves primarily to hydrate the ions present in solvent, resulting in a slight increase in aggregation number. This is the bound water. Additional water results in a dramatic increase in the aggregation number, which can only be understood by a reduction of the number of micelles and an increase in the radius of the water core [38]. These large aggregates are S_2 micellar solutions.

Finally, we note that the presence of water confers considerable oil solubility to amphiphiles, which are also quite water soluble. Thus the same amphiphiles are soluble in solutions which are essentially water containing little oil ranging to solutions which are essentially oil containing little water. It is these solutions which are the primary focus of this book. One can ask, for example, the following questions. What is the structure of those solutions containing both oil and water? What are the physical properties of these solutions? Are they of practical importance? How do these solutions differ as the molecular structure of the amphiphile is changed? These are the questions that will occupy our attention throughout the remainder of the book.

REFERENCES

1. C. R. Singleterry, J. Am. Oil Chem. Soc., 32:446 (1955).
2. S. Kaufman and C. R. Singleterry, J. Colloid Sci., 10:139 (1955).

3. A. Kitahara, T. Kobayashi, and T. Tachibana, J. Phys. Chem., 66:363 (1962).
4. M. Ueno and H. Kishimoto, Bull. Chem. Soc. Jpn., 50:1631 (1977).
5. A. Kitahara, Bull. Chem. Soc. Jpn., 30:586 (1956).
6. H. Gutmann and A. S. Kertes, J. Colloid Interface Sci., 51:406 (1973).
7. A. S. Kertes and H. Gutmann, in Surface and Colloid Science, Vol. 8 (E. Matijevic, ed.), Wiley, New York p. 194 (1976).
8. Y. C. Jean and H. J. Ache, J. Am. Chem. Soc., 100:6320 (1978).
9. Y. C. Jean and H. J. Ache, J. Am. Chem. Soc., 100:984 (1978).
10. K. Kon-No and A. Kitahara, Kogyo Kagaku Zasshi, 68:205B (1965).
11. S. Muto and K. Meguro, Bull. Chem. Soc. Jpn., 46:1316 (1973).
12. H. F. Eicke and H. Christen, Helv. Chim. Acta., 61:2258 (1978).
13. J. H. Fendler, E. J. Fendler, R. T. Medary, and O. A. El Seoud, J. Chem. Soc. Faraday Trans. 2, 69:620 (1973).
14. J. H. Fendler, E. J. Fendler, R. T. Medary, and O. A. El Seoud, J. Phys. Chem., 77:1432 (1973).
15. N. Muller, J. Colloid Interface Sci., 63:383 (1978).
16. L. A. Fucugauchi, B. Djermouni, E. D. Handel, and H. J. Ache, J. Am. Chem. Soc., 101:284 (1979).
17. H. F. Eicke, in Topics in Current Chemistry: Micelles, Vol. 87, Springer-Verlag, New York, p.91 (1980).
18. J. H. Fendler, Membrane Mimetic Chemistry, Wiley, New York, Chap. 3 (1982).
19. V. Soldatov, L. Oedberg, and E. Högfeldt, in Ion Exchange and Membranes, Vol. 2, Gordon and Breach, London, p. 83 (1975).
20. G. Zundel, Hydration and Intermolecular Interaction, Academic Press, New York, (1969).
21. E. Ruckenstein and R. Nagarajan, J. Phys. Chem., 84:1349 (1980).
22. W. A. Strobel and H. C. Eckstrom, J. Chem. Phys., 16:817 (1948).
23. W. Ostwald and R. Riedel, Kolloid Z., 69:185 (1934).
24. A. Kitahara, in Cationic Surfactants, (E. Jungerman, ed.), Marcel Dekker, New York, p. 289 (1970).
25. S. M. Nelson and R. C. Pink, J. Chem. Soc., 1744 (1952).
26. N. Pilpel, Nature London, 204:378 (1964).
27. K. Kon-No and A. Kitahara, J. Colloid Interface Sci., 35:636 (1971).
28. I. J. Heilweil, J. Colloid Sci., 19:105 (1964).
29. J. B. Perri, J. Colloid Interface Sci., 29:6 (1969).
30. H. Christen, H. F. Eicke, and M. Jungen, Helv. Chim. Acta., 56:216 (1973).

31. M. M. Davies, J. Am. Chem. Soc., 84:3625 (1962).
32. R. C. Little and C. R. Singleterry, J. Phys. Chem., 68:3453 (1964).
33. P. Debye and H. Coll, J. Colloid Sci., 17:220 (1962).
34. C. W. Brown, D. Cooper, and J. C. S. Moore, J. Colloid Interface Sci., 32:584 (1970).
35. J. David-Auslaender, H. Gutmann, A. S. Kertes, and M. Zangen, J. Solution Chem., 3:251 (1974).
36. A. Kitahara, Bull. Chem. Soc. Jpn., 31:288 (1958).
37. J. H. Fendler and E. J. Fendler, Catalysis in Micellar and Macromolecular Systems, Academic Press, New York, p. 314 (1975).
38. H. F. Eicke and P. Kivita, in Reverse Micelles (P. L. Luisi and B. E. Straub, eds.), Plenum Press, New York (1984).

4

The Phase Behavior and Properties of Solutions Containing Amphiphiles, Organic Liquids, and Water: Micellar Solutions

I. SIGNIFICANCE OF CONJUGATE PHASES

A. Isotropic Solutions Containing Both Oil and Water

Chapters 2 and 3 are devoted to solutions of amphiphilic compounds in water and in oil, respectively. Those systems for which water is the solvent and only traces or small quantities of hydrocarbon are present correspond to the state denoted as S_1 in Fig. 1.1. The systems so defined are isotropic, transparent micellar solutions. On the other hand, if the solvent is primarily a nonpolar oil and the solution contains traces or small quantities of water, inverted or reverse micelles predominate and according to the classification defined by Fig. 1.1, the state of the system is S_2.

Figure 1.1 also indicates that it is possible for a system to undergo a transformation from S_1 to S_2 through a series of intermediate states each of which is isotropic. In this chapter the properties of these intermediate systems, which contain substantial amounts of both oil and water as well as amphiphilic compounds, are considered. It will be seen that these solutions exhibit some rather remarkable properties which have considerable practical and scientific importance. For example, they govern the efficiency with which oil can be displaced from porous media by micellar solutions [1,2], they are intimately related to the stability of macroemulsions (see Section 4.II.F), and they determine the solvency of the amphiphilic compounds (see Chap. 7). All of these matter, and many others that have been reported [3] are evidently significant.

If the transformation from S_1 to S_2 proceeds continuously in an orderly fashion, one may want to know to what state in the sequence of intermediate states a particular isotropic solution of oil, water, and amphiphile corresponds. At first it might appear that the proportions of oil and water would define the extent to which the transformation has progressed. Thus one is tempted to assert that all isotropic solutions containing equal proportions of oil and water exhibit properties corresponding to each other. This is, however, far too simplistic a view. For example, the electrical conductivity of two different solutions both containing equal proportions of oil and water may differ markedly (see Fig. 4.34). One system may exhibit a conductivity that appears more oil-like than another solution containing identical proportions of oil and water. Thus it is possible to assert that the system, which has a conductivity more nearly like that of oil, has progressed to a state nearer S_2 than the system to which it is compared, even though both systems contain the same amounts of oil and water. Some measure other than the composition of the solution defining the position of states intermediate between S_1 and S_2 is clearly needed. There are several possibilities, including the electrical conductivity, but the one that is most convenient and the one adopted here relates to the phase behavior of the micellar solutions. The number and compositions of the various phases that coexist in equilibrium has proven to be an extremely valuable measure of the relative state of a micellar solution. Its primary value resides in the simplicity of the measurements that are required to characterize the system. Very often there is no need to determine the precise compositions of the equilibrium phases. Ample information may be available based simply on the measurement of the phase volumes or weights. It is, in fact, remarkable that such a simple experiment can be used for characterizing micellar systems and for comparing the efficiency of different amphiphilic compounds for a given application.

When oil, water, and surfactant are blended together and allowed to equilibrate, two or more phases may appear, and in many cases almost the entire inventory of surfactant will reside in one of the phases together with various proportions of oil and water. That phase containing the bulk of the surfactant is often called the micellar phase, although it may, from time to time, also be termed a microemulsion. The phase behavior of interest will involve at least three components, and often more. The phase behavior of three-component systems can, at fixed pressure and temperature, best be represented using a ternary diagram. These diagrams provide a simple perspective of phase behavior that is difficult to capture in any other way. Furthermore, phase boundaries of systems containing more than three components can also often be usefully shown in ternary diagrams. The difficulties of interpreting such a representation are described in a subsequent section. First, ternary systems are considered.

B. Ternary Systems of Amphiphile, Water, and
    Hydrocarbon

1. Degrees of Freedom

The phase rule is generally written as

$$p + f = N + 2 \qquad\qquad (4.1)$$

where p is the number of phases present in the system, N the number of independent constituents, and f the number of possible independent changes of state or degrees of freedom. A system is called
invariant, monovariant, bivariant, and so on, according to whether f
is zero, 1, 2, and so on. In a system composed of three components
and two phases, f is univariant at a fixed temperature and pressure.
This means that the mole or weight fraction of but one component in
one of the phases can be specified but all other compositions in both
phases are fixed. A ternary diagram showing a two-phase region is
depicted by Fig. 4.1. Any system whose overall composition lies
within the two-phase region will exist as two phases whose compositions are represented by the ends of the tie lines. In accordance
with the phase rule, the surfactant concentration in the phase labeled
M can be varied independently over a restricted range. For example,

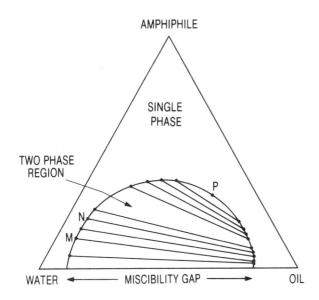

FIGURE 4.1 Ternary diagram representation of two-phase region.
The sloping lines are tie lines connecting conjugate phases.

the proportion of amphiphile can be increased so that the composition of one of the phases in the two-phase system moves from M to N; however, the other compositions in both phases are then fixed. Similarly, the proportion of the oil or the water in one of the phases determines the composition of both of the phases that coexist at equilibrium.

Also shown in Fig. 4.1 is a plait point or critical point, which is designated as P. The tie lines in the two-phase region surrounding this point are quite short, indicating that the two conjugate phases have nearly the same composition.

If three phases coexist, the system at constant temperature and pressure is, according to the phase rule, invariant. There then exists a region of the ternary diagram where systems whose overall composition fall within it divide into three phases having compositions which are invariant and represented by the corners of the tie triangle. Since a three-phase region must necessarily be bounded by two-phase regions [see Polatnik and Landau (4)], the boundaries of the three-phase region are tie lines in the adjacent two-phase regions and the region of three-phase invariant compositions must necessarily be triangular in form.

Figure 4.2 represents a ternary diagram showing the invariant triangular region. Any overall composition, such as point M lying within the triangle, called a tie triangle, will divide into three phases having compositions corresponding to the vertices A, B, and C of the triangle. The compositions A, B, and C are invariant in the sense that varying the position M, the overall composition, throughout the tie triangle will result in variations in the amounts of the phases A, B, and C but not in their composition.

The existence of a three-phase invariant region imposes certain restrictions on the topology of the phase diagram. For example, the system depicted by Fig. 4.2 has but one binary miscibility gap. The other two binaries, oil and amphiphile and water and amphiphile, are both miscible in all proportions. Since the three-phase region must be bounded by two-phase regions, two of the three bounding two-phase systems must terminate within the interior of the ternary diagram. The two bounding two-phase systems must therefore exhibit plait points. In Fig. 4.2 these two plait points are denoted as P_1 and P_2. These plait points are critical points because the compositions of the two phases in equilibrium with an overall composition within the two-phase region but near one of the plait points will have very nearly the same composition. Thus there are two critical points, P_1 and P_2, shown in Fig. 4.2.

If all three binary pairs exhibit a miscibility gap, the three bounding two-phase regions may all terminate on the two-phase boundaries. In this case a plait point would not exist. Figure 4.3 shows such a diagram.

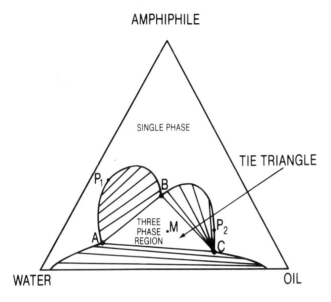

FIGURE 4.2 Ternary diagram showing a three-phase invariant triangular region.

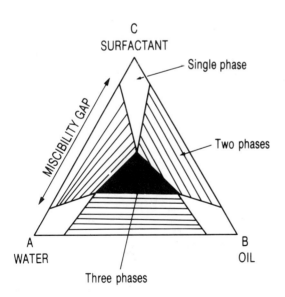

FIGURE 4.3 Three-phase region for ternary system having three binary miscibility gaps.

Figures 4.2 and 4.3 represent idealized phase diagrams which are seldom, if ever, observed when considering solutions of amphiphilic compounds. Figure 4.4 shows a system in which the surfactant is a commercial polyethoxylated nonylphenol [5]. This surfactant contains a mixture of different amphiphiles rather than being composed of a single molecular type [Eq. (2.90) expresses the molecular distribution]. Thus the system is not a true ternary, and can be represented only approximately using a ternary diagram.

The shaded triangle shown in Fig. 4.4 is a tie triangle, and a system having an overall composition represented by a point within this tie triangle will divide into three phases. One of the phases will have very little in it other than water since the vertex of the triangle is very close to the water vertex of the ternary diagram. This conjugate phase is sometimes called an excess water phase because it is in equilibrium with a phase containing substantial quantities of water, oil, and surfactant, which is a micellar solution or a microemulsion. Also in equilibrium with the same micellar solution is an excess oil phase. It is seen from Fig. 4.4 that while this phase contains some surfactant (more than the excess aqueous phase) and water, it is primarily oil and is often called an excess oil phase. Thus a composition within the tie triangle is composed of a micellar solution in equilibrium with excess oil and water phases.

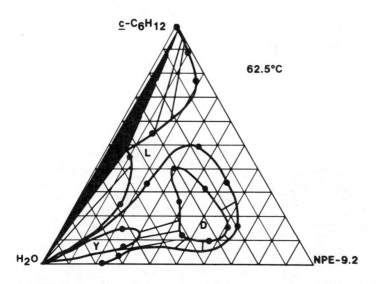

FIGURE 4.4 A phase diagram of the cyclohexane, i $C_9H_{19}C_6H_4$ $(OCH_2CH_2)_{9.2}OH$ (commercial polyethoxylated nonylphenol) and water system. The surfactant is designated as $NPE_{9.2}$. (After Ref. 5.)

The oil phase is often less dense than water or the micellar solution and has sometimes been called the upper phase. The micellar
solution frequently has a density intermediate between that of oil
and water and is called a middle phase since in a test tube it appears sandwiched between the upper excess oil phase and the lower
excess aqueous phase.

The tie triangle shown in Fig. 4.4 is bounded by three two-phase
regions having an appearance much like the idealized diagram (Fig.
4.2). The two-phase regions are in turn bounded by a single-phase
isotropic micellar solution designated L. The micellar solution represented by the surfactant-rich vertex of the tie triangle is intermediate between S_1 and S_2 because it contains substantial quantities
of both oil and water and thus cannot be labeled as either. The
properties of these intermediate isotropic solutions containing substantial proportions of oil and water are the subject of this chapter.

The features that distinguish this real system from idealized
systems depicted by Fig. 4.2 are the liquid-crystalline phases D and
Y, which appear at somewhat higher amphiphile concentrations. The
D phase is a lamellar phase, designated as G in Table 1.1 The phase
Y is composed of rods arranged hexagonally and corresponds to M_1
described in Table 1.1.

The liquid-crystalline phases generally appear at higher concentrations of amphiphile, although this need not be the case. The systems of greatest interest in this book are, therefore, primarily those
which contain small concentrations of amphiphile, since these represent in general the isotropic phases and are often of the greatest
applicability.

2.   The Transformation from S_1 to S_2

It will be seen that three-phase systems play an important role in
the characterization of micellar solutions, and therefore it is interesting to consider a sequence of phase diagrams which are characteristic of the transformation between S_1 and S_2 states. To help visualize this transformation, consider the series of partial phase diagrams
shown in Fig. 4.5. The chemical system is the one studied in Fig.
4.4, and in fact the curve in Fig. 4.5 for a temperature of 62.5°C
corresponds to the boundary separating the single-phase region L
from the two-phase regions shown in Fig. 4.4. The remaining details of the phase diagram have been omitted so that the phase
boundaries separating the two-phase region from the single-phase
region for several temperatures can be shown on a single ternary
diagram. The sharp point of intersection or point of the cusp represents that vertex corresponding to the composition of the micellar
solution which is in equilibrium with the excess water and oil phases.
Thus Fig. 4.5 shows the change in the composition of the micellar
solution as the temperature is increased. This is an important

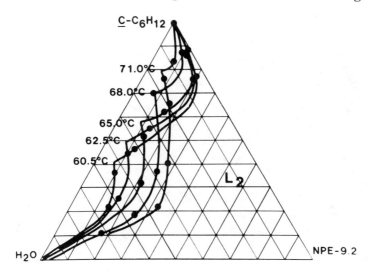

FIGURE 4.5   Isothermal cross sections of the aqueous-micellar solution and oleic-micellar solution phase boundaries. The system is the same as that shown by Fig. 4.4.   (After Ref. 5.)

diagram, for it suggests conditions by which the transformation from S_1 to S_2 may be carried out.   The composition of the micellar phase at the cusp is seen to evolve continuously from a point midway between the water and the cyclohexane vertices at 62.5°C to a point near the cyclohexane vertex at 71°C.   The isotropic system near the cyclohexane vertex contains amphiphile and cyclohexane but only a small quantity of water.   Clearly, this micellar system must be composed essentially of inverted micelles and therefore represents a S_2 state.   Thus the composition of the conjugate micellar solution is continuously transformed from a state that is intermediate between S_1 and S_2 to one that is S_2 by continuously increasing the temperature.

It is shown in Chap. 5 that there are a number of formulation variables other than the temperature which can be used to alter the composition of the micellar system so that it is transformed from one isotropic state to another.   Thus the trends depicted in Fig. 4.5 will be seen to be quite general.   By systematically changing one of the formulation variables, a micellar solution can be transformed from S_1 to S_2, and for a ternary system the intermediate states are characterized by the existence of a three-phase invariant region (tie triangle) for which the composition of the vertex corresponding to the micellar solution varies from the water-rich side of the triangular diagram to the oil-rich side.   The intermediate states are therefore characterized by the composition of the micellar solution, which is in equilibrium with excess aqueous and oleic phases. The existence of the conjugate phases in equilibrium with the micellar solution is an important restriction, since for this case the composition of the micellar solution is an invariant. Thus the surfactant-rich vertex of the tie triangle represents a unique point which serves to characterize the system. Micellar solutions containing about half water and half hydrocarbon appear at 65°C. The formulation at this particular temperature may be thought to be exactly intermediate between S_1 and S_2 states.

Figure 4.5 shows the evolution of the vertex representing the micellar solution conjugate to an excess oil phase and an excess water phase as the temperature of the system is raised. It is seen that the composition of the micellar solution becomes progressively richer in oil and correspondingly depleted of water. It is natural therefore to assert that the system approaches S_2 progressively with increasing temperature.

The existence of a three-phase invariant region is therefore anticipated to be one means whereby phase behavior can be used to characterize the position of a micellar system relative to the states S_1 and S_2. It is therefore interesting to inquire as to the origin of the three-phase region.

3. Origin of the Three-Phase Region

It is likely that a typical S_1 system will exhibit a phase diagram somewhat like Fig. 4.6. The two-phase region is shown to have a plait point positioned well toward the hydrocarbon side of the phase diagram. Thus for reasonable surfactant concentrations, a composition under the demixing curve will divide into a micellar solution in equilibrium with an excess oil phase. The oil phase will contain very little amphiphile and small quantities of water, whereas the aqueous phase will contain almost all of the amphiphile and some solubilized hydrocarbon. Such a system is designated as Winsor type I.

Compare Fig. 4.6 with Fig. 4.7, which is representative of S_2 systems and is designated as depicting type II behavior. This designation is used throughout the remainder of this book.

In Fig. 4.7 the tie lines slope in a different direction than they do in Fig. 4.6. This difference reflects a difference in the affinity of the amphiphile for water relative to oil. Under the conditions for Fig. 4.7, the amphiphile tends to partition primarily into oil for most overall compositions lying within the two-phase region. It is this preference for oil relative to water that confers the type II character on the system. Most overall compositions for the system depicted in Fig. 4.6 are S_1, whereas those of Fig. 4.7 are S_2. We might therefore ask how one phase diagram can be transformed systematically and continuously to the other without increasing the number of com-

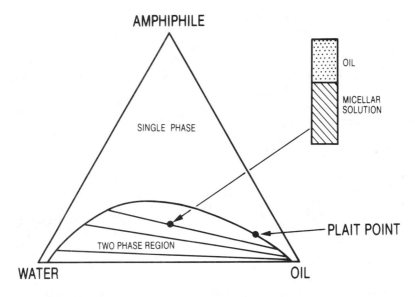

FIGURE 4.6   A typical S_1 system, also called Winsor type I system.

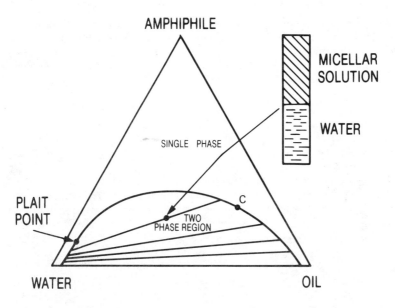

FIGURE 4.7   A ternary representation of type II phase behavior.

ponents. It will be seen in Chap. 5 that such a transformation can
be achieved by progressively changing the temperature, the surfactant lipophile, the surfactant hydrophile, or the structure of the
oil in a homologous series.

Evolution from type I to type II systems can occur through
an intermediate sequence of three-phase systems. These can be designated as type III. A type I phase diagram such as the one shown
by Fig. 4.6 can become a three-phase system at a critical tie line
which under slightly altered conditions broadens into a tie triangle
for which two micellar solutions, both water-rich, are in equilibrium
with the hydrocarbon phase. A type III system exhibiting a tie triangle having one very short leg is shown in Fig. 4.8. Of course
two vertices so closely spaced represent conjugate phases having
virtually the same composition. These two phases are therefore near
their critical composition, which exists at the point where the tie
line initially opens into a triangle. Thus the transition from type I
to type III phase behavior takes place at a critical tie line, and therefore all of the recent developments related to critical phenomena (see
Chap. 8) should apply to micellar solutions near the I–III transition.
Critical opalescence, ultralow tensions, and large-scale concentration
fluctuations are all observed near this transition. Similarly, as shown
in Fig. 4.8, type III behavior terminates and type II can initiate at
a critical tie line [6]. The sequence of states shown in Fig. 4.8 is
typical.

The importance of the particular micellar system containing equal
volumes of hydrocarbon and water and which is simultaneously in
equilibrium with both aqueous and hydrocarbon phases is now clear.
That system can be thought of as being an equal mixture of S_1 and
S_2 solutions. It is also called the optimum system or the optimum
middle phase or surfactant phase. The importance of this intermediate system cannot be overemphasized and it will play an essential
role in all of the subsequent discussion.

The term "middle phase" is often applied to describe the micellar
solution in a type III system. This micellar solution generally has
an intermediate density between oil and water and thus resides in
the middle separating oil on the top and water on the bottom. Figure
4.9 shows a photograph of a series of test tubes each containing equal
volumes of hydrocarbon and water and the same amphiphilic mixture.
The system shown contains more than three components, but the phase
behavior is typical. The only difference in the composition from tube
to tube is the quantity of added electrolyte, which increases from left
to right. In this case the micellar phase can easily be identified by
its slightly turbid or milky appearance. In the tubes on the left,
the micellar phase is the lower phase, and these represent type I
systems. An increase in the salinity causes the micellar phase to appear in the middle in equilibrium with two clear phases. This marks

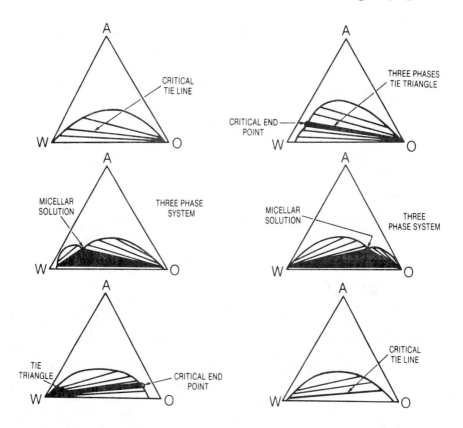

FIGURE 4.8 A sequence of ternary systems which are continuously transformed from type I to type III to type II. This is denoted as a I–III–II transformation. The existence of two critical tie lines is most notable.

the onset of type III behavior. A continued increase in the salinity will result in the micellar solution becoming the upper phase, which as shown by the tubes on the right, is in equilibrium with a clear lower phase.

In this discussion of three-component systems we have noted that the three-phase region can arise at a critical tie line, but this is not the only way that a three-phase region can form. It is also possible for three phases, all having virtually the same composition, to form around a single point in composition space. This single point is called a tri-critical point and represents another possible mechanism for the origination of a three-phase region.

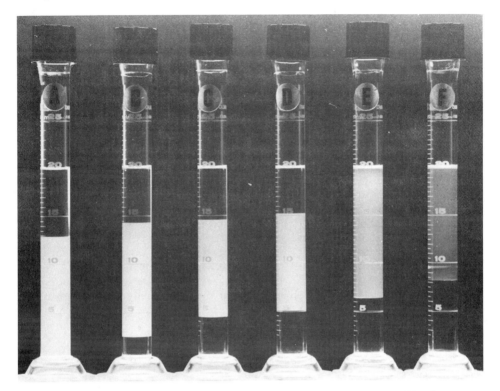

FIGURE 4.9 Phase changes of a micellar solution at low amphiphile
content when NaCl concentration increases: A: 56 g/l; B: 58 g/l;
C; 58.5 g/l; D: 59 g/l; E: 62 g/l; F: 65 g/l. 1.6% sodium dodecylsulfate; 2.4% 1-pentanol; brine/hexane = 1(vol); T = 25°C.

4. Microstructure of the Micellar Solutions

The microstructure of a type I system may be visualized to be swollen
micelles surrounded by water. Figure 2.4 shows various shapes for
micelles, and it is not difficult to imagine the hydrocarbon present in
a type I micellar phase to be dissolved in the interior of the micelles.
Furthermore, if enough hydrocarbon is solubilized, a spherical shape
is the most likely. This is because packing constraints which may
exist when forming micelles such that the lipophile density is liquidlike within the micellar interior need no longer be enforced. The presence of a hydrocarbon core practically ensures that spherical packing
is possible. This being the case, it is noted that dividing a given
total volume into spheres gives the maximum entropy and is therefore
preferred. It must be admitted that the entropic contribution is not
large, so that other factors may arise which will reduce the tendency
for spherical behavior. Nevertheless, experiments discussed in Section 4.III indicate that micellar solutions containing small quantities
of hydrocarbon are essentially spherical.

S_1 or type I systems may therefore be imagined to be oil droplets surrounded by a sheath of amphiphile which separates the oil
core from the continuous aqueous phase. A sketch of type I behavior depicting this microstructure is shown in Fig. 4.10a. It
should be stressed that the oil drops are generally small enough so
that the micellar solution is transparent and is considered to be a
single phase even though substantial regions of oil do exist. Similarly, type II systems may be visualized as swollen inverted micelles, as shown in Figure 4.10b. Again, experiments tend to confirm
the structure shown (see Section 4.III).

It is not possible to draw a similar picture of the microstructure
of a microemulsion containing substantial quantities of both water and
oil. It is known that there are large regions which are water-like,
separated by sheaths of amphiphile from similar large regions which
are oil-like. Scriven [7,8] has proposed that the equilibrium structure
must be bicontinuous; that is, both the aqueous regions and oil regions are continuous and the interface separating them has essentially a constant mean curvature, as is required by equilibrium. It
seems likely that Scriven's view is correct; however, because of
thermal fluctuations, which are always present, large extended regions are unlikely to persist as long as the interfaces remain fluid.
Thus it will be difficult to confirm or reject the Scriven hypothesis.
The issue of structure is pursued in some detail in Section 4.III and
in Chap. 8.

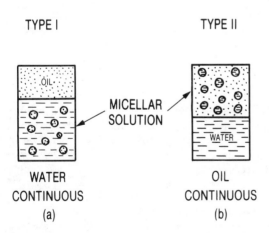

FIGURE 4.10 Sketch showing the likely microstructure of types I
and II systems.

C. Four-Component Systems

Most of the formulations of interest contain more than three components and their complete phase behavior cannot be represented using
a triangular diagram. It is not unusual, for example, to find that a
mixture of amphiphiles is preferred rather than a single compound.
Alcohols are often used, and although amphiphiles, they are sometimes called cosolvents or cosurfactants to emphasize that they have
poor solvency and are normally not applied alone. The oils are often
mixtures of a number of different organic components and the water
may contain electrolytes. For systems containing more than four
components there is no convenient complete method of representation,
but even for four-component systems the diagrams are, as will be
seen, often complex and difficult to visualize or interpret.

The phase behavior of four-component mixtures can be represented at constant temperature and pressure using a tetrahedral
figure which is a three-dimensional simplex. Figure 4.11 shows an
example of such a figure, and embedded within the tetrahedron are
three curves which are intended to illustrate three-phase behavior
[9]. Since any point within the tetrahedral volume represents the
composition of a four-component mixture, a point on one of the
curves shown represents a mixture of cosolvent, amphiphile, water,
and hydrocarbon. There are three curves, labeled $\overline{m_1 m_2}$, $\overline{u_1 u_2}$,
and $\overline{\ell_1 \ell_2}$. These three curves represent the loci of the compositions
of three mutually saturated phases. Thus a point on the curve
$\overline{m_1 m_2}$ represents the composition of a micellar solution in equilibrium
with an excess aqueous phase whose composition is represented by
a corresponding point on the $\overline{\ell_1 \ell_2}$ curve (ℓ denotes lower phase).
Similarly, both of these phases are in equilibrium, with an excess
oil phase having a composition on $\overline{u_1 u_2}$ (u denotes upper phase).
By connecting the three points, a tie triangle positioned within
the tetrahedral volume is then constructed. Just as for the three-
component ternary diagrams, any point on this tie triangle represents an overall composition that will divide into three phases
whose compositions are represented by the vertices of the triangle.

Thus the three-phase region can be thought to be composed of
a sequence of nonintersecting (no common edge or vertex) tie triangles positioned so as to constitute a volumetric region embedded
within the tetrahedral volume. This three-phase volume is bound
by the three curves $\overline{m_1 m_2}$, $\overline{\ell_1 \ell_2}$, and $\overline{u_1 u_2}$. This volume can originate or terminate in only one of the following ways:

1. Intersecting with a triangular face of a tetrahedron (including
    one of the faces of four-phase invariant volumes which may
    exist in a four-component system).

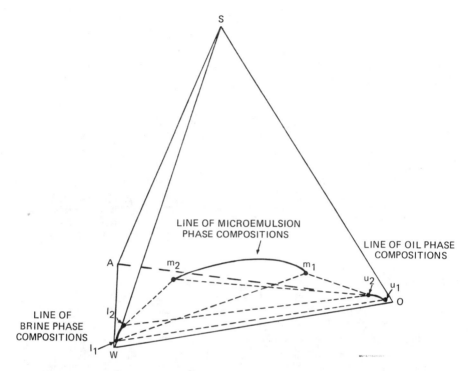

FIGURE 4.11 Sketches showing the locus of three-phase compositions and two tie triangles (dashed lines). (After Ref. 9.)

2. Decreasing to a line (critical tie line)
3. Shrinking to a point (tricritical point)

If one of the three ternaries composing the quaternary system forms a three-phase system, one triangular face of the tetrahedron will include a tie triangle, as shown by Fig. 4.11. The three-phase volume then intersects and terminates at this face. This means that all three curves $\overline{m_1m_2}$, $\overline{\ell_1\ell_2}$, and $\overline{u_1u_2}$ intersect the face of the tetrahedron. If two faces contain three-phase invariant systems, the loci of the three vertices ($\overline{m_1m_2}$, $\overline{\ell_1\ell_2}$, and $\overline{u_1u_2}$) may pass entirely from one surface of one of the faces to the other. The three-phase volume then arises at one surface and terminates at the other.

The three-phase volume can also terminate if the curve $\overline{m_1m_2}$ intersects either $\overline{\ell_1\ell_2}$ or $\overline{u_1u_2}$. When two curves intersect, the tie triangle clearly collapses to a line. This line is a critical tie line since the phases along two different curves take on the same composition at the point of intersection. If none of the ternary systems

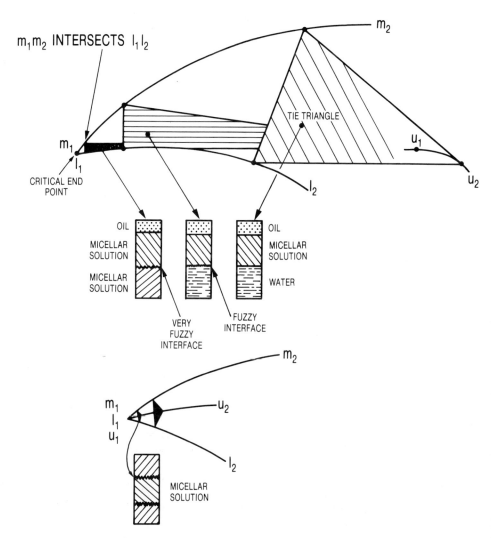

FIGURE 4.12   The origin of the three-phase region is determined by the curves showing loci of three-phase compositions.

represented by the faces of the tetrahedron exhibit three-phase
behavior, it is most likely that the three-phase volume does indeed
originate at a critical tie line and also ceases to exist at another one.
The remote possibility does exist, however, that all three curves
$\overline{m_1m_2}$, $\overline{\ell_1\ell_2}$, and $\overline{u_1u_2}$ converge to a single point. Such a point is
called a tricritical point. Three-phase regions can arise at a tricrit-
ical point.

The three-phase volume, when one exists, requires a large number of experiments to define its extent and the extent and nature of
the surrounding two- and single-phase regions [9-11]. Such a
large experimental program would ordinarily not be carried out except to delineate the phase behavior of systems that have considerable importance. Most often the study of the phase behavior is
prompted by a need to screen or compare various amphiphilic compounds which are candidates for a particular application. At this
point a fully detailed phase diagram delineating not only those regions for which three phases coexist, but also the two- and one-
phase regions as well as the liquid-crystalline phases and their
structure, is not required and the investigator may very well settle
for phase information within a subspace of the full phase diagram.
Often the subspace involves fixing a ratio of the surfactant to alcohol, which is maintained constant in the overall system. For example,
a series of test tubes may be filled with varying proportions of water,
hydrocarbon, and a mixture of amphiphiles where the mixture of
amphiphiles is a well-defined ratio of alcohol to surfactant and the
phase behavior of these mixtures observed. The overall compositions
contained in the series of test tubes can be represented as points
on a plane within the tetrahedral volume. Figure 4.13 shows a plane
having the distinguishing feature that each point on it represents a
composition having the same alcohol/surfactant ratio. Clearly, the
task of defining the phase behavior throughout a plane is markedly
simpler than defining it throughout the tetrahedral volume. This
gain in simplicity is offset to some extent by the disadvantage that
although the phase boundaries can be correctly delineated, the compositions of the conjugate phases cannot be shown since these compositions do not necessarily lie in the plane of investigation.

This difficulty can be illustrated by considering the intersection
of the plane which is being investigated with a three-phase volume.
Such an intersection is shown by Fig. 4.14. The curves denoted
$\overline{m_1m_2}$, $\overline{\ell_1\ell_2}$, and $\overline{u_1u_2}$ represent the loci of the three-phase compositions, so that a tie triangle has one vertex on each of the curves.
One such tie triangle is shown. Now the plane of investigation,
that is, the plane for which the surfactant-to-alcohol concentration
is a constant, is shown to intersect this triangle, and any overall
composition represented by a point on the line of intersection will
divide into three phases whose compositions are at the vertices of

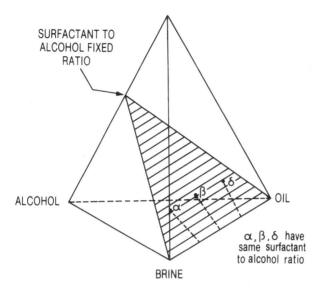

FIGURE 4.13 The plane containing all possible proportions of hydro-carbon to water blended with a definite proportion of alcohol to sur-factant.

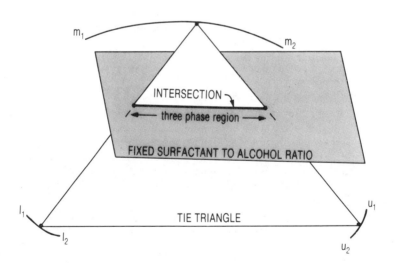

FIGURE 4.14 The intersection of a tie triangle and a plane of con-stant surfactant to alcohol concentration. The line of intersection represents three-phase systems.

145

the triangle. The vertices do not, in general, lie on the plane of investigation. Therefore, even though the overall composition of the system is represented by a point on the plane, the individual phases have different proportions of surfactant to alcohol than the value on the plane, and hence lie off the plane. Despite the fact that the equilibrium phases do not lie on the plane of fixed surfactant-to-alcohol concentration, an investigator would find any overall composition on the line of intersection to divide into three phases and thus will designate the line of intersection as part of a three-phase region. Furthermore, the plane of investigation will intersect other tie triangles and a three-phase region on the plane of investigation is determined. This three-phase region may appear as a lenticular region [10–12]. Figure 4.15 shows an example of such a lenticular region. Also shown are the adjacent two-phase regions in contact with the three-phase region, and the single-phase regions in contact with two-phase regions. The tie lines connecting conjugate phases in the two-phase region will in general not lie in the region of investigation. Thus tie lines are not and cannot be shown in Fig. 4.15.

Finally, it is noted that the point at the confluence of the single-phase, two-phase, and three-phase regions is a vertex of one of the tie triangles and therefore represents the composition of a micellar solution that is in equilibrium with excess aqueous and heptane phases even though the latter two compositions cannot be represented on the pseudoternary diagram. There are two such points shown in Fig. 4.15 (points labeled 1 and 10). These points at the vertices of the lenticular region will be shown to be of special significance in characterizing the micellar solution; however, it is first necessary to consider systems containing larger numbers of components.

D. Multicomponent Systems

Most systems of interest will include a number of components which exceeds four. Indeed, Fig. 4.15 shows a pseudoternary diagram which contains five components since the sodium chloride is an additional component; however, for this system the sodium chloride and the water partition into the various phases in very nearly [but not exactly (11,13,14)] a fixed ratio and can, with good accuracy, be represented as a single component.

As the number of components increases, the representation of the phase behavior becomes more difficult, ultimately requiring a computer model which can present to the experimenter different two- or three-dimensional views of the hypersurfaces separating the various regions of state. Furthermore, as the number of components increases, the number of experiments required to define the complete phase behavior becomes prohibitively large. Thus one must be able to characterize micellar solutions based on less than a complete phase diagram. For the five component system, Figure 4.15 provides an

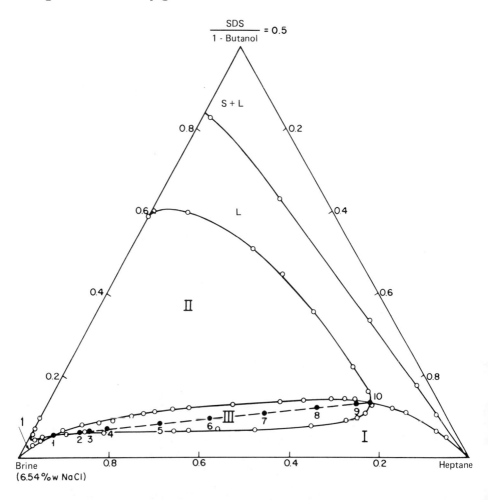

FIGURE 4.15 Pseudoternary phase diagram of the system sodium dodecylsulfate, 1-butanol, water, sodium chloride, and heptane. (After
Ref. 12.)

adequate picture for most purposes. Two of the three vertices of
the ternary diagram represent more than one component. The upper
vertex denotes a mixture of alcohol and surfactant that is thought
of as a single component, and can be called a pseudocomponent to
emphasize that it is not a true component. The water and sodium
chloride mixture is also a pseudocomponent. Ternary diagrams in
which one or more of the three vertices stand for more than one component will be called pseudoternary diagrams. The reader should be

certain to recognize that pseudoternary diagrams have quite different
structures than do ternary diagrams. One prominent difference in
comparing Fig. 4.15 with Fig. 4.8 is that the three-phase region of
a true ternary system must be a tie triangle. Such is not the case
for the three-phase region of a pseudoternary diagram, as shown
by Fig. 4.15. Pseudoternary representation other than the one
shown may be defined. Friberg and Buraczewska [15] used three
pseudocomponents, while Podzimek and Friberg [16] preferred to
use water/ionic surfactant as a pseudocomponent.

E. Characterization of Micellar Solutions

Types I, II, and III systems can be rather precisely defined for
true ternary mixtures consisting of amphiphile, oil, and water. In
this case a type III system can be defined as one for which a three-
phase tie triangle exists anywhere within the ternary diagram. Similarly, type I systems exhibit two-phase but not three-phase behavior,
and in the two-phase region the tie lines are sloped similar to those
shown in Fig. 4.6. In this case there is no three-phase region and
the two-phase region consists of a micellar solution in equilibrium
with an excess oil phase. Similarly, type II systems might, for a
ternary system, be classified as a system for which the multiphase
region is two-phase, generally yielding a micellar solution in equilibrium with an excess aqueous phase.

These definitions, although convenient for ternary systems, become impractical when more than three components are present. Thus
a more convenient but perhaps less precise operational definition is required. It is very useful to adopt the following definitions. A type III
system is one that exhibits three phases. This definition will apply
irrespective of the number of components present in the overall system so that the test is simple to apply. Unfortunately, this classification of a micellar system will depend, to some extent, on the surfactant concentration. This is easily seen by considering the ternary
system shown in Fig. 4.16. The formulations having overall compositions denoted by A, B, C, and D on the ternary diagram are to be
considered. These four formulations are distinguished by an increasing surfactant concentration. The number of phases is seen to vary,
however, and according to the definition, A and B are classified as
type III. Systems C and D are not. Thus even though a three-
phase region does exist and the composition of the micellar solution
at the vertex is invariant, the micellar system may or may not be
classified as type III.

Thus the simple definition offered here and used extensively in
subsequent chapters lacks precision. As shown by Fig. 4.16, it is
somewhat ambiguous. The definition does have the considerable advantage that it is readily applied to multicomponent systems which can
be characterized based on the results of a relatively few experiments

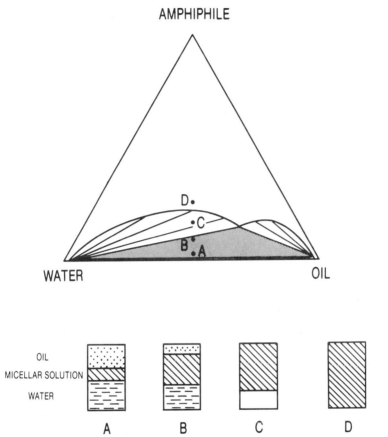

FIGURE 4.16 The number of phases in a system containing equal
proportions of oil and water depends on the overall composition.
Tubes A and B contain three phases, Tube C contains two phases,
and Tube D is composed of a single phase.

and as will be seen is quite useful especially when coupled with a
study which includes the effect of the surfactant concentration.

Following the definition of type III phase behavior, it is then
reasonable to define type I phase behavior as a system that exhibits
two phases, with a surfactant-rich aqueous phase in equilibrium with
an oil phase. System C in Fig. 4.16 exhibits two phases but is not
classified as type I since the micellar phase is in equilibrium with an
aqueous phase containing a small quantity of amphiphile. For type
I systems the micellar solution is in equilibrium with oil.

System C is classified as type II, which is defined as a two-phase system for which an aqueous phase is in equilibrium with a micellar solution.   Finally, system D is characterized as type IV which is defined as a single-phase micellar solution.   These classifications, types I, II, III, and IV, are due to Winsor [17,18].

Figure 4.17 shows a sequence of pseudoternary diagrams each different from the other in the amount of NaCl included in the formulation.   On each of the diagrams is shown a composition designated as A.   Thus A is representative of a sequence of tubes each containing the same quantities of heptane, surfactant, alcohol, and water but varying amounts of NaCl.   In the tube without salt, two

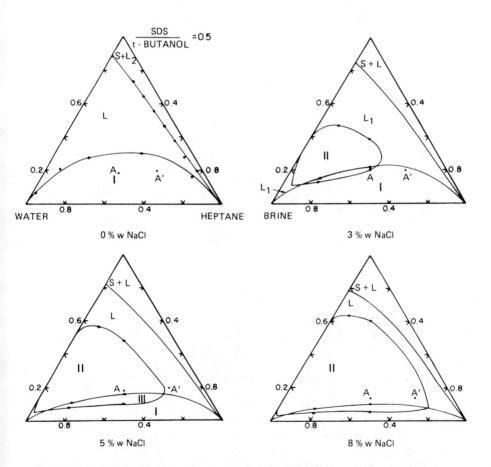

FIGURE 4.17   A series of pseudoternary diagrams shown for increasing sodium chloride concentration.   The surfactant (sodium dodecylsulfate) to n-butanol concentration is maintained constant.   (After Ref. 12.)

phases will be observed having an appearance quite similar to that seen in the tubes on the left of Fig. 4.9, in which a micellar solution is in equilibrium with an excess oil phase. This is because point A falls below the demixing curve in a type I region. For 3% NaCl, point A falls in the three-phase region and its appearance will be like that seen in the middle tubes. Finally, at high salt concentrations of 5%, point A falls in a type II system. Thus we see the I-III-II transition exemplified by Fig. 4.9 and representative of many systems [2,6,9,12,19,20].

A second system, designated as A', is also noted in Fig. 4.17. As the salt concentration is increased, the appearance of the phases will differ slightly from that shown in Fig. 4.9. At 5 wt % NaCl, for example, A' is within the single-phase region, and as seen in tube D of Fig. 4.18, the micellar phase fills the entire tube. No phase boundaries appear. This is type IV system in the Winsor notation

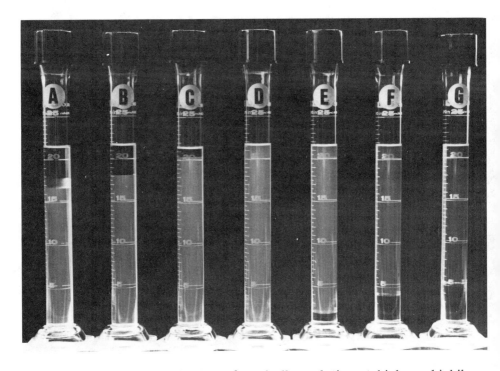

FIGURE 4.18 Phase changes of a micellar solution at high amphiphile content when the NaCl concentration increases: A: 19g/l, B: 21g/l; C: 22g/l; D: 26g/l; E: 28.5g/l; F: 30g/l; G: 42g/l. 3.8% sodium dodecylsulfate; 5.7% 1-pentanol; brine/hexane = 1 (vol); T = 22°C.

and the sequence of transitions resulting from increasing the salinity
is, when the overall composition is A', noted here as I to IV to II.

F. The R-Ratio

As defined in Chap. 1, the R-ratio measures the tendency for the
amphiphile to disperse into oil, divided by its tendency to dissolve
in water [see Eq. (1.19)]. If one of the tendencies far exceeds the
other, the interfacial region tends to take on a definite curvature.
If, for example, $R \gg 1$, then the cohesive energy between the amphiphile and the oil is much larger than that between the amphiphile and
the aqueous phase. The interfacial area in contact with oil tends to
be maximized and that with water minimized. Thus oil tends to become the continuous phase and the corresponding characteristic system is type II.

Similarly, for $R \ll 1$, the characteristic system is type I. The
interface is convex toward the oil and water tends to be the continuous phase. Moreover, it then follows that there is a correspondence
between type III behavior and $R = 1$.

The importance of identifying the relationship between the type
of phase behavior and the R-ratio will not become apparent until it
is desired to understand the way in which a variation in any one of
the parameters will influence the phase behavior. These matters are
discussed in Chaps. 5, 6, and 7.

Since, as has been noted, the classification of the phase behavior
lacks precision, the identification with the R-ratio is also imprecise
and at this point must simply be regarded as qualitative; however,
the R-ratio is a fundamental concept susceptible to quantitative definition. As noted in Chap. 1 and as detailed in Chap. 8, the phase
behavior of microemulsions cannot be described properly without introducing an intrinsic interfacial curvature. The R-ratio is a measure of that curvature.

G. Alternative Representations of Phase Behavior

Figure 4.19 illustrates yet another way of presenting the phase
boundaries [21]. Those shown are drawn for fixed ratios of water
to oil and surfactant to alcohol. In this case and for most similar
plots that appear in subsequent chapters, an equal volume of water
to oil is maintained. In accordance with the definitions presented in
Section 4.I.F, the number of phases that appear then determines the
system type. Types I, II, III, and IV all appear and are duly noted
in Fig. 4.19. In this representation type III behavior appears to be
the result of a superposition of types I and II systems. Although
this is no doubt a considerable simplification, it is one approach used
for modeling the thermodynamic behavior of type III systems (see
Chap. 8).

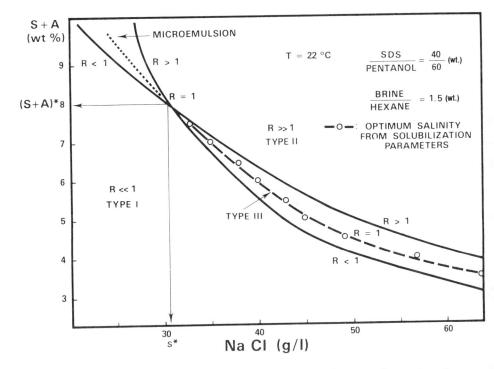

FIGURE 4.19 Typical phase boundaries at various surfactant and alcohol (S+A) concentrations as a function of salinity. The water–oil ratio is 1 by volume. The circles correspond to equal volumes of oil and brine solubilized in the micellar phase. (After Ref. 21.)

Upon addition of a sufficient quantity of amphiphile (surfactant plus alcohol), a single-phase (type IV) micellar system is found. From a practical viewpoint, it is interesting to note that the amount of amphiphile required to dissolve equal volumes of oil and water, creating a single-phase isotropic micellar solution, is minimum at a point along the curve denoted as R = 1. The R-ratio is unity when the micellar solution contains equal volumes of water and oil. Along the curve R = 1, the total volume of the micellar phase increases as the sodium chloride concentration is decreased. At a particular salt concentration, which is called the optimum salinity [2], the micellar phase "consumes" both the excess oil and the excess aqueous phase, leaving a single-phase type IV system. The minimum amount of amphiphile is denoted as (S + A)* to emphasize that it is an optimum (minimum). This quantity is a measure of the solvency of the amphiphilic mixture, as described in Chap. 7.

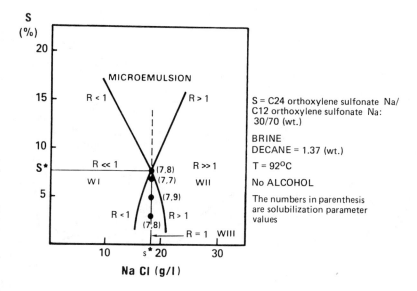

FIGURE 4.20 Typical phase boundaries at various surfactant (S) concentrations as a function of salinity and no alcohol. The water-oil ratio is 1 in volume. The circles correspond to equal volumes of oil and brine solubilized in the micellar phase. Note the vertical shape. (After Ref. 22.)

The phase diagram shown in Fig. 4.19 has a rotated appearance. The curve R = 1 is inclined relative to the abscissa. This is due primarily to the partitioning of the alcohol, an important issue discussed in some considerable detail in Chapter 5. Figure 4.20 shows a graph similar to that presented in Fig. 4.19 except that the amphiphile system contains no alcohol and the R = 1 curve is practically vertical [22].

Again as in Fig. 4.19, the minimum quantity of amphiphile to solubilize equal volumes of oil and water is found by following the R = 1 curve to the single-phase vertex. This is, as noted above, the optimum system and the minimum amount of amphiphile is designated as S*.

H. Transitions Involving Liquid-Crystalline Phases

The emphasis in the discussion thus far has been on I-III-II transitions that take place through a sequence of isotropic states. It must again be emphasized that in some cases, isotropic intermediate states between S_1 and S_2 may not be observed. For example, Winsor [23]

encountered anisotropic liquid-crystalline mesomorphous phases when
he added octanol to a mixture of an aqueous solution of undecane 3-
sodium sulfate and a paraffinic oil.   In fact, as noted in Section 1.III,
liquid-crystalline phases are often encountered.   An early review ar-
ticle by Ekwall et al. [24] shows a number of phase diagrams for
hydrocarbon/water/amphiphile blends in which large regions of different liquid-crystalline phases are found.   The progression of intermediate states is indicated in Table 1.1.

Shah and Hamlin [25] and Shah et al. [26] also reported the existence of intermediate anisotropic phases between S_2 and S_1 as water
is added to a mixture of hexadecane, hexanol, and potassium oleate.
Striking changes in the viscoelastic properties were found to correspond to the appearance and disappearance of birefringence.   Other
measurements, including proton NMR and electrical conductivity,
tended to support the sequence of liquid-crystalline phases proposed
by Ekwall et al. [24].

The preparation of systems that include liquid-crystalline phases
is often complicated by the persistence of coarse emulsions.   In some
cases these may last for months or perhaps even years.   Nevertheless,
the components must initially be intimately mixed to ensure that the
ultimate states are equilibrium states.   Thus in many cases long wait-
ing times are ultimately required to reach equilibrium.

The general trends for the coalescence of emulsions are discussed
in Section 4.II.F.   For types I and II systems, the isotropic amphiphilic phase begins to clear first.   For type III systems there generally exists one system for which phase separation occurs rapidly.
This is shown in Fig. 4.21.   The type III system contained in tube D
has already coalesced, whereas in the others, emulsions persist.

II. PHYSICOCHEMICAL PROPERTIES CHANGES
ACCOMPANYING THE I-III-II TRANSITION

A. Solubilization

The volumes of hydrocarbon and water dissolved within the micellar
solution are quantities of considerable practical interest since the
blending of oil and water to form a single thermodynamically stable
phase is often the purpose of amphiphiles.   That amount of water
and oil in a micellar solution is termed "solubilized" and the solubilized volumes are V_W and V_O, denoting water and oil, respectively.
Since V_W and V_O depend on the amount of amphiphile present, it is
perhaps better to consider them per unit amount of surfactant.   Thus
"solubilization parameters" [2] SP_O and SP_W have been defined for
oil and water, respectively, as

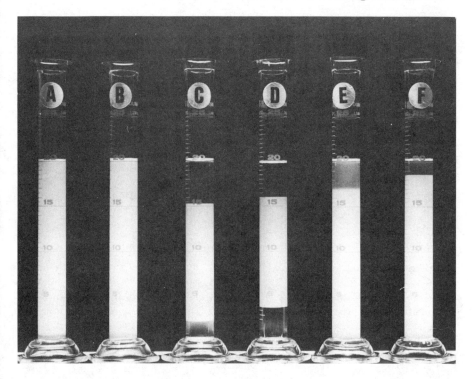

FIGURE 4.21 Shown are emulsions in various stages of coalescence.
The volume of emulsion (opaque portion) is a measure of its stability.
Systems A, B, C, E, and f corresponding to type I or II behavior
are relatively stable. The type III system D has already reached
its final equilibrium state. The NaCl concentration increases from
A to F. A: 56g/l; B: 58g/l; C: 58.5g/l; D: 59g/l; E: 62g/l; F:
65f/l. 1.6% sodium dodecylsulfate; 2.4% 1-pentanol; brine/hexane =
1 (vol); T = 25°C.

$$SP_o = \frac{V_o}{V_s} \quad \text{and} \quad SP_w = \frac{V_w}{V_s} \tag{4.2}$$

where V_s is the volume of surfactant contained in the micellar phase,
normally calculated excluding alcohol or other cosolvents whenever
they are present. Figure 4.22 shows a typical variation of the solubilization parameter as a function of increasing electrolyte concentration, which in this case promotes the classical I-III-II transformation.

SOLUBILIZATION
PARAMETERS
(cm^3 / cm^3)

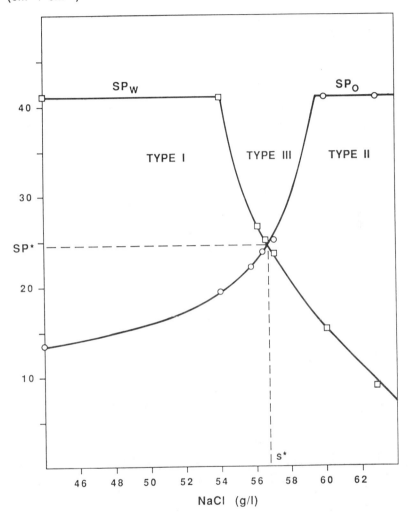

FIGURE 4.22 Solubilization parameters SP_O and SP_W of oil and brine as a function of salinity. The experiments correspond to a cut at 4% (S+A) in Fig. 4.19.

At low salinity the volume of oil dissolved in the aqueous micellar phase is very small. An increase in the electrolyte concentration results in an increase in the solubilization of oil. For type I systems, very little water is in the organic phase and virtually all the water is contained in the micellar phase. SP_W is therefore essentially constant as long as type I behavior is maintained. When the type III regime is reached, water is expelled from the micellar phase and SP_W decreases. At the same time, the oil uptake in this phase continues to increase up to the point where all the oil present in the system is solubilized. A type II system is then obtained and further increase in the salinity results in a constant value of SP_O, while SP_W is seen to decrease regularly. It is interesting to point out that in general no abrupt discontinuity in the physical properties of the micellar solution is observed in the SP_O curve when the type I/type III boundary is crossed, nor in the SP_W curve at the type III/type II boundary. (Viscosity may be an exception; see Section 4.II.E.). This indicates further that the structure of the micellar phase changes gradually during the I-III-II transition, from S_1 to S_2.

The SP_O and SP_W curves intersect inside the three-phase domain. At that point the micellar phase contains equal amounts of water and oil and the corresponding solubilization parameter is noted SP*. It will be seen later that SP* is one parameter that characterizes the solubilizing power of the surfactant (see Section 7.I). According to the definition given in Section 4.I, R = 1 when $SP_O = SP_W = $ SP*. Measurements such as those depicted by Fig. 4.21 have been repeated at various surfactant and alcohol concentrations and are the results that are used to locate the R = 1 curve shown in Fig. 4.19.

Another representation of the phase volumes has been found to be useful [27]. It permits one to visualize the evolution of the interfaces as one of the formulations is systematically changed. An example is provided by Fig. 4.23. The volume fractions of the aqueous, micellar, and oil phases, respectively, are plotted against salinity. A vertical cut in such a diagram gives a direct indication of the relative phase volumes and the location of the interfaces, as well as the range of salinity for which three phases are observed (see Fig. 4.9).

The change in phase volumes illustrated by Figs. 4.22 and 4.23 is typical. There are, however, exceptions and the results may depend on which formulation variable is used to promote the I-III-II transformation. Another way of stating this same reservation is to note that in a five-component system such as the ones studied in Figs. 4.22 and 4.23, one can, by systematically varying the concentration of a component or the ratio of two components, penetrate the three-phase region along many different paths. Some of the paths may not display the sequence of phase volumes illustrated by Figs. 4.22 and 4.23.

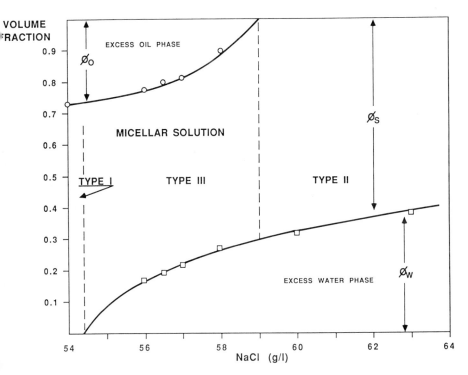

FIGURE 4.23   Volume fraction of the phases corresponding to the
experiment described in Fig. 4.22.

One of the interesting examples of a somewhat different phase-
volume sequence is illustrated by Fig. 4.24 when the overall composition is located at a point beneath the critical tie line.   At first a
micellar solution is shown to be in equilibrium, with an excess oil
phase and an excess aqueous phase.   The three systems shown, u,
v, and w, contain slightly different proportions of oil, water, and
surfactant, but since all three lie within the tie triangle, the phases
have identical compositions.   By changing one of the formulation variables, the middle-phase micellar solution, having a composition at
point u, appears to grow and to incorporate the oleic phase.   This
is the typical behavior depicted by Fig. 4.23.   On the other hand,
if the overall composition is at a point below the critical tie line, the
micellar solution will appear to shrink in size.   Figure 4.25 shows an
example of this "unusual" phase behavior [28].
        The important point is that the phase behavior, which is called
here "typical" or usual, is not necessarily always observed.   In a
multidimensional space composed of multiphase phase regions, a rich

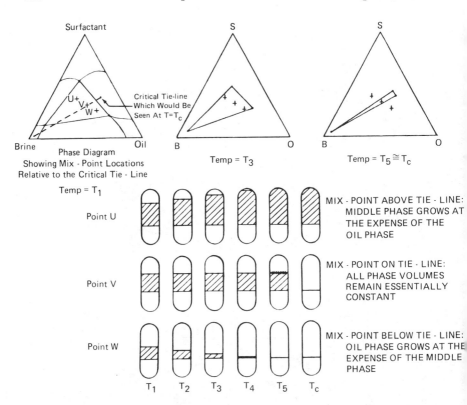

FIGURE 4.24 An "unusual" volume sequence different from those shown by Figs. 4.22 and 4.23 resulting from overall composition being located at a point beneath a critical tie line.

variety of phase sequences is obviously possible, depending on the path followed when one formulation variable is systematically changed [6].

B. Interfacial Tensions

In a number of applications low interfacial tensions between oil and water are required and surfactants are used to achieve the desired state. The low-tension states have their origin in two different mechanisms. By adsorbing at the interface between oil and water, the amphiphile reduces the interfacial tension essentially by rendering the transition more diffuse. This mechanism is the prominent one as the concentration of the amphiphile is increased until micells form in either the oil or the water phase. At this point further decrease in the interfacial tension by this mechanism is difficult because the

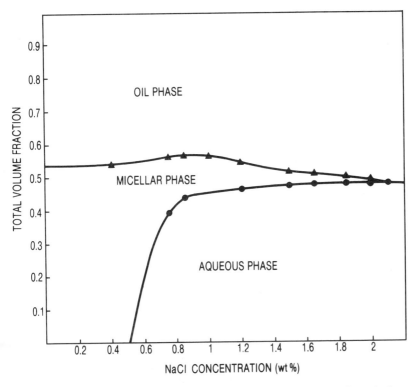

FIGURE 4.25 "Unusual" phase behavior of mixtures of n-dodecane and brine containing 2% (wt) Petrostep 465, 1% (wt) isobutanol and 1% (wt) pentanol-1. Petrostep 465, supplied by Stepan Chemical Co., is a petroleum sulfonate of broad equivalent weight distribution, with an average of 465. (After Ref. 28.)

chemical potential of surfactant is essentially constant almost independent of concentration.

The existence of ultralow interfacial tensions is well known and definitely correlated to phase behavior, in particular to the existence of type III behavior. It is not, therefore, strictly a function of surfactant adsorption. Another mechanism must exist which helps to create the low interfacial tensions, and this mechanism must in some way be rooted in the phase behavior of amphiphilic systems. Certainly, as two phases approach a critical endpoint, the interfacial tension between them vanishes. Thus ultralow interfacial tensions may be thought to originate at critical points; however, what distinguishes micellar systems from other solutions which are molecular dispersions is that the low interfacial tensions tend to persist over a wider range of compositions in the case of micellar solutions. This is attributed to the fact that the chemical potential of the components in a micellar system varies slowly with composition; that is, the chemical environments of the oil, water, and amphiphile remain essentially the same even though there may be a substantial variation in their proportions. Thus water molecules tend to be surrounded by water molecules which are collectively shielded from oil molecules by the interfacial surfactant layer. Water tends to be bulk water and oil tends to be bulk oil. Of course, this is a simplistic viewpoint and the chemical environment of the components of a micellar solution does vary to some extent when the overall composition is changed.

As discussed in Section 8.II, which deals with critical phenomena, the interfacial tension is predicted by critical scaling theory to remain small as long as deviation of the chemical potential of the components does not change markedly from their value at the critical point.

It is therefore expected that those micellar solutions which solubilize the largest quantities of oil and water will exhibit the lowest interfacial tensions, since it is precisely these solutions for which the chemical environments will be least sensitive to changes in composition. Large quantities of water and oil imply large regions of "bulk-like" oil and water. It is, therefore, not surprising to find a correlation between phase behavior and interfacial tension.

Shinoda et al. have reported interfacial tensions [27] and phase behavior [29] for a nonionic surfactant as a function of temperature. The phase behavior observed for a system containing 5% surfactant and various water/cyclohexane ratios is given in Fig. 4.26. Shinoda's original phase nomenclature, which dealt primarily with emulsified systems, has been replaced by that used here. The equivalence has been acknowledged by Shinoda [27,30] and is discussed further in Section 4.II.F.

The volume fractions of oil, water, and surfactant phases, as well as the interfacial tension between the respective phases, are plotted against temperature in Fig. 4.27. These data correspond to a path along the vertical cut C'C in Fig. 4.26. Figure 4.27 shows that when temperature is raised, the interfacial tension γ_{mo} between the micellar solution and the excess hydrocarbon phase decreases (type I systems), whereas the interfacial tension γ_{mw} increases between the micellar solution and the excess aqueous phase (type II systems). The two curves appear to intersect inside the three-phase region, but because of the narrowness of this region, further detailed investigation was not carried out.

Using the sessile drop technique, Healy et al. [31] have been able to investigate the interfacial tension behavior within the three-

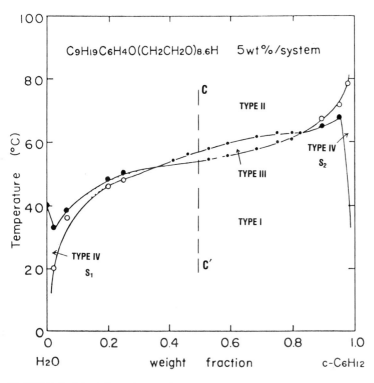

FIGURE 4.26 The phase diagram of water and cyclohexane containing 5 wt% polyoxyethylene (8.6) nonylphenylether as a function of temperature. (After Ref. 29.)

phase regime. A typical result is given in Fig. 4.28. γ_{mo} is seen to decrease when the electrolyte concentration is raised, while γ_{mw} increases. Again, no abrupt change, within experimental uncertainty, is observed at the I-III- or III-II transition. The two curves intersect inside the three-phase region; the corresponding system has been designated as optimum [2] and the value of the interfacial tension at the intersection where $\gamma_{mo} = \gamma_{mw}$ is denoted as γ^*. In the example shown by Fig. 4.28, γ^* is very low, on the order of 10^{-3} dyn/cm.

Figure 4.28 shows the change in solubilization parameters observed over the same range of salinities as the interfacial tension. The general features of the interfacial tension curves compared to that of the solubilization parameter curves suggest that a direct correlation exists between the amount of oil (or water) solubilized in a surfactant phase and the interfacial tension which it displays against

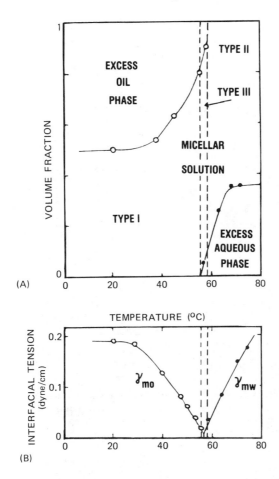

FIGURE 4.27   The effect of temperature on the volume fractions of water, oil and amphiphilic phases (A) and the interfacial tension between these phases (B).   The systems correspond to the cut C'C in Fig. 4.26.   (After Ref. 27.)

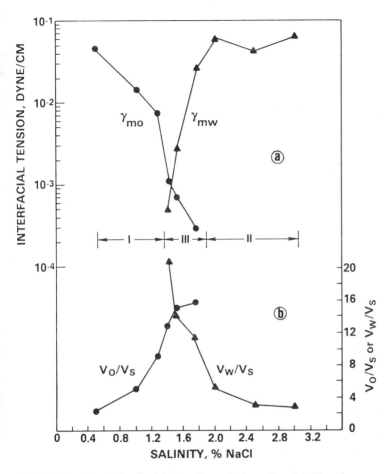

FIGURE 4.28   Interfacial tensions (a) and solubilization parameters
(b) as a function of salinity for the system 3% 63/37 MEA C12 OXS/
TAA, 48.5% oil, 48.5% water, x% NaCl.   MEA C12 OXS is a commercial dodecylorthoxylene sulfonate, monoethanolamine salt.   TAA
stands for tertiary amyl alcohol.   The oil is a 90/10 mixture of
Isopar M and heavy aromatic naphta.   (After Ref. 2.)

the excess oil (or water) phase: the higher the solubilization, the lower the interfacial tension. Empirical equations have been given to relate interfacial tension and solubilization. Healy and Reed [2] have proposed

$$\log \frac{\gamma_{mo}}{\gamma'_{mo}} = \frac{a}{m_o SP_o + 1} \qquad (4.3)$$

$$\log \frac{\gamma_{mw}}{\gamma'_{mw}} = \frac{b}{m_w SP_w + 1} \qquad (4.4)$$

where a, m_o, γ'_{mo}, b, m_w, and γ'_{mw} are universal parameters. Similarly, Nelson [32] proposed

$$\log \gamma_{mo} = \frac{a'}{1 + b'SP_o} - c' \qquad (4.5)$$

$$\log \gamma_{mw} = \frac{a'}{1 + b'SP_w} - c' \qquad (4.6)$$

where a', b', and c' are parameters. Typical values are 4.80, 0.1, and 5.4, respectively. According to Nelson, these equations yield, for a given solubilization parameter, a similar value for γ_{mo} but a substantially lower value for γ_{mw} than the equations reported above. Salter [33] has measured more than 200 different interfacial tensions and concluded that a substantial correlation does exist between interfacial tension and the phase volume.

Theoretical interpretations defining this relationship have been proposed. Dealing with spherical water external (S_1) or oil-external (S_2) micellar solutions, Robbins [34] developed a model that relates interfacial tension within the micelle to SP_w or SP_o, respectively. Miller et al. [35] proposed that the low tensions are associated with the presence of relatively large micelles or similar entities in the interfacial region. Following the Cahn and Hilliard theory [36], which relates the interfacial tension between immiscible liquid phases in a binary system to the thickness of the interface, they found that the order of magnitude of the interfacial tension between a micellar solution and an excess phase is given by

$$\gamma = \frac{B}{a_0^2} \qquad (4.7)$$

where B is a constant and a_0 is the micelle radius.

Recently, Huh [37], using a layer model for the micellar phase in a type III system, derived the following equations:

$$\gamma_{mo} \approx \frac{A \cos(\pi\phi_1/2)}{96\pi\tau^2 SP_o^2} \tag{4.8}$$

$$\gamma_{mw} \approx \frac{A \cos(\pi\phi_2/2)}{96\,\pi\tau^2 SP_w^2} \tag{4.9}$$

where A is a constant, τ is the apparent thickness of surfactant layer at the oil/brine interface, and ϕ_1 and ϕ_2 are, respectively, the volume fractions of oil and brine in the system. If $\phi_1 = \phi_2$ and if the system is taken at optimum, $SP_o = SP_w = SP^*$ and $\gamma_{mo} = \gamma_{mw} = \gamma^*$. One therefore has

$$\gamma^* = \frac{C}{(SP^*)^2} \tag{4.10}$$

where C is a constant. This relationship has been verified under a variety of conditions [38–41]. Equation (4.10) allows an easy and reasonably accurate estimation of the interfacial tensions from simple solubilization measurements.

The correlation that exists between the interfacial tensions and the phase volume is to some extent anticipated based on critical scaling theory (see Section 8.II). Although Eqs. (4.3), (4.4), and (4.10) do appear to fit measurements representative of a wide variety of systems, critical scaling theory does not demand that all interfacial tensions fall on the same curve; that is, the values of the parameters need not be universal.

The interfacial tension γ_{ow} between the excess aqueous and organic phases in type III systems has been investigated [10,42–46]. When three phases i, j, and k coexist at equilibrium, thermodynamic arguments show that the interfacial tensions between the phases obey the relation [47]

$$\gamma_{ij} \leq \gamma_{ki} + \gamma_{kj} \tag{4.11}$$

The equality corresponds to Antonov's rule and implies that a small amount of k spreads to form a film separating phases i and j. The equality holds automatically at a critical point. When the inequality applies, a small amount of k forms a lens at the interface between phases i and j.

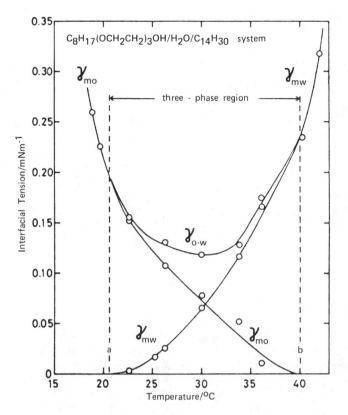

FIGURE 4.29 The interfacial tensions between O–W, O–M and M–W as a function of temperature. Surfactant: $C_8H_{17}(OCH_2CH_2)_3OH$; tetradecane/water = 1 (wt). The systems investigated correspond to line a-b in Fig. 7.12. (After Ref. 46.)

As regards mixtures of oil, water, and amphiphile, Kunieda and Shinoda [46] reported that for type III systems, $\gamma_{ow} < \gamma_{mo} + \gamma_{mw}$ (Fig. 4.29). This inequality was also confirmed by the presence of a stable lens of micellar phase at the water/oil interface in some systems.

Temperatures a and b in Fig. 4.29 correspond to critical end temperatures (lower and upper, respectively) for the W and M phases, and M and O phases, respectively. Thus very low interfacial tensions γ_{mo} and γ_{mw} are observed near temperatures a and b.

Antonov's rule certainly applies at the lower and upper critical temperatures. This is trivially true, for example, at the lower critical temperature since $\gamma_{mw} = 0$ and $\gamma_{mo} = \gamma_{ow}$ because the composition of the phases m and w are indistinguishable. A similar argument applies at the upper critical temperature. γ_{ow} is seen to display a minimum inside the three-phase region. At optimum, $\gamma_{mo} = \gamma_{mw} = \gamma^*$. Thus, at optimum, $\gamma_{ow} < 2\gamma^*$. Clearly, the minimum in γ_{ow} becomes ultralow when the two critical end temperatures approach each other. If they coincide, γ_{ow} becomes zero since the three coexisting phases become identical (tricritical point). In that case the width of the three-phase region also becomes zero (see Section 4.II.C). This has been supported by the correlation reported by Kunieda and Shinoda [46], between γ_{ow} and the difference between the two critical end temperatures for a number of systems containing nonionic surfactans.

Working on mixtures of brine (NaCl), dodecane, pentanol-1, and sodium p-octylbenzenesulfonate at constant temperature, Bellocq et al. [42] have identified two critical endpoints and have reached similar conclusions: namely, γ_{ow} is close to γ_{mo} or γ_{mw}, depending on which critical point is considered. They also reported that inside the three-phase region, γ_{ow} remains equal to the largest of the two other interfacial tensions, in agreement with the results of Pouchelon et al. [43] for a different system. This implies that in these cases, the inequality still applies. On the other hand, Bellocq et al. [42] and Seeto et al. [44] have studied alcohol, hydrocarbon, water and sodium chloride systems. Based on interfacial tension measurements, they found that Antonov's rule applies to this system within experimental error. In this latter case however, from observations of spreading behavior, possible deviations from the rule were detected.

C. Width of the Three-phase Region

When one of the formulation variables is systematically changed (this is sometimes called a scan), the range of values over which type III behavior is observed has been found to depend on the system investigated. Figure 4.30 shows that the range of salinity over which the system exhibits type III behavior increases when the number of carbon atoms of the hydrocarbon (n-alkane series) increases and the surfactant and alcohol concentrations are kept constant [48]. Further investigation of a number of different systems shows that a simultaneous increase in salinity and alkane carbon number results in a decrease of the solubilization [49,50] (see Chap. 7). More generally, it has been observed that high solubilization of oil and brine at optimum is obtained when the three-phase region is narrow, and vice versa, whatever the variables under consideration. Schechter, Wade, et al. [39–41,51] have related the optimal solubilization parameter SP* to the width of the three-phase region defined as

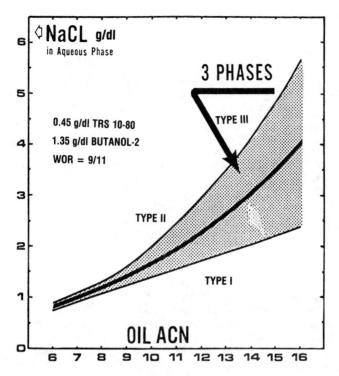

FIGURE 4.30 Phase map showing the effect of simultaneous changes
of sodium chloride concentration and number of carbon atoms of
n-alkane (ACN) on the type III region. TRS 10-80 is a commercial
petroleum sulfonate (Witco Chemical Company). (After Ref. 48.)

the range of alkane carbon number ACN over which type III systems
are observed as follows:

$$SP^* \cdot \Delta ACN = d \tag{4.12}$$

where d is a constant characteristic of the class of surfactant under
consideration. Values of d are 5.5, 24.7, and 40.3 for alkylbenzene-
sulfonates, α-olefin sulfonates, and ethoxylated oleyl sulfonates,
respectively.

In the case of ethoxylated alkylphenols, the optimal solubilization
parameter has been found to correlate with the reciprocal of the
range of surfactant hydrophile-lipophile balance (ΔHLB) yielding
type III behavior, as illustrated in Fig. 4.31. HLB is defined (as
usual) as the weight percentage of ethylene oxide in the surfactant

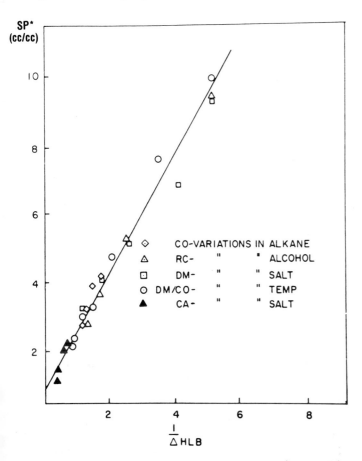

FIGURE 4.31 Solubilization parameter at optimum SP* correlated
with the reciprocal of the width of the three-phase region expressed
in HLB units. The data have been obtained with ethoxylated alkylphenols of various lipophiles under a variety of conditions of alkane
carbon number, salinity, temperature, and alcohol concentration.
The water-oil ratio is 4 (wt). The surfactant concentration is
0.035 molar. (After Ref. 39.)

molecule. The points on Fig. 4.31 correspond to systems representing a variety of alkyl chains for various alkane carbon numbers,
alcohol concentration, temperature, and salinity.

A correlation between the width of the three-phase region and
the interfacial tension can be anticipated from critical scaling theory.
Figure 4.29 provides an illustration of the relationship. The nearer

the two critical endpoint temperatures are positioned to one another, the smaller will be the interfacial tensions at the optimum. When the two critical endpoints correspond, the system is termed tri- critical [52–55], which is defined as a point at which three conju- gate phases become identical simultaneously. The three interfacial tensions at this point vanish simultaneously.

D. Electrical Conductivity of Micellar Phases

The electrical conductivity of ordinary emulsions (macroemulsions) is used to decide which of the immiscible phases (water or oil) is the continuous one. This measurement is often decisive. Water-in-oil emulsions yield oil-like solution electric conductivities, and vice versa. Intermediate conductivities do not normally exist. Micellar solutions, on the other hand, do exhibit intermediate states and the electrical conductivities attending the I-III-II transformation often decrease gradually. A number of measurements of the electrical conductivities of micellar solutions have been reported [56–60]. These studies have been carried out primarily in the hope of elucidating the structure of the intermediate micellar phases. This aspect is discussed in greater detail in Section 4.II.

The most remarkable feature to be noted is that as the propor- tions of oil relative to water are gradually changed, the electrical resistivity also gradually changes. There is not, as one might ex- pect, an abrupt transition from water-to-oil continuous systems. Figure 4.32 shows the electrical resistivity associated with the I-III-II transformation. For the case shown the resistivity is low at a point near the I-III boundary and increases progressively. This increase in resistivity corresponds to the increase in oil con- tent of the micellar solution.

A second feature of electrical transport in micellar solutions con- cerns the point at which the inversion from water to oil continuity takes place as the water/oil ratio is systematically changed or as the system undergoes a I-III-II transformation. For type I systems the R-ratio is less than unity and the interface tends to be convex toward the oil phase. Because of this intrinsic bending tendency, water continuous systems tend to resist the transformation to oil continuity even though the proportion of oil to water may be large. Thus type I systems tend to retain a conductivity characteristic of the aqueous phase even at relatively large oil/water ratios. On the other hand, type II systems (R > 1) are expected to reflect an electrical conductivity characteristic of oil even at rather small oil/ water ratios. Type III systems yield intermediate behavior.

Figure 4.33 shows three pseudoternary diagrams representing type I, II, and III phase behavior. The difference between the three systems is the alcohol/surfactant ratio. Increasing this ratio promotes the I-III-II transition. Also shown on each diagram is a

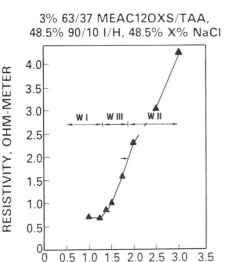

3% 63/37 MEAC12OXS/TAA,
48.5% 90/10 I/H, 48.5% X% NaCl

FIGURE 4.32 The change in electrical resistivity of the amphiphilic phase along the I–III–II transition on increasing the electrolyte concentration. The system is the same as that described in Fig. 4.28. (After Ref. 31.)

path representing a sequence of compositions in the single-phase portion of the diagram labeled scan A. Along this path the amount of amphiphile (surfactant plus alcohol) is held fixed while the oil/water ratio is steadily increased. The electrical conductivity of micro-emulsions having compositions on these paths can be measured. They are reported in Fig. 4.34. At the highest alcohol/surfactant ratio (A/S = 1.5) the system is type II, as shown by Fig. 4.33. Thus R > 1 and the interface is convex toward the aqueous phase. Based on the results shown in Fig. 4.34, the transition to oil continuity appears complete at a composition of roughly 30 wt % n-decane, since at this point the electrical conductivity is approximately equal to that of oil. Conversely, for A/S = 0.65, the system is type I. At approximately 60 wt % decane the conductivity is reduced to nearly that of the oil phase. The difference between the two systems is the R-ratio. Thus systems for which R < 1 tend to remain water continuous. This intrinsic curvature of the interfacial region is an important feature of micellar solutions and is incorporated into the thermodynamic model which is considered in Section 8.III.

    The points at which the electrical conductivity suddenly increases as shown in Fig. 4.34, and similar results found by other

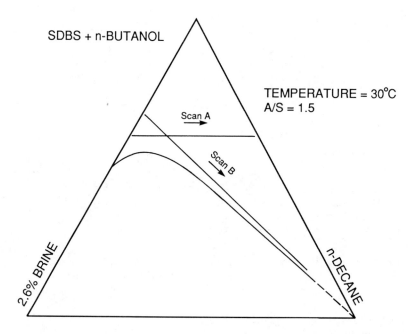

FIGURE 4.33 Pseudoternary diagrams showing the phase boundaries of the sodium dodecyl benzene sulfonate-n-butanol-n-decane-brine system at various alcohol to surfactant ratios, A/S. (After Ref. 58.)

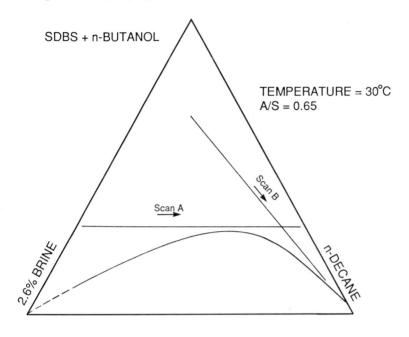

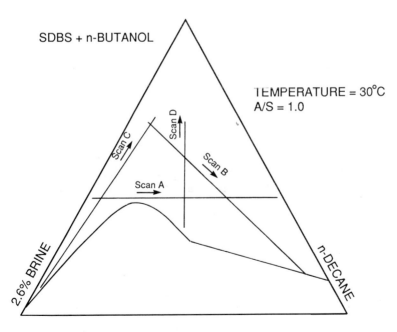

FIGURE 4.33 (Continued).

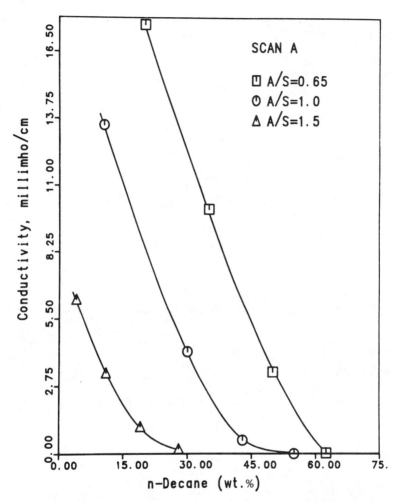

FIGURE 4.34   Electrical conductivities of those micellar solutions composed along scan A of the three phase diagrams given in Fig. 4.33.   (After Ref. 58.)

investigators [56,57,61], have been interpreted as a percolation
threshold [62]. In percolation modeling, sites composing a medium
or lattice are randomly assigned values, usually on a binary system.
Nearest-neighbor sites of equal value define bonds between sites.
Sets of connecting bonds define clusters. On an occupation basis,
as the probability of occupation increases, the probability that
large clusters form also increases. Above some critical occupation,
there exists a nonzero probability that a randomly chosen site will
belong to an infinite cluster which will percolate the lattice [63–65].

It is seen in Fig. 4.34 that the electrical conductivities began to
rise rapidly at a certain composition which can be thought of as the
percolation threshold. At this point it is reasonable to imagine a
continuous pathway of water extending through the micellar solution.
Lagues et al. denote this as the onset of phase inversion [56,62].
In Chap. 8 a different thermodynamic model representing the inversion process is proposed. Curves similar to those shown in Fig.
4.34 are predicted.

The results shown in Figs. 4.32 and 4.34 are typical as long as
the micellar solution viscosities remain small. If, however, liquid
crystals, gels, or even high-viscosity Newtonian fluids exist along
any one of the scans, the electrical conductivity may exhibit both
maxima and/or minima, reflecting the changes in the ion mobility
caused by variations in viscosity. Winsor [66] has studied a system
in which the G phase shown in Fig. 1.1 appears. His results show
maxima and minima reflecting the large viscosities associated with the
G phase. Clausse et al. [59] have observed similarly complex behavior of electrical conductivities with increasing water content. In
addition to correlating the results with solution viscosity, they have
observed a relationship with the type of phase behavior exhibited
by micellar solutions [59].

E. Viscosity Changes of the Micellar Phase

In contrast to conductivity which varies gradually, the viscosity of
the micellar phase has been found in some cases to display discon-
tinuities in slope during the I-III-II or I-IV-II transitions. However,
as long as only isotropic systems are considered, the viscosity
changes in the vicinity of these boundaries are not large. The viscosity does not vary by more than one order of magnitude.

Figure 4.35 shows that upon varying salinity, the surfactant
phase viscosity exhibits two maxima, occurring roughly at the type I/
type III and type III/type II boundaries. Between these two peaks,
inside the three-phase regime, a minimum in viscosity is observed.
As Healy et al. state [31], this behavior is due to the changes in
the surfactant phase composition as shown by its volume changes
(Fig. 4.35, upper curve).

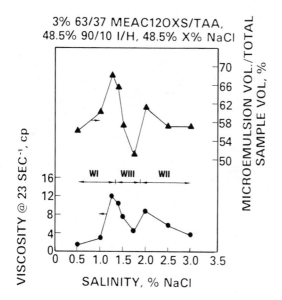

FIGURE 4.35 Changes in viscosity and volume of the micellar phase along the I–III–II transition on increasing the electrolyte concentration. The system is the same as that described in Fig. 4.28. (After Ref. 31.)

The same pattern of viscosity behavior has been observed by Bennett et al. [67], who also indicated that some micellar mixtures near the viscosity peak may display non-Newtonian behavior. Investigating a different system, Salter [68] reported, on varying the salinity, only one peak located at the onset of type II system rather than two peaks as shown in Fig. 4.35. He also pointed out the dependence of the viscosity on the composition of the phase. In contrast, Thurston et al. [69] found a viscosity maximum at the I/III boundary but not at the II/III boundary, again emphasizing the importance of the system composition.

All of these results show that although almost discontinuous changes in the slope of the viscosity may be observed, they do not yield wide variations in viscosity, the upper/lower value ratio usually being on the order of 2 or 3. Jones and Dreher [70] reported viscosity measurements on the single-phase micellar solutions (type IV) observed between type I and II when the alcohol concentration varies in the system. They found either a maximum or a minimum of the viscosity, depending on the system investigated.

Similarly, Bourrel et al. [50] investigated the behavior of the viscosity of a micellar phase in relation to the phase boundaries.

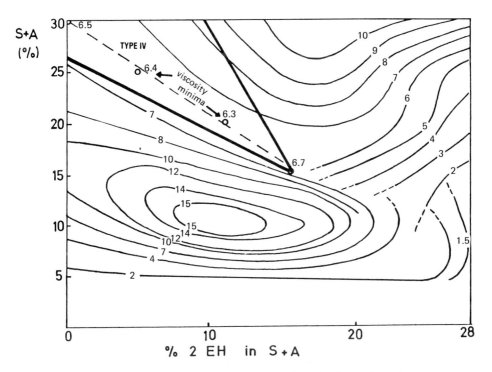

FIGURE 4.36 Isoviscosity contours superimposed on the phase
boundaries obtained by varying the proportions of hydrophobic alcohol (2-ethylhexanol = 2EH) in the alcohol mixture. Surfactant
S = Petrostep 465 (Commercial petroleum sulfonate). Alcohol A =
Isobutanol + 2-ethylhexanol. Oil = Octane. Water: 0.05 wt % NaCl
S/A = 72/28 (wt). Water/oil = 1 (wt). Temperature = 30°C. The
numbers indicate the value of the viscosity in cp. (After Ref. 50.)

Two alcohols (isobutanol and 2-ethylhexanol) were used in various
proportions to achieve the phase transition, the surfactant/alcohol
ratio being held constant. The results are shown in Fig. 4.36 in
the form of isoviscosity contours mapped on the phase diagram. Two
regions of maximum viscosity correspond to type I and type II systems. In the single-phase domain, when passing from S_1 to S_2 solutions, the viscosity exhibits little variation, although a valley of
shallow minima exists in this region. This result, obtained under
conditions for which the I-III-II transition is achieved with relatively small changes in composition compared to the cases discussed
above in which the salinity is changed, indicates that all of the
microstructures are almost certainly liquid-like and readily deform-

able. In contrast, for those cases when a G phase is interposed between S_1 and S_2, there is a very large (often many hundred-fold) increase in viscosity [17].

F. The Stability of "Ordinary" Emulsions

Ordinary emulsions or macroemulsions must be distinguished from micellar solutions or microemulsions, which are the primary focus of this book. Macroemulsions are composed of drops of one liquid phase interspersed within a second immiscible liquid phase. The drops will ultimately cream, that is, rise or fall, depending on whether they are less or more dense than the continuous phase. The larger drops will tend to move faster than the smaller ones, and collisions and coalescence will ultimately occur. The initially disperse system of drops will ultimately appear as a second phase, separated from the initially continuous phase by a single interface.

Microemulsions or micellar solutions, on the other hand, are composed of submicroscopic structures which are so small that Brownian motion tends to keep them suspended, and coalescence of the drops leads to an increase rather than a decrease in free energy, as is evidently the case for ordinary emulsions. Thus the free energy of the micellar solution is a minimum. The system is thermodynamically stable.

One primary difference between a micellar solution and a macroemulsion can be thought to be drop size. The size of emulsion drops is frequently orders of magnitude larger than the size of the submicroscopic regions composing a micellar solution. It is important to keep this difference in mind; however, the fundamental difference is thermodynamic stability. Ordinary emulsions are not thermodynamically stable.

Micellar solutions coexist in equilibrium with immiscible excess phases. Thus it is clearly possible to blend a micellar phase and a conjugate excess phase mechanically to create a macroemulsion which will ultimately coalesce to reconstitute the original two phases. When such experiments are carried out, a number of interesting observations can be made.

The micellar solution tends to be the continuous region[†] [71,72] and the excess phase the discontinuous one. This is true over a wide range of relative phase volumes of the two phases. Thus even a large volume of the excess phase when blended with a small volume

---

[†]This observation is apparently well known to any chemist who has worked with surfactants for even a relatively short period. The references cited are some of the more recent ones which restate the rules and provide evidence.

of micellar solution will probably result in drops of the excess phase being dispersed into the micellar phase. The mechanism that dictates the more rapid coalescence of two drops of micellar phase as compared to two drops of excess phase, thereby yielding a continuous micellar phase, is that of surface tension gradients [73-75]. When two drops approach one another, the liquid film of continuous phase separating the two drops will become thinner. If the film becomes thin enough, it will become unstable and coalescence will occur. Any mechanism that slows the rate of thinning will therefore reduce the rate of coalescence. As the liquid film thins, any surfactant molecules in the interface separating the drop from the liquid film will tend to be displaced by the movement of the liquid film. If this displaced surfactant is slow to be replaced, large surface tension gradients develop which tend to retard the film-thinning process [73]. It is not difficult to see that the displaced surfactant can be replaced more readily if the surfactant is in the drop phase than when it is in the continuous phase. Thus continuous phases containing surfactant develop larger surface tension gradients and hence smaller coalescence rates than when the surfactant is inside the drop. Clearly, when two immiscible liquids are stirred, the faster coalescing drops tend to become the continuous phase.

Thus one immediate correlation with phase behavior is seen. In a macroemulsion the micellar phase is likely to be the continuous one. There are other correlations that can be observed. A sequence of different micellar solutions in equilibrium with excess phases (two-phase systems) can all be mechanically blended simultaneously to create series of macroemulsions [76]. The time required for a certain fraction of the excess phase to coalesce can readily be measured and compared, as illustrated in Fig. 4.21. Figure 4.37 shows on a different system the time required for one-half of the original excess phase volume to separate from the emulsion as a function of the surfactant hydrophile-lipophile balance (HLB)[†]. There are two maxima in stability, corresponding, in the case at hand, to the boundaries separating type I from type III and type II from type III. Results obtained on different systems by Salager et al. [71] have shown that the maxima in emulsion stability do not necessarily coincide with the I/III and III/II boundaries. In Fig. 4.37, the two maxima are separated by a deep minimum. The maximum in emulsion stability corresponds to a slow appearance of the separated excess phase and may be attributed to a retardation of any one or more of the three consecutive steps that take place: creaming, colliding, and coalescence.

[†]For polyethylene oxide nonionic surfactants such as those used to conduct the experiments reported in Fig. 4.37, HLB is simply defined as one-fifth of the weight fraction of the ethylene oxide content of the molecule.

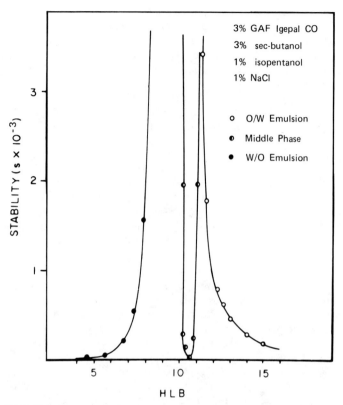

FIGURE 4.37 Emulsion stability in terms of time for one-half the clear phase to appear versus HLB of a series of Igepal CO (GAF Corp.) ethoxylated nonylphenols of known HLB. The concentrations refer to the aqueous phase. Aqueous phase/nonane = 4 (by volume). (After Ref. 76.)

It is believed that the maxima in emulsion stability are due to a slowing of the droplet rise at the I/III/ and II/III boundaries caused by an increase in viscosity of the micellar phase (see Section 4.II.E) and a decrease in the density difference between the drops and the micellar phase. Hazlett and Schechter [77] have measured the coalescence rate in a shear field and found that the coalescence efficiency does not increase markedly at either the I/III or II/III boundaries. A modified stability ratio defined to be proportional to the number of collisions between droplets required to yield a single coalescence is shown in Fig. 4.38. A large value of the modified stability ratio is indicative of a stable emulsion. Smaller values indicate rapid coalescence.

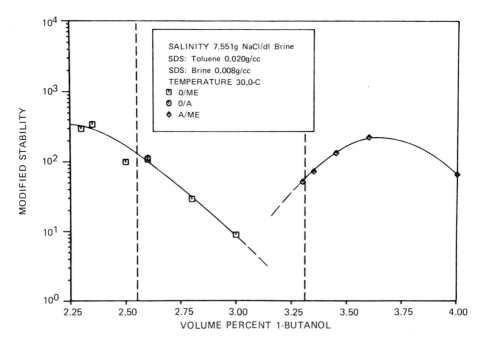

FIGURE 4.38 Coalescence in micellar system composed of sodium dodecyl sulfate, toluene, butanol, and brine. The systems studied include oil/micellar solution (O/ME), brine/microemulsion (A/ME), and brine/oil (A/O). (After Ref. 77.)

The particular system studied exhibits a I/III boundary at approximately 2.5 vol % butanol and a II/III boundary at about 3.3 vol %. These boundaries are indicated in Fig. 4.38 by vertical dashed lines. There is no increase in the modified stability ratio at these boundaries. Thus the maxima in emulsion stability shown in Fig. 4.37 must be attributable to a reduced creaming and/or collision rates; it is not due to a decrease in the coalescence efficiency.

It is also instructive to note that the modified stability ratio decreases by several orders of magnitude within the three-phase region, indicating that emulsions formed between the micellar solution and one of the excess phases will be unstable. This corresponds to the point of minimum emulsion stability shown in Fig. 4.37. Thus the deep minimum is a result of greatly increased coalescence efficiency [78] and does not correspond to an increase in the collision frequency. By maintaining a constant shear rate, the collision frequency is essentially fixed for the studies plotted in Fig. 4.38.

Thus Figs. 4.37 and 4.38 both reflect a profound relationship which correlates emulsion stability to surfactant phase behavior.

Although this correlation has been shown to exist for a wide range of systems [71,80–83], it has not been widely used by practitioners interested in selecting amphiphiles to break emulsions, despite the fact that it would seem reasonable to employ phase behavior as a guide.   However, Shinoda and his co-workers [72,79] have long used emulsion stability as a measure of the transition between S_1 and S_2 states.   The "phase inversion temperature" (PIT) used extensively by this group is that point at which the emulsion inverts.

The rapid coalescence which is seen to occur within the three-phase region is not predicted by the usual theories related to emulsion stability [84].   A mechanism of percolation has been suggested by Hazlett and Schechter [77].   This mechanism envisions that as the composition of the micellar solution becomes richer in the excess phase, which is of course also the discontinuous phase, a path may span from one drop to another, through which the discontinuous phase may move.   This causes a rapid coalescence at film thicknesses which are far greater than that required for film rupture by the usual mechanisms.   Figure 4.39 depicts the process.   Shown is a cloud of very tiny drops.   These represent the solubilized discontinuous phase in the micellar solution.   Depicted also is a continuous path or cluster of these tiny drops, leading from one macroscopic drop to another.   These are the drops of the discontinuous phase.   Rapid coalescence is thought to occur by flow through the discon-

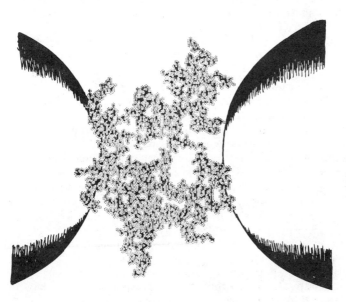

FIGURE 4.39   Destabilization of thin films by percolation.

tinuous path of tiny drops. This corresponds to the percolation
phenomenon described in Section 4.II.D as it relates to the electrical conductivity of micellar solutions.

Rapid breaking of emulsions is associated with type III systems
for R = 1. It is known that exceedingly stable emulsions are associated with liquid-crystalline structures that surround the interface
[85–87]. The nature of these liquid crystals has not been widely
investigated, but practitioners are aware of their importance in creating highly stable emulsions. Thus it is possible, but certainly not
yet established, that systems for which R = 1 may represent both the
most stable and the least stable emulsions, depending on whether or
not liquid-crystalline phases form.

As mentioned previously, the alcohol concentration has a strong
effect on the rate of phase separation: the higher the concentration,
the more rapid the rate. This is due, as discussed further (Section
5.V.C), to the increase in fluidity of the C region when alcohol
adsorbs at the water/oil interface. Thus when the system tends to
form anisotropic liquid-crystalline phases, the addition of alcohol
may reduce emulsion stability.

Other factors that contribute to the enhancement of interface
fluidity (and are also operative in destroying the liquid-crystalline
phases), such as temperature or surfactant branching will also tend
to promote quicker phase separation.

G. Contact Angle Variation

Reed and Healy [88] have investigated the contact angle on both
high- and low-energy substrates with the phases in equilibrium.
The contact angle was inferred based on the shape of sessile drops.
It is by convention measured through the more dense phase. Thus
in the case of a type III system, for a drop of micellar phase against
oil, the contact angle is therefore measured through the micellar
phase. For a micellar phase against water, the contact angle is
measured through the water. A similar convention is used in regard
to moving interfaces. They are called advancing when the more
dense fluid moves with respect to substrate it has contacted (the
cell fluid is first equilibrated with the substrate); otherwise they
are called receding. Difficulties in contact angle measurements and
reproducibility were encountered by Reed and Healy because very
low interfacial tensions result in flat sessile drops and also because
of the sensitivity of phase behavior to temperature. Contact angles
were thus reported with a precision ranging from ±5 to ± 12° [88].

Within this accuracy, the steady values of contact angles did
not show any effect of the nature of the substrate or of the cleaning procedure. Furthermore, advanced and receded contact angles
were found pairwise the same, hence hysteresis-free.

Receded contact angles were also measured as a function of salinity for the system whose composition is given in the caption of Fig. 4.28. As salinity increases, the I-III-II transition is observed and the contact angles were measured with the more dense fluid filling the cell and the less dense fluid composing the drop. The results are shown in Fig. 4.40.

The contact angles are independent of substrate within ±10° with the exception of θ_{wm}^r on Teflon at high salinities. This atypical behavior on Teflon is not interpreted. The solid lines in Fig. 4.40 intersect at 1.7% NaCl in the three-phase region and can thus be used to define an optimal salinity for contact angle. This salt concentration is of approximately the same value, which is found to be optimum based on equal solubilization of oil and water.

The results presented above are interpreted by assuming that the substrate that governs contact angle behavior is a consequence of surfactant adsorption, the effective substrate being determined

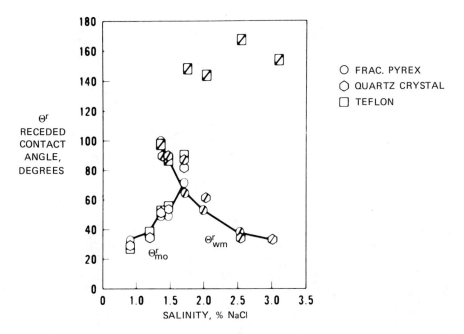

FIGURE 4.40 Receded contact angles on three substrates for the system described in Fig. 4.28. θ_{mo}^r (open symbols) refers to the contact between substrate-micellar phase and excess oil phase. θ_{wm}^r (barred symbols) refers to the contact between substrate-micellar phase and excess aqueous phase. The more dense fluid fills the cell. (After Ref. 88.)

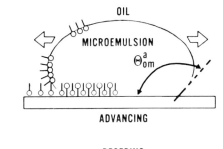

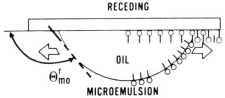

FIGURE 4.41 Model of effective substrate for interpretation of advancing (θ^a_{om}) and receding (θ^r_{mo}) contact angle results. (After Ref. 88.)

by the last contacting fluid. Figure 4.41 illustrates a model surfactant adsorption for the micellar phase of type I and III systems on high-energy surfaces. In the upper drawing, the substrate is first equilibrated with excess oil (which also contains surfactant) and monolayer adsorption occurs, while in the lower drawing the substrate is first equilibrated with the micellar phase and bilayer adsorption occurs.

It is thus concluded that advanced and receded contact angles are the same since molecular configuration of surfactant at all interfaces is invariant to expansion or retraction of either drop shown in Fig. 4.41.

III. MICROSTRUCTURE OF MICELLAR SOLUTIONS

Type I and type II systems are depicted in Fig. 4.10 as being dispersions of spherical drops. For type I systems, the aqueous phase is imagined to be the continuous one, whereas type II systems are depicted as oil continuous. In this section the experimental evidence supporting this simple microstructure is examined. It is seen that in the extremes of large water/oil or oil/water ratios, the micellar solution can accurately be portrayed as a dispersion of drops. Most results indicate that these drops are small, being on the order of 10^2 Å in radius. They have generally been found to be sharply

distributed about the mean value (monodisperse), although polydispersity in some systems has been reported. Thus for micellar
solutions, which are primarily S_1 or S_2 states, the microstructure is
apparently that of small drops; however, the structure of the intermediate states such as those associated with type III systems are as
yet unresolved issues. Phase behavior represented on a pseudoternary diagram as shown in Fig. 4.15 suggests that type III systems
are "mixtures" of type II and type I systems. This is the approach
taken in constructing a thermodynamic model (see Section 8.III) to
represent the states intermediate between S_1 and S_2; however, it
does not seem reasonable to think that spherical drops are representative of the actual form of the interface when the micellar solution
is composed of substantial quantities of both oil and water. As noted
in Section 1.I, Scriven [7,8] has proposed that the interfacial layer
separating water from oil is bicontinuous.

Experiments to determine microstructure have considered primarily
"dilute" solutions, that is, those containing relatively small quantities
of dispersed phase. Light scattering, neutron scattering, component
diffusion, dialysis, osmotic pressure, sedimentation, nuclear magnetic
resonance, and fluorescence have all been applied to provide insight
into the microstructure of the micellar solutions.

Although the dimensions of swollen micelles or inverted micelles
are described here, it should not be thought that these are permanent, unchanged immutable structures. They are not static entities.
Whatever shape is represented in the paragraphs that follow, it
should be regarded as an average value.

A. Light Scattering

For unpolarized light Brunetti et al. [89] state that the intensity
$I(\theta)$ of the light scattered at an angle θ from a suspension of particles in excess of that from the continuous phase is given by

$$I(\theta) = (1 + \cos^2\theta)Kv_m\phi S(q) \qquad (4.13)$$

where q is the wave transfer vector, defined as $4\pi\tilde{n}\sin(\theta/2)/\lambda_0$,
$I(\theta)$ is the intensity, and

$$K = \frac{2\pi^2\tilde{n}^2}{\lambda_0^4}\left(\frac{d\tilde{n}}{d\phi}\right)^2$$

In these equations θ is the scattering angle, v_m the volume of a
micelle, ϕ the micelle volume fraction, $\tilde{n}$ the refractive index, and
λ_0 the wavelength in vacuum of the incident light. $S(q)$ is a structure factor related to the interactions between drops [90] which in

the limit as $q \to 0$ is related to the osmotic pressure, π_o, by compressibility relationship [89]

$$S(\phi) = \frac{kT}{v_m} \left(\frac{\partial \pi_o}{\partial \phi} \right)^2 - 1 \qquad (4.14)$$

The osmotic pressure can be expanded into a power series in ϕ (virial expansion) valid for small ϕ,

$$\pi_o = \frac{\pi kT}{v_m} \left(1 + \frac{B\phi}{2} + \cdots \right) \qquad (4.15)$$

where B is a virial coefficient. From Eqs. (4.13), (4.14), and (4.15),

$$\frac{\phi}{I} = \frac{1}{Kv_m} \left(1 + \frac{B\phi}{2} \right) \qquad (4.16)$$

Based on this expression, it is seen that by measuring the intensity at a given scattering angle and plotting ϕ/I versus ϕ, then $1/Kv_m$ is determined as the intercept of the plot extrapolated to $\phi \to 0$ and $B/2v_m K$ is found as the slope of the plot at small ϕ. From these two values v_m and B are thus determined. Values so determined are reported in Table 4.1 with $v_m = 4\pi R^3/3$.

The results presented in Table 4.1 show that for the systems studied, primarily S_2 micelles, the drop radii vary over a narrow range and are approximately 60 Å. A close examination of the data reveals that these radii tend to increase with increasing proportions of water relative to amphiphile. This seems to be a rather general trend. The radii do not appear to be very sensitive to the chain length of the alcohols which are used in conjunction with sodium dodecylsulfate. Three different alcohols are compared.

The virial coefficient, B, is negative for attractive interactions between micelles and positive for repulsive ones. The values of B given in Table 4.1 show a strong dependence on the chain length of the alcohol even though the radii do not. The virial coefficient ranges from negative to positive for the sequence pentanol, hexanol, heptanol.

A number of other investigators [91–93] have used light-scattering techniques to obtain information about micellar sizes. The range of radii reported for systems other than the ones given in Table

TABLE 4.1 Micellar Radii and Virial Coefficient for the System Water/Sodium Dodecyl Sulfate (SDS)/Alcohol[a]/Dodecane

Composition (mol %)	C_5	C_5	C_6	C_6	C_6	C_6	C_7	C_7
Water	9.71	17.48	8.10	13.20	16.76	14.04	18.46	23.93
Dodecane	68.57	56.06	71.56	63.56	58.86	62.53	55.78	48.11
SDS	5.58	6.86	4.66	5.69	5.78	4.03	6.73	6.89
Alcohol	16.41	19.61	15.62	17.55	18.60	19.38	19.01	21.03
Drop radius (Å)	50 ± 6	62 ± 7	47.6 ± 1	64 ± 4	66 ± 1	70 ± 2	62 ± 2	67 ± 3
Virial coefficient	-23 ± 3	-27 ± 3	-0.5 ± 0.5	-4.4 ± 0.5	-6.1 ± 1	-8.8 ± 2	5 ± 2	4 ± 3

[a]C_5, 1-pentanol; C_6, 1-hexanol; C_7, 1-heptanol.
Source: Ref. 89.

4.1 is in general agreement with the values reported. S_2 systems
exhibit radii ranging from 50 to 100 Å. It must be recognized, however, that as the amount of water is decreased, keeping the amphiphile concentration constant, the radii will generally correspondingly
decrease, and ultimately, in the absence of water, the small aggregates described in Chap. 3 will be observed. Thus while the values
of droplet radii that have been measured do not vary over a wide
range, it seems clear that this is in part related to the limitation of
the experimental methods to dilute solutions and to the systems selected for study rather than being representative of all possible
micellar systems. Light scattering is best applied to a system where
multiple scattering is a minimum.

B. Dilution of Micellar Systems

The determination of the micellar volume, v_m, based on light scattering requires, according to Eq. (4.16), variation of ϕ; that is, the
micellar solution must be diluted. The assumption that is made during the dilution process is that v_m is independent of ϕ. If this is
not the case, a plot of ϕ/I versus ϕ will not actually be a straight
line, even for small ϕ. It is self-evident that the dilution of a micellar solution can only be accomplished when the system is so dilute
that droplet-droplet interactions are weak (normally less than 15% by
volume dispersed phase [94]), and requires the addition of fluid
having precisely the same chemical composition as that of the continuous phase. Unfortunately, it is not easy to sample the continuous
region directly to determine its chemical composition. The difficulty
can be appreciated by examining the systems used for the study
summarized in Table 4.1. All of the systems studied include both
alcohol and SDS. The continuous region is evidently the dodecane;
however, the alcohol will be present in the interfacial region, in the
dodecane and in the water. The solution to be added to dilute the
micellar solution will thus not be simply dodecane, but must include
that amount of the alcohol and trace quantities of water and SDS
which are present dissolved in the continuous dodecane phase. Thus
it becomes most important to determine that fraction of the total
amount of alcohol which resides within the dodecane.

One procedure that can be used to determine the fraction of the
alcohol residing in the continuous region entails "titrating" along the
phase boundary of a demixing curve [20,91,94–98]. To ascertain
the composition of the oil region in a S_2 system, a small quantity of
oil is added to a micellar solution which is composed so that it lies
at a point on the phase boundary separating single-phase solutions
from multiphase regions. The addition of the small quantity of oil
(the continuous phase) will produce a turbid solution which given
sufficient time will separate into distinct phases. This process can
best be understood by referring to Fig. 4.7. Imagine a point on

the phase boundary, as, for example, that represented by point C. On adding a small quantity of oil to this solution, a small quantity of water is expelled since the overall system composition is moved to a point within the two-phase region. The solution will initially be turbid because the water expelled will initially be present in the form of small drops. This solution can be titrated to clarity by adding alcohol containing traces of water and perhaps surfactant. The concept is that, if the ratio of alcohol to oil is the same as that of the continuous phase, the net effect is to dilute the micelles, that is, to cause them to be more widely separated from each other, without, however, changing the volume of a drop.

If the process of adding oil and subsequent titration with alcohol produces a linear relationship that intersects the origin, it seems reasonable to assume that the ratio is representative of the continuous region and that the micelles do not change volume or shape during the dilution process. Figure 4.42 shows the result of titrating a system composed of sodium dodecyl sulfate (SDS), dodecane, pentanol, and water [99]. The slope of the line shown in this figure is, according to the arguments expressed above, representative of the ratio of pentanol to dodecane comprising a continuous region. It is important to note that this ratio is quite different from the overall composition.

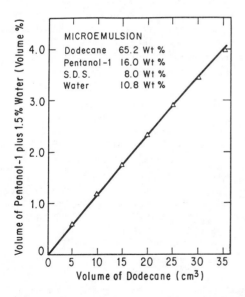

FIGURE 4.42   The amount of pentanol required to clarify the turbidity created by the addition of oil.   (After Ref. 99.)

The validity of the titration procedure can be assessed using dialysis [99]. The essential concept is quite simple. It is supposed that if a small chamber initially containing pure oil is separated from a micellar solution by a membrane having pores which are small enough to reject the micelles but large enough to permit the free passage of alcohol, water, and surfactant which are molecularly dispersed within the continuous oil region, the rate of buildup to equilibrium levels of the concentration of these molecularly dispersed components will be rapid and can therefore be measured. The membrane does not reject all micelles, and their concentration will slowly increase in the small oil compartment. The important feature is that the buildup of micelles takes place on a time scale which is much longer than that characterizing the buildup of molecularly dispersed components. Results based on dialysis are in good agreement with those shown in Fig. 4.42.

C. Neutron Scattering

Small-angle neutron scattering studies have been carried out to elucidate the size of micelles [100–103]. The wavelength of a cold neutron beam is small, approximately 5 Å, and is therefore well suited for investigating micellar microstructure. Furthermore, by replacing ordinary water with deuterium, the scattering density of the aqueous core of S_2 micelles can be greatly enhanced and facilitate determination of the core radius. Results reported by Kotiarchyk et al. [103] indicate that for the water/AOT/decane system, the drop radius increases with increasing water/amphiphile ratio. Radii similar in magnitude to those shown in Table 4.1 are reported. It is also reported that the drop radius increases with increasing temperature at a fixed amphiphile/water ratio. This is consistent with the trends that are expected based on consideration of the natural curvature of the interface. This is discussed in Section 8.II.

Since neutron scattering experiments can be carried out over a wide range of values of the wave transfer vector, both the average micellar radius and the dispersion of the radii about the average can be determined without diluting the micellar solution. For the water/AOT/decane system, Kotiarchyk et al. [103] report that there is substantial dispersion about the average value. A standard deviation of about 25% of the average radius was reported.

D. Quasi-elastic Light Scattering

The temporal fluctuations that give rise to a broadening in the frequency of scattered light contain information on the motion of droplets suspended in the continuous phase [104]. In particular the translational diffusion coefficient D_T can be calculated from the line width (defined as the half-width at half-maximum) of the scattered light. Since there are well-known relationships between D_T and the droplet size for simple shapes (e.g., the Stokes-Einstein equation), the hydrodynamic radius can therefore be determined.

The fluctuations are conveniently described by time-dependent correlation functions. The autocorrelation function is defined as follows:

$$C(\tau) = \langle I(0)I(\tau) \rangle = \lim_{T \to \infty} \frac{1}{T} \int_0^T I(t)I(t - \tau) \, dt \qquad (4.17)$$

where $I(t)$ is the time-dependent scattering intensity. Since the random movement of the droplets is correlated over a time small compared to a characteristic fluctuation time, the product of intensities is large for small times and is equal to $\langle I^2 \rangle$. Since a particle is subject to Brownian motion, it moves within the fluid and becomes uncorrelated at long times. Thus autocorrelation function decay is related to the Brownian motion of the particles, which is characterized by D_T, the Brownian diffusion coefficient. D_T therefore appears as the principal parameter of interest. Jakeman and Pike [105] show that

$$C(\tau) = \langle I^2 \rangle (1 + b \exp -2\Gamma\tau) \qquad (4.18)$$

where $\Gamma = D_T q^2$ and b is an experimental constant. It is assumed in deriving Eq. (4.18) that the average number of droplets in the observed scattering volume can be large but not large enough so that there is substantial interaction between droplets. Droplet-droplet interactions would impede their random motion, an effect not included in Eq. (4.18). Bellocq et al. [106], Graciaa et al. [107], and Kahler and Prager [108] have considered more concentrated solutions by taking into account interactions.

Once D_T is determined, if the shape of the scatterer is known or assumed, the droplet size can be determined. For example, the Stokes-Einstein equation for a sphere is

$$D_T = \frac{kT}{6\pi\mu a} \qquad (4.19)$$

where k is the Boltzmann constant, μ the viscosity of the continuous phase, and a the micellar radius.

A number of investigators [109–113] have applied this technique to microemulsions. Both type I and type II systems have been studied. The drop radii measured by this technique are hydrodynamic radii. The results reported by Gulari et al. [113] range to 600 Å and the trends are similar to those reported in Table 4.1. Increasing the volume of the discontinuous phase while maintaining the concentration of the amphiphile constant tends to result in larger drops. It was thought that the surface area remains constant, implying that the area per amphiphile is therefore effectively constant.

Gulari et al. [113] also reported a substantial increase in drop size with increasing temperature for a system composed of heptane, AOT, and water. The same trend was observed based on the neutron scattering [103]. It was observed that at the highest temperatures studied, phase separation takes place and the increasing drop size was attributed to the approach of the system to the phase boundary [113].

E. Sedimentation

Early work by Mathews and Hirschhorn [114] and Bowcott and Schulman [95] utilized the ultracentrifuge to measure the drop size in micellar solutions. More recently, an extensive study has been carried by Hwan [115]. He reported phase behavior, interfacial tensions, and drop size obtained by sedimentation. The surfactants used in Hwan's study were predominantly dodecyl and nonyl orthoxylenesulfonates and the oil studied was composed of 9 parts by volume paraffinic to 1 part by volume aromatic refined oils but otherwise complex mixtures of unknown composition. The results are summarized in Table 4.2. Shown are the sedimentation coefficients as a function of sodium chloride concentration for both of the surfactants studied. The nonyl orthoxylene sulfonate (C_9OXS) surfactant has an optimum salinity of approximately 5 wt % NaCl (this varies slightly as a function of the overall surfactant concentration), while the corresponding value for the dodecyl orthoxylenesulfonate ($C_{12}OXS$) is 1.75 wt % NaCl. Results are reported for both type I and type II systems. Type III surfactant phases were also subjected to centrifugal forces and near the optimum salinity patterns consistent with both water-continuous and oil-continuous systems were observed simultaneously, indicating that the system cannot be considered to be strictly oil or water continuous.

The sedimentation coefficient listed in Table 4.2 is calculated by measuring the rate of movement of the boundary between clear phase and the phase containing droplets starting initially with a uniform dispersion. Specifically it is defined by the following equation

$$S = \frac{dx}{dt} \frac{1}{\omega^2 x} \qquad (4.20)$$

where dx/dt is the velocity of the sediment, ω the angular velocity of the centrifuge, and x the distance between the boundary

Properties of Solutions Containing Amphiphiles

TABLE 4.2 Micelle Size Calculated by Extrapolating the Sedimentation Coefficient to Zero Surfactant Concentrations[a]

Salinity (wt %)	$S^0 \times 10^{13}$ (sec)	Drop radius (Å)
Nonyl orthoxylene sulfonate (C_9OXS)		
Micellar solution type		
I 1	1.2	23
2	3.2	33
3	12.1	63
Optimum salinity 5		
II 8	20.2	94
9	16.2	84
10	13.8	54
Dodecyl orthoxylene sulfonate ($C_{12}OXS$)		
0.5	9.6	55
0.8	30.0	90
1.0	39.2	101
Optimum salinity 1.75		
2.5	44.0	144
3.0	26.0	110
3.5	10.8	71

[a]Oil is Isopar M and Heavy Aromatic Naphtha (Exxon trade names) mixed 9 parts to 1 part by volume.

and the axis of rotation. Generally, the drop radius can be calculated by extrapolating to zero drop concentration and applying Stokes' law to each individual drop, or the radius can be obtained based on empirical corrections to Stokes' law to account for the presence of many drops. This has been discussed by Hwan [115]. In any case the calculation based on Stokes' law yields a hydrodynamic radius which should be comparable to that obtained by quasi-elastic light scattering.

The results show that for the systems studied, the water-in-oil micellar solutions have larger hydrodynamic radii than do the oil-in-water solutions, and that the drop radii increase in size as the optimum salinity is approached. Indeed, for the longer-chain surfactants, radii in excess of 100 Å are found. Thus one may expect the effective drop radius to continue to increase as type III systems are approached. It is, however, difficult to interpret the sedimentation results observed with type III systems.

Hwan reported the systems studied to be monodisperse. Dvolaitzky et al. [102b] have evaluated the polydispersity of a microemulsion using sedimentation techniques.

F. Other Techniques

The dielectric properties of micellar solutions have been measured
by several investigators [116-119]. These measurements should contribute to our understanding of the micellar structure since it seems
reasonable to expect that a relationship between dielectric dispersion
and micellar structure should exist. This, however, is a subject of
current research and will not be discussed further here. Nuclear
magnetic resonance [120,121] and spin-label techniques [122] have
also been used to obtain information about the microstructure of
micellar solutions.

IV. OPTIMAL STATES

A. Recounting of Their Special Characteristics

A number of interesting and important properties associated with
type III behavior, especially at the particular conditions for which
the micellar phase contains equal volumes of oil and water, have been
noted in Section 4.II. The systems, for which $SP_W = SP_O = SP^*$,
are designated as optimal because a number of properties attain either
a maximum or a minimum under practically the same conditions. For
example, the interfacial tension between the excess oil and excess
water phases attains a minimum, the coalescence rate of ordinary emulsions is a maximum, the oil recovery efficiency of a micellar solution is
a maximum, the quantity of amphiphile required to solubilize equal
quantities of oil and water is a minimum, and the contact angles of the
excess oil-micellar solution and the excess water-micellar solution are
equal, and many other characteristics are all singular at about the
same conditions. Thus the states for which $SP_O = SP_W = SP^*$, which
corresponds to $R = 1$, are, with good reason, thought of as special
states and hence are designated as being optimal systems.

B. Locating the Optimum

The technique for finding the optimum state given an oil that is to
be solubilized together with water generally consists of systematically
varying one of the formulation variables, X, while holding all of the
other variables constant. The formulation variables are enumerated
and their influence is described in detail in Chap. 5, but based on
the discussion presented in Section 4.I and 4.II, they clearly include
temperature and salinity. A criterion is then selected which will
identify the optimum state. Most often this criterion is based on

solubilization and that state for which $SP_W = SP_O$ is optimum. Under
these conditions $X = X^*$; that is, the optimum state for the particular system is characterized by a single number which can be called
the optimum. For example, X is frequently the electrolyte concentration and X^* may be called the optimal salinity. Of course, X^* is
special only because the salinity is the formulation variable which
has systematically been changed while holding all others constant.
In fact, it is not sufficient to cite an optimal salinity without precisely
defining the values of all of the other formulation variables; furthermore, if the electrolyte concentration is fixed and the temperature
varied, the optimum would be represented by X^*, which would now
represent an optimal temperature. Thus although in the chapters to
follow we shall frequently cite an optimal salinity, for example, simply
because it is that variable which is changed, it is important to recognize that all the formulation variables taken together form an optimal
state. There is actually not a single formulation variable having a
higher status than that of any other formulation variable.

Solubilization is not the only criterion that can be used to locate
the optimal state. Others can and have been applied. The point
where the interfacial tensions between the micellar phase and the excess water and excess hydrocarbon phases are equal is one example
[2]. The center of the three-phase range [48] or the point at which
the coalescence rate of ordinary emulsions is highest [123] are both
valid operational criteria. It is remarkable that the values of X^*
found by applying any one of these criteria do not differ greatly
from one another. Often they are the same to within experimental
error. Although there exists no proof that the X^* corresponding to
the various criteria must be identical, experience has taught that
this is a very good assumption and is certainly reliable in searching
for the optimal state.

Having selected a formulation variable that can feasibly be
changed, and a suitable criterion, then based on a knowledge of the
influence of the formulation variable, the search for the optimum can
be made systematic. This is discussed in Chaps. 5 and 6.

It is well to recognize that an optimum state with $R = 1$ does not
always exist. In some cases, for example, varying one of the formulation variables will proceed from S_1 to S_2 through a series of liquid-
crystalline states such as indicated by Fig. 1.1. Shah and Hamlin
[25] found, for example, that by steadily increasing the hexanol
concentration in a potassium oleate/water/decane system, the intermediate states correspond to the sequence of liquid crystals indicated
by Fig. 1.1

If liquid-crystalline intermediate states appear, the chemical composition of the system or the temperature has to be changed in some
way so as to find the desired sequence of isotropic states. This may
require experimentation. Guidance is provided by the discussion in
Chap. 5. The trends are discussed in terms of the R-ratio. Since,

however, this concept is qualitative rather than quantitative, it is not possible to predict just what changes will produce success.   Only the trends are defined.

In other cases, isotropic intermediate states which are highly viscous are observed and again optimization is not possible.   As noted in Chap. 5, each amphiphile is associated with a range of alkanes that will yield normal isotropic behavior.   Thus it is not possible to insist that a certain amphiphile can always be made to match a particular oil by, say, varying the electrolyte concentration.   In some cases liquid-crystalline mesophases and gels result rather than fluid isotropic states.   Experienced researchers have recognized these limitations and know what changes to try when unusual phase behavior is encountered.   Our aim in the following chapters is to demonstrate that the usual and the unusual behavior are both readily understood based on consideration of the R-ratio.

REFERENCES

1. R. N. Healy and R. L. Reed, Soc. Pet. Eng. J., 14:491 (1974).

2. R. L. Reed and R. N. Healy, in Improved Oil Recovery by Surfactant and Polymer Flooding (D. O. Shah and R. S. Schechter, eds.), Academic Press, New York (1977).

3. L. M. Prince, in Microemulsions, Theory and Practice (L. M. Prince, ed.), Academic Press, New York, p. 21, (1977).

4. L. S. Polatnik and A. I. Landau, Phase Equilibria in Multicomponent systems, Holt, Rinehart and Winston, New York, p. 247 (1964).

5. D. H. Smith, J. Colloid Interface Sci., 108:471 (1985).

6. K. E. Bennett, C. H. K. Phelps, H. T. Davis, and L. E. Scriven, Soc. Pet. Eng. J., 21:747 (1981).

7. L. E. Scriven, Nature, 263: 123 (1976).

8. L. E. Scriven, in Micellization, Solubilization, and Microemulsions, Vol. 2, (K. L. Mittal, ed.), Plenum Press, New York, p. 877, (1977).

9. S. J. Salter, paper SPE/AIME 7056 presented at the 5th SPE/AIME Symposium on Improved Methods for Oil Recovery, Tulsa (Apr. 1978).

10. A. M. Bellocq, D. Bourbon, and B. Lemanceau, J. Colloid Interface Sci., 79:419 (1981).

11. A. M. Bellocq, J. Biais, B. Clin, A. Gelot, P. Lalanne, and B. Lemanceau, J. Colloid Interface Sci., 74:311 (1980).

12. J. van Nieuwkoop and G. Snoei, J. Colloid Interface Sci., 103: 400 (1985).

13. A. W. Adamson, J. Colloid Interface Sci., 2:261 (1969).

14. W. C. Tosch, S. C. Jones, and A. W. Adamson, J. Colloid Interface Sci., 3:297 (1969).

15. S. Friberg and I. Buraczewska, in Micellization, Solubilization, and Microemulsions, Vol. 2 (K. L. Mittal, ed.), Plenum Press, New York, p. 791 (1977).

16. M. Podzimek and S. Friberg, J. Disp. Sci. Tech., 1:341 (1980).

17. P. A. Winsor, Solvent Properties of Amphiphilic Compounds, Butterworth, London (1954).

18. P. A. Winsor, Trans. Faraday Soc., 44:376 (1984).

19. C. E. Blevins, G. P. Willhite, and M. J. Michnick, Soc. Pet. Eng. J., 21:581 (1981).

20. J. G. Dominguez, G. P. Willhite, D. W. Green, in Solution Chemistry of Surfactants (K. L. Mittal, ed.), Academic Press, New York, p.673 (1979).

21. M. Bourrel and C. Chambu, Soc. Pet. Eng. J., 23:327 (1983).

22. M. Bourrel, C. Chambu, and F. Verzaro, Proceedings of the Second European Symposium on EOR, Technip, Paris, p. 39 (Nov. 1982).

23. P. A. Winsor, J. Phys. Chem., 56:391 (1952).

24. P. Ekwall, L. Mandell, and K. Fontell, Molecular Cryst. Liq. Cryst., 8:157 (1969).

25. D. O. Shah and R. H. Hamlin, Jr., Science, 171:483 (1971).

26. D. O. Shah, A. Tamjeedi, J. W. Falco, and R. Walker, Jr., AIChE J., 18:1116 (1972).

27. H. Saito and K. Shinoda, J. Colloid Interface Sci., 32:647 (1970).

28. K. S. Chan and D. O. Shah, paper SPE 7896 presented at the SPE Symposium on Oilfield and Geothermal Chemistry, Houston (Jan. 1979).

29. K. Shinoda and H. Kunieda, J. Colloid Interface Sci., 42:381 (1973).

30. K. Shinoda and H. Saito, J. Colloid. Interface Sci., 26:70 (1968).

31. R. N. Healy, R. L. Reed, and D. G. Stenmark, Soc. Pet. Eng. J., 16:147 (1976).

32. R. C. Nelson, Soc. Pet. Emg. J., 22:259 (1982).

33. S. J. Salter, paper SPE 12036 presented at the 58th SPE Annual Technical Conference and Exhibition, San Francisco (Oct. 1983).

34. M. L. Robbins, paper SPE/AIME 5839 presented at the 3rd SPE/AIME Symposium on Improved Methods for Oil Recovery, Tulsa (Mar. 1976).

35. C. A. Miller, R. N. Hwan, W. J. Benton, and T. Fort, J. Colloid Interface Sci., 61:554 (1977).

36. J. W. Cahn and J. E. Hilliard, J. Chem Phys., 28:258 (1958).

37. C. Huh, J. Colloid Interface Sci., 71:408 (1979).

38. L. N. Fortney, "Criteria for Structuring Surfactants to Maximize Solubilization of Oil and Water," M.A. thesis, The University of Texas at Austin (May 1981).

39. A. Graciaa, L. N. Fortney, R. S. Schechter, W. H. Wade, and S. Yiv, Soc. Pet. Eng. J., 22:743 (1982).

40. Y. Barakat, L. N. Fortney, R. S. Schechter, W. H. Wade, and S. Yiv, Proceedings of the Second European Symposium on EOR, Technip, Paris, p. 11 (Nov. 1982).

41. Y. Barakat, L. N. Fortney, R. S. Schechter, W. H. Wade, S. Yiv, and A. Graciaa, J. Colloid Interface Sci., 92:561 (1983).

42. A. M. Bellocq, D. Bourbon, and B. Lemanceau, J. Disp. Sci. Tech., 2:27 (1981).

43. A. Pouchelon, J. Meunier, D. Langevin, D. Chatenay, and A. M. Cazabat, Chem. Phys. Lett., 76:277 (1980).

44. Y. Seeto, J. E. Puig, L. E. Scriven, and H. T. Davis, J. Colloid Interface Sci., 96:360 (1983).

45. A. M. Bellocq, D. Bourbon, B. Lemanceau, and G. Fourche, J. Colloid Interface Sci., 83:427 (1982).

46. H. Kunieda and K. Shinoda, Bull. Chem. Soc. Jpn., 55:1777 (1982).

47. B. Widom, J. Chem. Phys., 62:1332 (1975).

48. J. L. Salager, J. C. Morgan, R. S. Schechter, W. H. Wade, and E. Vasquez, Soc. Pet. Eng. J., 19:107 (1979).

49. M. Bourrel, A. M. Lipow, W. H. Wade, R. S. Schechter, and J. L. Salager, paper SPE 7450 presented at the 53rd SPE Annual Technical Conference and Exhibition, Houston (Oct. 1978).

50. M. Bourrel, C. Chambu, R. S. Schechter, and W. H. Wade, Soc. Pet. Eng. J., 22:28 (1982).

51. Y. Barakat, L. N. Fortney, C. Lalanne-Cassou, R. S. Schecter, W. H Wade, and S. Yiv, paper SPE/DOE 10679 presented at the

3rd SPE/DOE Symposium on Enhanced Oil Recovery, Tulsa (Apr. 1982).

52. R. B. Griffith and B. Widom, Phys. Rev., A8:2173 (1973).

53. M. Kahlweit and R. Strey, J. Phys. Chem., 90:5239 (1986).

54. M. Kahlweit, R. Strey, and P. Firman, J. Phys. Chem., 90: 671 (1986).

55. D. H. Smith, paper SPE/DOE 14914 presented at the 7th SPE/ DOE Symposium on Enhanced Oil Recovery, Tulsa (Apr. 1986).

56. M. Lagues, J. Phys. Lett., 40:L331 (1979).

57. M. Lagues and C. Sauterey, J. Phys. Chem., 84:3503 (1980).

58. A. Lam and R. S. Schechter, "A Study of Diffusion and Electrical Conductivity in Microemulsions," J. Colloid Interface Sci., in press.

59. M. Clausse, J. Heil, A. Zradba, and L. Nicolas-Morgantini, Jornadas del Comité Español de la Detergencia, Tensioactivos y Afines, Barcelona (Mar. 1986).

60. J. van Nieuwkoop and G. Snoei, J. Colloid Interface Sci., 103: 417 (1985).

61. B. Lagourette, J. Peyrelasse, C. Boned, and M. Clausse, Nature, 281:60 (1979).

62. M. Lagues, R. Ober, and C. Taupin, J. Phys. Lett., 39:487 (1978).

63. M. F. Sykes and J. W. Essam, Phys. Rev., 133:(1A), A310 (1964).

64. M. F. Sykes, D. S. Gaunt, and M. Glen, J. Phys. Math. Nucl. Gen., A9(10):1705 (1976).

65. S. Kirkpatrick, Rev. Mod. Phys., 45(4):574 (1973).

66. P. A. Winsor, Trans. Faraday Soc., 46:762 (1950).

67. K. E. Bennett, H. T. Davis, C. W. Macosko, and L. E. Scriven, paper SPE 10061 presented at the 56th SPE Annual Technical Conference and Exhibition, San Antonio (Oct. 1981).

68. S. J. Salter, paper SPE 6843 presented at the 52nd SPE Annual Technical Conference and Exhibition, Denver (Oct. 1977).

69. G. G. Thurston, J. L. Salager, and R. S. Schechter, J. Colloid Interface Sci., 70:517 (1979).

70. S. C. Jones and K. D. Dreher, Soc. Pet. Eng. J., 16:161 (1976).

71. J. L. Salager, L. Quintero, E. Ramos, and J. Anderez, J. Colloid Interface Sci., 77:288 (1980).

72. K. Shinoda and H. Kunieda, in Encyclopedia of Emulsion Technology, Vol. 1, Basic Theory (P. Becher, ed.), Marcel Dekker, New York, p. 337 (1983).

73. G. D. M. McKay and S. G. Mason, J. Colloid Sci., 18:674 (1963).

74. D. T. Wasan, J. J. McNamara, S. M. Shaw, and K. Sampath, J. Rheol., 23:181 (1979).

75. H. Sonntag, J. Netzel, and B. Uterberger, Spec. Discuss. Faraday Soc., 1:57 (1970).

76. M. Bourrel, A. Graciaa, R. S. Schechter, and W. H. Wade, J. Colloid Interface Sci., 72:161 (1979).

77. R. D. Hazlett and R. S. Schechter, "Stability of Macroemulsions in Microemulsion Systems," Colloids Surf., in press.

78. R. M. Flumerfelt, A. B. Catalano, and C. H. Tony, Surface Phenomena Enhanced Oil Recovery (D. O. Shah, ed.), Plenum Press, New York, p. 571 (1981).

79. K. Shinoda and H. Arai, J. Phys. Chem., 68:3485 (1964).

80. L. M. Baldauf, R. S. Schechter, W. H. Wade, and A. Graciaa, J. Colloid Interface Sci., 85:187 (1982).

81. F. S. Milos and D. T. Wasan, Colloids Surf., 4:91 (1982).

82. J. L. Salager, M. Minana-Perez, M. Perez-Sanchez, M. Ramirez-Gouveia, and C. I. Rojas, J. Disp. Sci. Tech., 4:313 (1983).

83. A. Graciaa, Y. Barakat, R. S. Schechter, W. H. Wade, and S. Yiv, J. Colloid Interface Sci., 89:217 (1982).

84. R. D. Hazlett and R. S. Schechter, "Droplet Coalescence Near Liquid/Liquid Critical Points," Colloids Surf., in press.

85. S. Friberg and L. Mandell, J. Am. Oil Chem. Soc., 47:149 (1970).

86. S. Friberg, L. Mandell, and M. Larsson, J. Colloid Interface Sci., 29:155 (1969).

87. D. T. Wasan, S. Shah, N. Aderangi, M. S. Chan, and J. J. McNamara, Soc. Pet. Eng. J., 18:409 (1978).

88. R. L. Reed and R. N. Healy, Soc. Pet. Eng. J., 24:342 (1984).

89. S. Brunetti, D. Roux, A. M. Bellocq, G. Fourche, and P. Bothorel, J. Phys. Chem., 87:1028 (1983).

90. W. G. M. Agterof, J. A. J. van Zomeren, and A. Vrij, Chem. Phys. Lett., 43:363 (1976).

91. (a) A. Graciaa, J. Lachaise, A. Martinez, M. Bourrel, and C. Chambu, C. R. Acad. Sci., B282:547 (1976); (b) A. Graciaa, L. Lachaise, A. Martinez, and A. Rousset, C. R. Acad. Sci., B285:295 (1977).

92. A. A. Calje, W. G. M. Agterof, and A. Vrij, in Micellization, Solubilization, Microemulsions, Vol. 1 (K. L. Mittal, ed.) Plenum Press, New York (1977).

93. A. M. Cazabat and D. Langevin, J. Chem. Phys., 74:314 (1981).

94. A. M. Cazabat, C. R. Acad. Sci., B296:1389 (1983).

95. J. E. Bowcott and J. H. Schulman, Z. Elektrochemie, 59:283 (1955).

96. A. Graciaa and J. Lachaise, J. Phys. Lett., 39:1235 (1978).

97. A. Graciaa, "Physical Chemical Study of Microemulsions," Ph.D. dissertation, University of Pau, France (1978).

98. A. M. Cazabat, J. Phys. Lett., 44:1593 (1983).

99. A. Graciaa, J. Lachaise, M. Bourrel, R. S. Schechter, and W. H. Wade, J. Colloid Interface Sci., in press.

100. C. Cabos and P. Delord, J. Appl. Crystallogr., 12:502 (1979).

101. C. Cabos and P. Delord, J. Phys. Lett., 41:L455 (1980).

102. (a) C. Taupin, J. P. Cotton, and R. Ober, J. Appl. Crystallogr., 11:613 (1978); (b) M. Dvolaitzky, M. Guyot, M. Lagues, J. P. Le Pesant, R. Ober, C. Sauterey, and C. Taupin, J. Phys. Chem., 69:3279 (1978).

103. M. Kotiarchyk, S. H. Chen, and J. S. Huang, J. Phys. Chem., 86:3273 (1982).

104. R. Pecora, J. Chem Phys., 40:1604 (1964).

105. E. Jakeman and E. R. Pike, J. Phys. A1., 2:411 (1969).

106. A. M. Bellocq, G. Fourche, P. Chabrat, L. Letamendia, J. Rouch, and C. Vaucamps, Opt. Acta., 27:1629 (1980).

107. A Graciaa, J. Lachaise, P. Chabrat, L. Letamendia, J. Rauch, C. Vaucamps, M. Bourrel, and C. Chamber, J. Phys. Lett., 38:253 (1977).

108. E. W. Kahler and S. Prager, J. Colloid Interface Sci., 86:359 (1982).

109. S. Qutubuddin, C. A. Miller, G. C. Berry, T. Fort, and A. Hussam, in Surfactants in Solution, Vol. 3 (K. L. Mittal and B; Lindman, eds.), Plenum Press, New York (1984).

110. A Vrij, E. A. Nieuwenhuis, N. M. Fijnant, and W. G. H. Agterof, Faraday Discuss. Chem. Soc., 65:101 (1978).

111. A. M. Cazabat, D. Langevin, and A. Pouchelon, J. Colloid Interface Sci., 73:1 (1980).

112. C. Hermansky and R. A. Mckay, J. Colloid Interface Sci., 73:324 (1980).

113. E. Gulari, B. Bedwell, and S. Alkhafaji, J. Colloid Interface Sci., 77:202 (1980).

114. M. B. Mathews and E. Hirschhorn, J. Colloid Interface Sci., 8:86 (1953).

115. R. N. Hwan, "A Mechanism for Ultralow Interfacial Tension in Systems Containing Microemulsions: Theoretical Considerations and Experiments with Ultracentrifuge," Ph.D. dissertation, Carnegie-Mellon University (1978).

116. M. Clausse, P. Sherman, and R. J. Sheppard, J. Colloid Interface Sci., 56:123 (1976).

117. J. Peyrelasse, C. Boned, P. Xans, and M. Clausse, C. R. Acad. Sci., B284:235 (1977).

118. J. Peyrelasse, V. E. R. MacClean, C. Boned, R. J. Sheppard, and M. Clausse, J. Phys., D11:L117 (1978).

119. T. A. Bostock, M. H. Boyle, M. P. MacDonald, and R. M. Wood, J. Colloid Interface Sci., 73:368 (1980).

120. J. Biais, B. Clin, P. Lalanne, and B. Lemanceau, J. Chem. Phys., 74:1197 (1977).

121. A. M. Bellocq, J. Biais, B. Clin, P. Lalanne, and B. Lemanceau, J. Colloid Interface Sci., 70:524 (1979).

122. C. Ramachandran, S. Vijayan, and D. O. Shah, J. Chem. Phys., 84:1561 (1980).

123. A. Graciaa, Y. Barakat, R. S. Schechter, W. H. Wade, and S. Yiv, J. Colloid Interface Sci., 89:217 (1982).

5

Methods for Promoting Phase Changes

I. GENERAL

Interesting and important physical properties of micellar solutions are associated with a S_1-to-S_2 transition. Examples have been cited in Chap. 4 illustrating that this progression of states can be observed when the concentration of either electrolyte or some alcohols is increased. In this chapter the mechanisms underlying this transition are systematically explored using the R-ratio as the basis for understanding. R, as defined by Eq. (1.21), represents a ratio of cohesive energies. The numerator can be considered to be a measure of the interaction of the C layer with O while the denominator represents that of the C layer with W. These two interactions are considered separately in this chapter. Those factors that are predominately hydrophilic in nature are taken up first (Section 5.II). Section 5.III considers lipophilic interactions. The proposition developed in these two sections is that those changes that tend to increase R promote I-III-II transitions. Based on data presented in this chapter, it seems clear that this proposition is generally valid and is most helpful in formulating microemulsions.

In some cases it is not really possible to separate changes in hydrophilic interactions from changes in lipophilic ones. Some variables, such as temperature or alcohol concentration, affect all three regions simultaneously. It is not possible, for example, to change the alcohol concentration in the O region without also changing it correspondingly in the ohter two regions. Such changes influence the numerator and the denominator of the R-ratio simultaneously (Section 5.IV).

TABLE 5.1   Factors Increasing the R-Ratio

	Section discussed
Hydrophilic interactions	
Increased electrolyte concentration	5.II.A
More lipophilic counterion	5.II.B
Increased hydrogen ion concentration	5.II.C
Less hydrophilic hydrophile	5.II.D
Decreased polarity of W region	5.II.E.
Lipophilic interactions	
Decreased oil molecular weight	5.III.A
Increased length of lipophile	5.III.B
Increased branching of lipophile	5.III.B.
Simultaneous interactions	
Higher long-chain alcohol concentration	5.IV.A
Lower short-chain alcohol concentrations	5.IV.A
Lower temperature (anionics)	5.IV.B
Higher temperature (nonionics)	5.IV.B
Entropic effects: Decreased water/oil ratio	5.V

Table 5.1 summarizes those factors influencing the R-ratio and identifies the section of this chapter in which each factor is discussed. The reader interested in formulating a micellar solution subject to the variation of a restricted number of formulation variables can refer to the applicable sections. The others can be omitted without difficulty.

One can also achieve a I-III-II transformation by decreasing the water/oil ratio. As discussed in Section 5.V, this change is one that cannot be strictly understood in terms of the R-ratio. The mechanism is related primarily to entropic changes in the system rather than to an alteration in the ratio of cohesive energies. To fully understand the forces that are acting, a thermodynamic analysis is required. In Section 5.V the need for a thermodynamic analysis is argued. The analysis is, however, deferred until Chap. 8.

In Section 1.IV it is noted that for some formulations liquid crystals rather than isotropic micellar solutions will form [1]. The conditions under which the transition from liquid crystalline to isotropic solutions occurs can be understood qualitatively in terms of the R-ratio. This important concept is discussed in Section 5.VI.

Section 5.VII is devoted to complex, but subtle issues which are not readily understandable without the aid of a model. For example, almost all commercial surfactants are mixtures and therefore form mixed micelles having a composition which will be different from that in the excess phases (Section 2.VII). This has an impact on the phase behavior, which can be predicted using the pseudophase model described in Section 5.VII.

II. HYDROPHILIC INTERACTIONS

The hydrophilic interactions appear in the denominator of the R-ratio. The cohesive energies involved are A_{cw}, A_{ww}, and A_{hh}, as shown in Eq. (1.21). These terms can be changed by altering any one of a number of variables which are described in this section. Any change that tends to increase the denominator of R is expected to promote a II-III-I transformation.

A. Inorganic Electrolytes

For ionic amphiphiles, the cohesive energy most profoundly changed upon the addition of inorganic electrolyte is A_{hh}. This term is related to the free energy associated with the formation of an electrically charged C layer. This free energy has been calculated (Section 2.VI) and shown to be a factor opposing the formation of micelles. The form of the equation expressing this free energy does not change in the presence of oil, so that starting with Eq. (2.18), the following result is easily obtained:

$$\overline{\Delta f}_{el} = \frac{8\pi^2 e^2 \Gamma_s^2 \lambda}{\varepsilon} \tag{5.1}$$

where $\overline{\Delta f}_{el}$ is the electrostatic free energy per unit of C-layer area. If the temperature dependence of ε and λ can be ignored[†], then

$$A_{hh} = -\overline{\Delta f}_{el} = \frac{-B}{I^{1/2}} \tag{5.2}$$

where B is a positive constant and I is the ionic strength. Equation (5.2) shows that A_{hh} for an ionic amphiphile is a negative number which increases as the ionic strength is increased.

[†]Since λ and ε do depend on temperature, there are entropic contributions to $\overline{\Delta f}_{el}$, and to be quantitative, only that portion which is enthalpic should be equated to A_{hh}.

A second but perhaps smaller effect of added electrolyte relates to the cohesive energy, A_{cw}. The interaction of the ionic head group with polar water is changed by virtue of the common-ion effect, which for an anionic surfactant Y^- can be represented by the equation

$$Y^- + X^+ \rightleftharpoons XY \tag{5.3}$$

Increasing X^+, the counterion concentration, favors the formation of uncharged XY. The net result is a reduction in the charge per unit of area and a decrease in A_{cw}.

Equation (5.3) essentially reflects counterion binding (see Section 2.IV) and the tendency for counterions to associate strongly with charged amphiphilic groups in the C layer. More refined treatments of this mechanism have been reported [2–8] but are not discussed here.

Increasing the electrolyte concentration then tends to decrease the denominator of R through both A_{cw} and A_{hh}. Since it is generally not possible to separate the influence of these cohesive energies, the grouping $(A_{cw} - A_{hh})$ will be considered simultaneously when dealing with the effect of electrolyte in ionic amphiphiles. Figure 4.9 shows the expected I-III-II transition with increasing sodium chloride concentration.

Multivalent cations are known to associate more strongly to anionic amphiphile micelles than univalent cations [7–10], thus further decreasing the ionization in accordance with Eq. (5.3). Calcium sulfonates, for example, are known to be oil soluble, a property essential to various industrial applications, such as lubrication. Adding calcium chloride, or more generally, multivalent cations, to a system containing anionic surfactant results in a larger reduction of $(A_{cw} - A_{hh})$ than is obtainable by an equivalent amount of sodium chloride. The difference is not explainable simply by considering the ionic strength and its influence on the double layer, [Eq. (5.2)]. Specific binding must also be taken into account (see Section 5.VII.F).

Another effect of added electrolytes is to modify the structure of water by disrupting hydrogen bonds. This mechanism is most readily observed using nonionic surfactants where ionization and hence long-range coulombic interactions are absent. Increasing the electrolyte concentration is thus expected to yield I-III-II transitions even when nonionic rather than ionic surfactants are used, since the A_{cw} in the R-ratio is decreased. Based on this simple view, those cations that are the most highly hydrated should, for a given concentration, be the most effective in reducing A_{cw}. This is demonstrated to be the case in Section 6.III.C.

Figure 5.1 shows a typical phase diagram obtained using a nonionic surfactant (an ethoxylated nonylphenol) [11]. On increasing the NaCl concentration, a I-III-II transformation is observed, as expected.

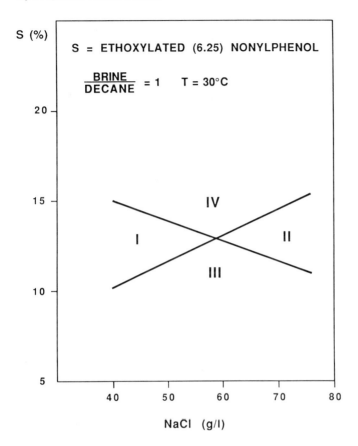

FIGURE 5.1   The phase behavior of ethoxylated nonylphenol (6.25 moles of ethylene oxide) at various surfactant concentrations as a function of salinity.   Note the large width of the type III and type IV regions, compared to Fig. 4.19 or 4.20.   (After Ref. 11.)

Note that the range of salinities yielding type III behavior is quite large compared to that observed using anionic surfactants (compare Fig. 4.20, for example).

Cationic surfactants also experience a decrease in $(A_{cw} - A_{hh})$ upon the addition of electrolyte, as illustrated by the phase changes shown in Fig. 5.2.   The mechanism is the same as that described for anionic surfactants.

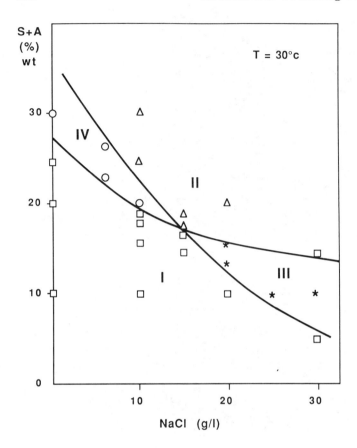

FIGURE 5.2 The effect of sodium chloride on the phase behavior of a commercial cationic surfactant Noramium MC 50 supplied by Ceca S.A. $(RN(CH_3)_3 \ Cl; \ R = coprah)$ noted S, containing 50% isopropanol and 25% water. Active material/hexanol-1 = 1 (wt); hydrocarbon: toluene-xylene cut; water/oil = 1(wt). (After T. Outassourt and M. Bourrel unpublished results.)

B. Organic Counterions

Although the CMC of anionic amphiphiles is practically unchanged if Na^+ is substituted for K^+ [12], it is decreased, often substantially, upon the substitution of an organic cation. These then become more strongly bound to, or included in, the C layer than are Na^+ ions. It is therefore expected that a given micellar system containing organic counterions will be displaced toward S_2 as compared to the same systems containing inorganic counterions at the same ionic strength. This is the observed trend [13].

In some cases the organic counterion can itself be considered to be an amphiphile. Mixtures of cationic and anionic surfactants could, for example, be considered to be a blend consisting of an organic counterion together with an ionic amphiphile where the role of organic counterion is, for example, assigned to that amphiphile present in excess. Such systems are interesting and appear again in Section 6.V, where detailed results are presented and discussed.

C. pH Effect with Carboxylated Surfactants

Carboxylic (fatty) acids are weak acids, and thus the dissociation of their salts is strongly pH dependent. The nonionized acid form, R-COOH, where R stands for any alkyl radical, is known to be oil soluble (oleic acid is a typical example). The dissociation equilibrium can be described simply by the reaction

$$R\text{-}COO^- + H^+ \rightleftharpoons R\text{-}COOH \tag{5.4}$$

where $R\text{-}COO^-$ is the ionized amphiphilic species. Decreasing the pH favors the formation of the uncharged species R-COOH. This results in a reduction of the polarity of the C region and thus in a decrease in the interaction energies $(A_{cw} - A_{hh})$. Decreasing pH is thus expected to yield a I-III-II transition. Conversely, starting with a carboxylic acid, an increase in pH is expected to produce a II-III-I transition. The observed trend is often, but not always, in this expected direction, since there are competing factors.

Adding NaOH, for example, alters the hydrophilic interaction in at least two ways. The cohesive energies $(A_{cw} - A_{hh})$ tend to increase because the proportion of dissociated surfactant increases. However, NaOH is also an electrolyte and its addition will tend to decrease $(A_{cw} - A_{hh})$. Adding NaOH may therefore promote either transition, depending on which of the two competing tendencies is dominant. Nelson [14] and Miller et al. [15] have both noted the existence of this conflict. Figure 5.3 shows the phase behavior of a micellar system containing oleic acid plotted as a function of the NaOH concentration. The two possible trends are seen. At 2 g/dl NaCl, for example, the II-III-I sequence appears on initially adding NaOH, but at higher NaOH concentrations, a type III system again appears, indicating the initiation of a I-III-II sequence. Thus at low NaOH concentrations, $(A_{cw} - A_{hh})$ increases due to surfactant dissociation faster than it decreases because of the added counterion. At high NaOH concentrations, the inverse is true.

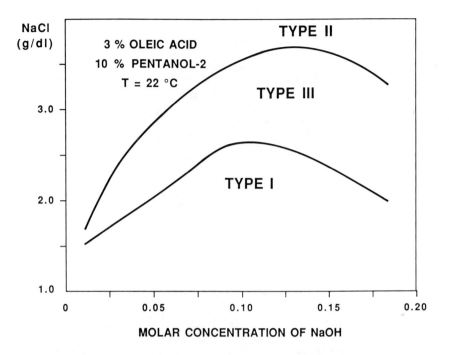

FIGURE 5.3   Effect of NaCl and NaOH concentrations on the phase behavior of systems containing 3 vol % oleic acid and 10 vol % pentanol-2. The hydrocarbon is decane and the water to oil ratio (volume) is one. (Adapted from Ref. 15.)

D. Surfactant Hydrophiles

1.   Nonionic Surfactants

Compounds containing varying quantities of ethylene oxide condensed with, for example, oleic acid, oleic alcohol, alkyl phenols, or sorbitan monooleate, form an important class of commercially available nonionized amphiphilic agents used as detergents, emulsifiers, and for the preparation of soluble oils.

For these compounds $RX(CH_2CH_2O)_mH$ (where X is usually O, N, or S and R an alkyl, acyl, or aryl radical, etc.), the interaction of the C layer with the W region is due primarily to hydrogen bonding between the water molecules and the ether oxygen atoms of the polyethylene oxide chain. The hydrophilic character of the surfactant may thus be very conveniently increased or decreased by variation of m. An illustration of the phase changes accompanying such a variation is provided by Fig. 5.4 [16]. m has been varied continuously by mixing two ethoxylated nonylphenols, characterized by m = 6.0

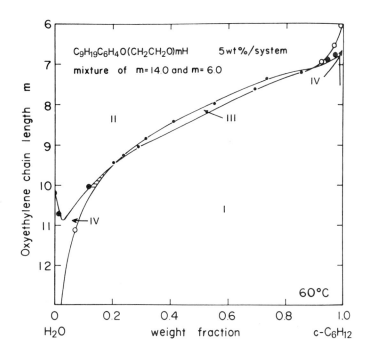

FIGURE 5.4   The effect of the average oxyethylene chain length on
the phase diagram of water plus cyclohexane containing 5wt% surfactant mixtures of $C_9H_{19}C_6H_4O(CH_2CH_2O)_6H$ and $C_9H_{19}C_6H_4O$-
$(CH_2CH_2O)_{14}H$.   (Adapted from Ref. 16.)

and 14.0.   (These amphiphiles are commercial products and thus represent the usual Poisson distribution of ethylene oxide units; 6.0
and 14.0 are thus mean values for each surfactant.)   The manner of
calculating the resulting value of m when the two surfactants are
mixed was not given by the authors, but it is likely that a linear
mixing rule was assumed.

Figure 5.4 shows that for water/cyclohexane ratios between 4 and
0.25 a I-III-II transition is obtained upon decreasing m, which is
equivalent to decreasing A_{cw}.   For water/cyclohexane ratios outside
this range, diminishing m yields the sequence I-IV-II.

Comparison of Figs. 5.4 and 4.26 provides a striking illustration
of the similarity resulting from a decrease in the ethylene oxide number of the amphiphilic agent and to an increase in temperature.   Both
diagrams exhibit the same overall features; that is, the same phase
sequence is obtained.   This is because by decreasing m or increasing
T, the A_{cw} term is decreased.   Based on the R-ratio the same qualitative behavior is to be expected.

2. Ionic Surfactants

For comparably composed C layers containing amphiphilic salts with
similar lipophiles but with different hydrophiles, the values of the
R-ratio will be determined by the relative hydrophilic solvent affinities of the different ionic groupings.

Little information is available for classification of the various
ionic hydrophiles with respect to their influence on phase behavior.
From Klevens, [17] solubilization measurements of n-heptane in aqueous solutions of dodecane-1-sodium sulfonate, dodecane-1-sodium carboxylate, dodecane-1-sodium sulfate, and dodecyl ammonium chloride,
Winsor [1] inferred that the hydrophilic solvent affinity decreases in
the order $-SO_3Na > -CO_2Na > -SO_4Na > -NH_3Cl$.

A more comprehensive comparison of the solvent affinities of the
$-CO_2Na$, $-SO_4Na$, and $-NH_3Cl$ groups was obtained by Winsor [13]
using corresponding undecane-3 derivatives and hexanol-1 to achieve
the phase transition I-G-II in systems containing cyclohexane, water,
and the amphiphilic salt under consideration. The results confirm
the sequence of decreasing hydrophilic solvent interaction $-CO_2Na >$
$-SO_4Na > -NH_3Cl$, the difference between the last two being the more
pronounced. The differences between the $-CO_2Na$ and $-SO_4Na$ groups
do not appear significant, especially in view of the difficulty of comparing the ranges of appearance of the G phase.

More recently, Lipow [18] has investigated the effect of hydrophilic structure on the phase behavior of surfactant-brine-oil mixtures by comparing the electrolyte concentration required to obtain a
type III system. Although the experiments were carried out using
various amphiphile alkyl chain lengths and alcohol concentrations, the
following order of decreasing hydrophilicity can be inferred:

$$-N(CH_3)_3Cl \gg -SO_3Na > -SO_4Na > -CO_2Na > -NH_3Cl$$

The difference in the $-SO_3Na$, $-SO_4$, and $-CO_2$ groups is not large
(see Fig. 6.12).

The classification of the various hydrophilic headgroups can be
made quantitative by comparing them at optimum behavior. Using a
definition of equal solubilization parameters at optimum (see Section
4.IV), the phase behavior of sodium octadecyl sulfate ($C_{18}SO_4Na$),
sodium stearate ($C_{17}CO_2Na$), sodium octadecyl phosphate ($C_{18}PO_4Na_2$),
and octadecyltrimethylammonium chloride [$C_{18}N(CH_3)_3Cl$] has been investigated as a function of the sodium chloride concentration. All
other conditions are identical.

Figure 5.5 shows the range of salinities for which type III behavior is observed and indicates the optimal salinity (equal volumes
of oil and brine dissolved in the micellar phase) for each surfactant.
With $C_{18}PO_4Na_2$, the I-IV-II transition is obtained instead of I-III-II.
In this case the salinity corresponding to the center of the type IV

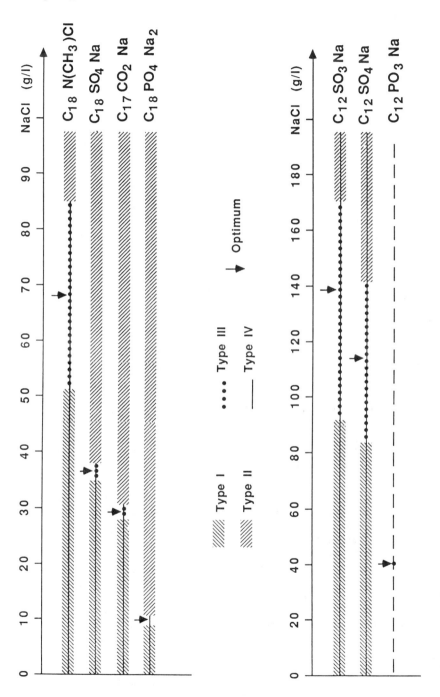

FIGURE 5.5 Comparison of the hydrophilicity for systems containing 2wt% active surfactant, 3wt% butanol, octane, and brine. The water to oil ratio by volume is one and the temperature is 50°C. (After A. Boix, F. Verzaro, and M. Bourrel, unpublished.)

region is taken as the optimal salinity. Figure 5.5 shows that for
the octadecyl series, the order of decreasing hydrophilic solvent interaction is $-N(CH_3)_3Cl$ > $-SO_4Na$ > $-CO_2Na$ > $-PO_4Na_2$. It must
be emphasized that the interaction with water of the cationic $C_{18}N$-
$(CH_3)_3Cl$ is sensitive to the Cl^- anion concentration, not to the Na^+
concentration like the anionic surfactants under consideration. It is
thus possible that the classification of $C_{18}N(CH_3)_3Cl$ in the series
would be different if sodium salts other than sodium chloride were
used. Sodium sulfate, for example, may be expected to affect the
phase behavior of $C_{18}N(CH_3)_3Cl$ systems to a larger extent than sodium chloride.

An estimation of the hydrophilicity of the sulfonate and phosphonate groups can be obtained by comparing results found using sodium
dodecyl sulfonate ($C_{12}SO_3Na$), sodium dodecyl sulfate ($C_{12}SO_4Na$),
and sodium dodecyl phosphonate ($C_{12}PO_3Na$). These results are
shown in Fig. 5.5. The larger optimal salinities obtained with the
dodecyl series as compared to the octadecyl series are due to the
lower interaction energies of these surfactants with oil. This point
is discussed in the following section.

Figure 5.5 shows that the hydrophilicity of the $-SO_3Na$ group is
slightly higher than that of $-SO_4Na$. Thus the final sequence of decreasing hydrophilic solvent interaction is

$$-N(CH_3)_3Cl \gg -SO_3Na > -SO_4Na > -CO_2Na \gg -PO_4Na_2$$

The results of the dodecyl series show that

$$-SO_3Na > -SO_4Na \gg -PO_3Na$$

The corresponding values of optimal salinities are reported in Table
5.2, together with the purity of the surfactants and the values of
solubilization parameters at optimum SP* which are discussed in Section 7.V.A.

For the case of the carboxylates, the question arises of defining
the alkyl chain length. In the preceding results, all the carbon atoms
of the molecule have been counted as belonging to the hydrophobic
moiety. The first carbon carrying the oxygens might be considered
as part of the hydrophilic head and not counted as a part of the lipophile. In that case $C_{18}CO_2Na$ should be compared with other octadecyl
surfactants studied instead of the $C_{17}CO_2Na$ which was actually used.
This substitution will result in a lower optimal salinity but should not
change the hydrophilic sequence.

In addition to those hydrophiles considered above, a number of
other hydrophilic ionic moieties are available. When a surfactant is
required to function in concentrated electrolyte solutions, high solvency is maintained by increasing the hydrophilicity rather than reducing the length of the lipophile. An example of amphiphiles with

TABLE 5.2 Purity and Optimal Salinities[a] of the Surfactants Investigated in Fig. 5.6

Surfactant	$C_{18}N(CH_3)_3Cl$	$C_{18}SO_4Na$	$C_{17}CO_2Na$	$C_{18}PO_4Na_2$	$C_{12}SO_3Na$	$C_{12}SO_4Na$	$C_{12}PO_3Na$
Active material (wt %)	93	98.5	95	98	99	100	96
Inorganic (wt %)	4.5 (NaCl)	1.5 (Na_2SO_4)	—	—	—	—	—
Water + unreacted material (wt %)	2.5	—	5	2	1	—	4
Optimal salinities NaCl(g/liter)	68	37	30	10	140	115	40
SP* (cm^3/cm^3)	8	24.3	21	>30	8.2	7	—

[a]The optimal salinities include the inorganic electrolytes contained in the surfactants.

increased hydrophilicity are the ethoxylated ionics discussed next. Other examples are the α-olefin sulfonates, disulfonates, or α-sulfo-methyl esters.

3.   Ethoxylated Ionic Surfactants

Typical representatives of this class of amphiphilic compounds are the sulfated polyoxyethylenated alcohols $R(OCH_2CH_2)_mSO_4^-M^+$, where m usually varies between 1 and 3.   Other ionic groups commonly encountered are $-CH_2CO_2^-$, $-CH_2SO_3^-$, and $-CH_2CH_2SO_3^-$.

Although the CMC values decrease upon introducing ethylene oxide units between the lipophilic chain and the ionic headgroup [19], the hydrophilic interaction of these ethoxylated ionic surfactants increases with ethylene oxide content.   As a result, this type of amphiphilic compound exhibits an improved tolerance [20,21] to electrolyte in general and multivalent cations in particular.

Figure 5.6 shows the results observed for sodium oleyl ethylene oxide sulfonates $[C_8H_{17}-CH=CH-C_8H_{16}-(OCH_2CH_2)_mSO_3Na]$, with m = 0, 1, 2, and 3 [22].   The optimal salinity is given for various n-alkanes, referred to by their alkane carbon number, ACN.   For a given ACN, the optimal salinity does not increase linearly with m. The first ethylene oxide adduct does not change the optimal salinity significantly and can be considered as presenting more or less balanced affinities for oil and water.   For the different hydrocarbons studied, the introduction of a second ethylene oxide unit yields an increase in optimal salinity of about 1.5% independent of the ACN. As m increases from 2 to 3, the interaction energies with water increase and the R-ratio decreases.   Thus a II-III-I transformation is promoted by increasing m.

4.   Mixtures of Surfactants

Mixing two (or more) surfactants to obtain intermediate states between S_1 and S_2 is a widely used technique and yields generally satisfactory results.   In the case of nonionics, it is especially convenient to blend compounds of the same lipophilic moiety but differing by their ethylene oxide content to obtain the behavior of amphiphilic species of intermediate ethylene oxide numbers (EON) (see Fig. 5.4).   Simple linear mixing rules will be acceptable only when the EON of the two surfactants blended do not differ too widely.   Generally, surfactants that differ by less than 2 EON units satisfy a linear mixing rule.

One of the difficulties that arises when the number of EON units differs markedly is that selective partitioning of certain of the amphiphilic molecules between oleic and aqueous phases takes place, thereby greatly complicating the phase behavior.   This complexity is discussed further in Section 5.VII.D.

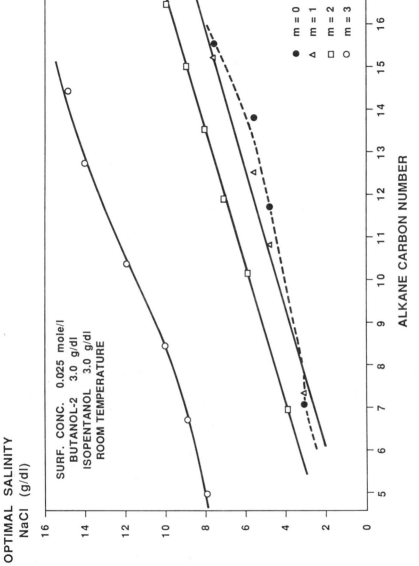

FIGURE 5.6 Optimal salinity as a function of ACN for oleyl ethylene oxide sulfonates. m indicates the number of oxyethylene units. The concentrations refer to the aqueous phase. (After Ref. 22.)

E. Other Polar Solvents

The hydrophilic interaction can also be changed by replacing water with another polar solvent immiscible with hydrocarbon. All the terms in the denominator of the R-ratio are altered, and some of the changes tend to increase R, while others decrease it. The ultimate summation of these individual effects is not predictable.

Winsor [1,23] reported that replacing water by ethanediol (ethylene glycol) in a type II system produces a type I system. He proposed that this transformation was due to a decrease in A_{ww} and/or A_{cw}. The surfactants used by Winsor were ionic (sodium sulfo-dioctyl succinate, sodium tetradecyl sulfate, or undecane-1 ammonium chloride). Because the dielectric constant of ethanediol ($\varepsilon/\varepsilon_0 = 37.7$ at 20°C) is less than that of water ($\varepsilon/\varepsilon_0 = 81.1$ at 20°C), Eq. (5.1) shows electrical free energy to decrease, thereby increasing A_{hh}. This change must tend to increase the R-ratio, as does decreasing A_{cw}. Thus (to explain the trend observed by Winsor) the decrease in A_{ww} must be the dominant effect when water is replaced by ethanediol.

Winsor also noted that the usual type of phase behavior obtained upon addition of electrolyte and dodecanol-1 was observed when water is replaced by ethanediol; however, the tendency to form gel phases was greatly reduced.

Using nonionic surfactant (ethoxylated nonylphenol), Verzaro et al. [24] reported a different behavior; namely, replacing water in a type I system by ethanediol yields a type IV or a type III system. Figure 5.7 shows the results for a commercial ethoxylated nonylphenol having an average ethylene oxide number of 6. The polar phase is a mixture of water and ethanediol, the proportions of which are varied. The trend I-III-II differs from that observed for ionic surfactants implying that the predominant term is, in this case, A_{cw}.

Based on these observations, it is seen that predicting the variation in the R-ratio when water is replaced by other polar solvents is difficult. This difficulty is made even more evident in Chap. 7, where it is shown that replacing water by 1,3-propanediol or 1,4-butanediol promotes a phase sequence opposite to that found for ethanediol.

III. LIPOPHILIC INTERACTIONS

The discussion so far has focused on those changes which influence primarily the cohesive energies that exist between the W and C regions. These are terms that appear in the denominator of the R-ratio. In this section variation of cohesive energies in the numerator is considered. These can be modified in a number of different ways. The most obvious approaches include varying the molecular weight of the oils or altering the lipophile. These are considered first.

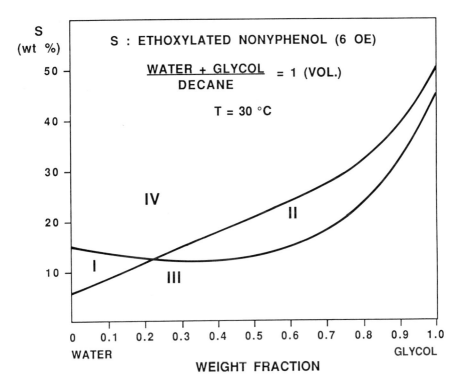

FIGURE 5.7 The effect of replacing water by ethylene glycol on phase behavior. (After A. Sanchez and M. Bourrel, unpublished results.)

A. Hydrocarbon Molecular Weight or Type

Increasing the molecular weight of a homologous series of hydrocarbons increases A_{oo} (promoting separation of O) more rapidly than A_{co} (promoting miscibility). This can be understood within the context of the argument developed in Section 1.V.

   For a given surfactant, Figure 5.8 depicts the qualitative variation of the numerator of R as a function of ACN. From Eqs. (1.14), (1.15), and (1.16) it can be shown that the numerator of R varies with ACN according to

$$A_{co} - A_{oo} - A_{\ell\ell} = c \sqrt{A_{\ell\ell}} \; ACN - a(ACN)^2 - A_{\ell\ell} \qquad (5.5)$$

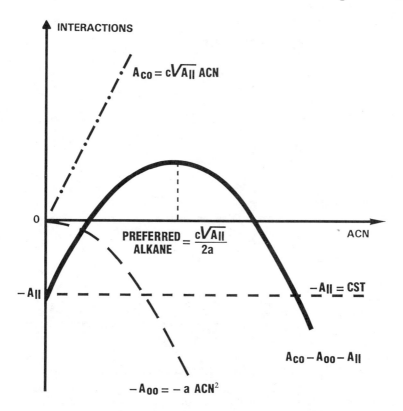

FIGURE 5.8   Qualitative variation of the cohesive energies A_{co}, A_{oo}, $A_{\ell\ell}$ and of the sum $A_{co}-A_{oo}-A_{\ell\ell}$ with alkane carbon number. a and c are held constant.   For a fixed lipophilic chain length, it is seen that the numerator of the R-ratio goes through a maximum for $ACN = c\ \sqrt{A_{\ell\ell}}/2a$.

It can be seen that the numerator does not vary monotonically with ACN but exhibits a maximum.   This result suggests that the trends observed when increasing the molecular weight of the hydrocarbon (oleic phase) will not necessarily be monotonic, especially if very small molecular weights such as propane and butane are included in the study.

Only few investigations of the phase behavior of systems containing alkanes of a very low number of carbon atoms have been reported. The largest body of data corresponds to hydrocarbons whose molecular weights are apparently situated beyond the maximum in the numerator of R shown in Fig. 5.8.   In this region, increasing ACN decreases the numerator of R since A_{oo} increases faster than A_{co} and thus

promotes the separation of oil out of the micellar phase. This results
in the II-III-I transition.

It is important to note that upon starting with small ACN (less
than 5 or 6, generally) a different trend is predicted; namely, increasing ACN should promote a I-III-II transition. At low ACN, A_{co}
increases more than A_{oo}, yielding a decrease in R. For small ACN
there is a second important factor. The lateral interaction between
the surfactant lipophiles, $A_{\ell\ell}$, greatly exceeds A_{co}, resulting in amphiphile precipitation, gel-like behavior, or liquid crystals (see Section 5.VI).

Homologous series of organic liquids other than n-alkanes, such
as alkylbenzenes or alkylcyclohexanes [25–27], have also been investigated. They all show that increasing hydrocarbon molecular weight
within the same series promotes II-III-I transition. Furthermore,
mixtures of paraffinic and aromatic oils have shown that increasing
the aromatic content of the oil promotes the I-III-II transition [25,26],
as would be expected from the higher solvency of aromatics for the
surfactant lipophile.

The shape and size of the hydrocarbon molecule are also parameters that must be taken into account when dealing with its interaction with the C region. Hydrocarbon molecules will in general in-
terpenetrate and mix with the lipophilic portions of the amphiphilic
molecules. The effective area per hydrophile in the C layer is usually considerably greater than the cross-sectional area of the lipophilic
part of the amphiphile. The interpenetration of the lipophilic chains
of the amphiphile with the hydrocarbon can therefore occur without
necessarily changing the effective area per polar group. This will
depend on the shape and polarity of the hydrocarbon.

Spegt et al. [28] have carried out an x-ray diffraction study of
liquid-crystalline phases occurring at 82°C in the ternary systems
sodium myristate/water/ethylbenzene and sodium myristate/water/n-
octadecane. Their main conclusion was that the hydrocarbon may
be incorporated into the lamellar G phase in a wide range of proportions without changing the gel structure. With octadecane dissolved
in the lamellar phase, the thickness of the bimolecular leaflet for a
given water/soap ratio apparently increases linearly with the hydrocarbon content, while the effective area per hydrophile remains constant. On the other hand, when ethylbenzene is dissolved in the
lamellar phase, the effective area per amphiphile increases with ethylbenzene content. Thus the molecular structure of the oleic phase can
be expected to be an important factor determining the phase behavior
and solvency of amphiphilic compounds. The increase in area per hydrophile with increase in ethylbenzene content is, for example, probably due to the localization of polarizable ethylbenzene molecules between the headgroups and their molecular packing so that separation
is thereby increased. This effect is not appreciable with the less

polarizable $\underline{n}$-octadecane molecules, which are probably contained within the hydrocarbon region of the micelles, where they exert little effect on the area occupied by the hydrophiles. Furthermore, it is likely that the $\underline{n}$-octadecane molecules may orient more or less perpendicularly to the oil/water interface, between the soap lipophiles; whereas, as has been noted, the ethylbenzene molecules may not accommodate such an arrangement. These same considerations relate to the solubilization of hydrocarbons within the isotropic micellar phase and have important consequences on the solubilizing power of these systems. Solubilization is discussed in Chap. 7. A further illustration of the presence of hydrocarbon molecules in the C region is provided by the fractionation of oil mixtures in micellar systems (see Section 5.VII.G).

B. Surfactant Tail Structure

1. Length of the Lipophile: Mixtures of Surfactants

The effect of this structural parameter can be examined using essentially the same arguments as applied to describe the influence of molecular weight of the oil on the R ratio. If $A_{\ell\ell}$ is assumed to increase as n^2, where n is the lipophilic chain length, as might be expected from Eq. (1.4), $A_{co} - A_{oo} - A_{\ell\ell}$ is given by

$$A_{co} - A_{oo} - A_{\ell\ell} = d \sqrt{A_{oo}} \, n - bn^2 - A_{oo} \qquad (5.6)$$

where d, b, and A_{oo} are constants for a given oil. Figure 5.9 shows that upon increasing n, the numerator first increases, goes through a maximum, and then decreases. The position of the maximum is a function of a number of variables, but most particularly, it depends on the molecular structure of the oleic phase. It is expected that the value of n at the maximum, for example, decreases when the ACN is decreased. For most hydrocarbons of practical interest the trend on increasing n is I-III-II, which means that R is an increasing function and therefore A_{co} in general increases faster than $A_{\ell\ell}$.

One method for increasing n continuously is blending two different surfactant molecules having different length of lipophilic chains. A result is shown by Fig. 5.10. The phase behavior is shown for various proportions of a binary mixture of alkyl orthoxylene sulfonates which have lipophilic chain lengths of 12 and 24 carbon atoms, respectively. The average chain length, n, shown along the ordinate is a mole fraction average of the two surfactants. It is seen that for a given oil, say decane, increasing n yields the expected I-III-II transformation. A similar trend is shown for hexadecane, with the transition requiring a larger n (longer surfactant tail) to complete.

Unlike the case of nonionics, the partitioning of ionic surfactant mixtures between equilibrium excess phases is often negligible; most

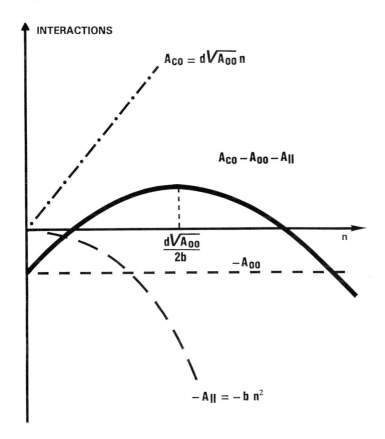

FIGURE 5.9 Variation of A_{co}, A_{oo}, $A_{\ell\ell}$ and of $A_{co}-A_{oo}-A_{\ell\ell}$ with n, number of carbon atoms of the surfactant lipophile, under the geometric mean assumption. d and b are constants. The hydrocarbon is maintained constant. The numerator of R goes through a maximum for $n = d\sqrt{\bar{A}_{oo}}/2b$.

of the surfactant is concentrated in the micellar phase and only traces are found in the excess phases. Linear mixing rules can thus generally predict the behavior of mixtures of ionic amphiphiles [25, 30]. The exceptions concern the mixtures of surfactants differing widely by the length of their lipophilic portions, for example, or by the number of hydrophilic groups (monosulfonates and disulfonates, for example). In those cases the highly hydrophilic species partition preferentially into water (see Section 7.I.G). A detailed documented discussion of the fractionation of mixtures of sulfonates has been reported by Salter [31].

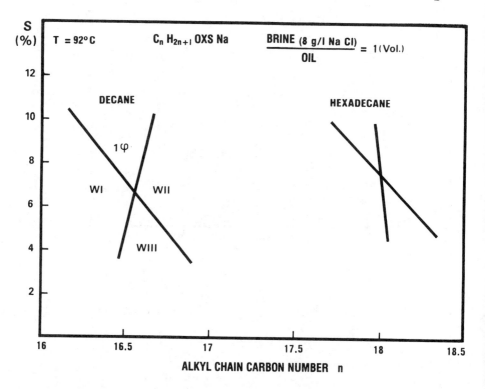

FIGURE 5.10 The effect of blending sodium C_{12} orthoxylene sulfonate and sodium C_{24} orthoxylene sulfonate on phase behavior for two alkanes. n refers to the average number of carbon atoms of the surfactant lipophile. (Adapted from Ref. 29.)

Most of the reported phase behavior studies for micellar systems correspond to that part of the curve shown in Fig. 5.9 for which the numerator of R increases with increasing n. As noted previously, on the descending part of this curve (large n) the lateral interactions $A_{\ell\ell}$ increase rapidly and hence tend to correspond to systems for which the amphiphile precipitates and separates as a solid, a liquid crystal, or a gel (see Section 5.VI).

2. Structure of the Lipophile

For a given alkyl chain and hydrophilic head, surfactants may differ, for example, by the point of attachment of the polar group. One effect of displacing the polar group from the terminal position to a central position is to increase the area occupied by the surfactant in C, resulting in a decrease of $A_{\ell\ell}$ and for ionic amphiphiles, an increase of A_{hh}. Both of these trends increase the value of R and

yield I-III-II transitions [26,32,33]. This trend has been shown by
Winsor [34] to apply when intermediate liquid-crystalline phases form.
Winsor found that the volume of cyclohexanol required to produce a
G lamellar phase transition in a type I system decreased on displacing
the SO_4Na group from the -1 to the -7 position along the tetradecane
chain [1]. This can be interpreted as meaning that ionic amphiphiles
with near-midchain attachment of the ionic group have under similar
conditions a larger value of R than those systems for which the ionic
group has a near-endpoint attachment (see Section 7.V.B).

The area per amphiphile in the C region is also increased by
adding side groups (methyl, ethyl, etc.) to the main alkyl chain, or
by adding aromatic rings, introducing double bonds, and so on. The
impact of these factors on the reduction of the lateral interactions
can, to some extent, be appreciated by considering their effect on
the physical state of hydrocarbons. For example, squalane is a liquid
at room temperature because of the presence of lateral CH_3 groups
which prevent crystallization, whereas the linear paraffin of equivalent molecular weight is a solid under similar conditions. Similarly,
8-octadecene is liquid at room temperature, as opposed to n-octadecane, which is a crystalline solid. The influence of the lipophile
structural parameters has been recently investigated [26,32,33] and
the results are in agreement with trends predicted here based on the
R ratio; namely, branching tends to increase R and promote I-III-II
transformations.

IV. FACTORS INVOLVING SIMULTANEOUS
CHANGES IN LIPOPHILIC AND HYDROPHILIC
INTERACTIONS

A. Addition of Organic Compounds Exhibiting Amphiphilic Character

As noted in Section 1.I, there are many different molecular structures
that exhibit amphiphilic character. These all tend to aggregate to
some extent in aqueous solution and to reduce the surface tension.
There is no clear distinction between molecules which can be considered as surfactants and those which have sometimes been called cosolvents or cosurfactants. In this section the organic additives given
the greatest emphasis are alcohols, which of course exhibit hydrophilic character through the OH group and lipophilic character through
the hydrocarbon chain. The only justification for considering alcohols
as a separate organic additive rather than a surfactant is that alcohols
are seldom, if ever, used alone because of their poor solvency.

It should be understood that the discussion here regarding alcohols also applies to compounds of similar characteristics, such as ke-
tones, amines, and so on. Because of their amphiphilic character,
some of the alcohol molecules will be dissolved by incorporation into
the C region, their polar group lying among the ionic groups and

neighboring water molecules and their hydrocarbon groups lying among the hydrocarbon groups of the amphiphilic salt within the C region. However, all of the alcohol molecules will not reside within the C region. A portion will be distributed between the O and W regions of the solution according to its relative hydrophilic, amphiphilic, or lipophilic character. Methanol, for example, will be largely concentrated in the W region, while a large part of dodecanol-1 will partition into the O region. On the other hand, alcohols exhibiting balanced interactions with oil and water are expected to partition more favorably to the C region. The C_3 or C_4 alcohols are, as seen from Fig. 5.11 [35], nearly balanced. This figure shows precisely the same type of phase map as was found in Figs. 4.19 and 4.20, except that in Fig. 5.11, no amphiphile other than alcohol is used. The

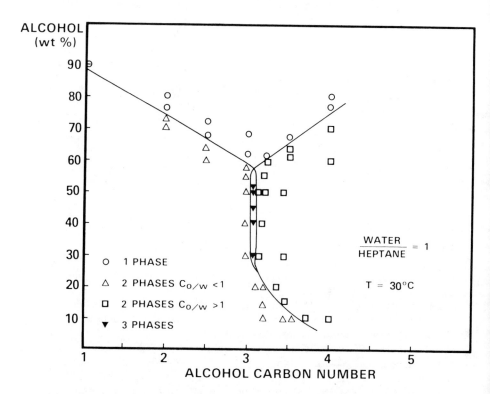

FIGURE 5.11 A phase diagram for systems containing alcohol but no other amphiphilic compounds. The similarity with phase maps of amphiphiles having greater solvency is striking. Noninteger alcohol carbon numbers refer to mixtures. $C_{o/w}$ is the alcohol concentration in the oil phase divided by that in the aqueous phase. (After Ref. 35.)

phase map is obtained by varying the lipophilic tail length of the alcohol. Noninteger carbon numbers have been obtained by mixing, on a mole-fraction basis, the two alcohols adjacent to the desired value.

For the case considered in Fig. 5.11, an alcohol having a carbon number of 3.08 yields the minimum of the single phase boundary, and therefore this alcohol can be considered optimum, that is, the alcohol for which R = 1. Furthermore, three-phase systems are observed over a very narrow range of alcohol carbon numbers, thus emphasizing the similarity between alcohols and surfactants and the difficulty of distinguishing between an amphiphile and cosolvent in any precise way.

The sensitivity of the phase behavior to the alcohol carbon number (i.e., the narrow width of the type III region in Fig. 5.11) can be understood as follows: If the ratio R is approximated by $R \approx A_{co}/A_{cw}$, A_{co} is assumed to be proportional to the number n of carbon atoms of the amphiphile (alcohol) and A_{cw} is considered independent of n, then

$$R \approx \frac{\beta n}{A_{cw}} = 1 \qquad (5.7)$$

for a system taken at optimum (β is a constant). If n is changed by Δn, then R entails a change,

$$\Delta R \approx \frac{\Delta n}{n} \qquad (5.8)$$

since $A_{cw} = \beta n$ at optimum. A variation of one carbon atom in the amphiphile chain ($\Delta n = 1$) therefore yields a change in R roughly inversely proportional to the chain length. It follows that starting from an optimized system, a given change in amphiphile tail length will more readily produce phase change with short amphiphiles than with long ones.

Adding alcohol to a surfactant/water/oil system entails a change in interactions energy per unit area of interface of the C layer with the O and W regions. This change depends on the nature of the alcohol and on its interfacial concentration. This can be illustrated by simple calculations in which A_{co} and A_{cw} are viewed as the sum of the surfactant and alcohol interactions per unit area of interface. If x_a and x_s are the interfacial mole fractions of alcohol and surfactant, respectively, whose interaction energy with oil per molecule of surfactant or alcohol, respectively, are a_{co}^a and a_{co}^s, the A_{co} term can be written, if S_a and S_s are the respective interfacial molecular areas of alcohol and surfactant, as follows:

$$A_{co} = \frac{x_a a^a_{co} + x_s a^s_{co}}{x_a S_a + x_s S_s} \tag{5.9}$$

Similarly,

$$A_{cw} = \frac{x_a a^a_{cw} + x_s a^s_{cw}}{x_a S_a + x_s S_s} \tag{5.10}$$

where a^a_{cw} and a^s_{cw} are, respectively, the interaction energy per molecule with water of the alcohol and surfactant molecules. Neglecting the changes in $A_{\ell\ell}$, A_{hh}, A_{oo}, and A_{ww}, R can be written as

$$R \approx \frac{a^s_{co} - (a^s_{co} - a^a_{co})x_a}{a^s_{cw} - (a^s_{cw} - a^a_{cw})x_a} \tag{5.11}$$

since $x_a + x_s = 1$. For a given surfactant, a^s_{co} and a^s_{cw} are constants; furthermore, generally, $a^s_{co} > a^a_{co}$ and $a^s_{cw} > a^a_{cw}$. Thus, increasing the proportion of alcohol in the C layer decreases both the numerator and the denominator of the R-ratio. If the effect of various alcohols is to be compared, a^a_{cw} can be taken as a constant and a^a_{co} as increasing with the number of carbon atoms of the alcohol. It follows that increasing the C layer composition of those alcohols for which $a^a_{co} > a^a_{cw}$ will tend to increase the R-ratio, thus promoting the I-III-II transition, while the alcohols for which $a^a_{co} < a^a_{cw}$ will produce the II-III-I transition on increasing their concentration in the C region. Both trends have been reported in the literature for both anionic [36,37] and nonionic surfactants [38].

The results shown in Fig. 5.11 can be understood by noting that $a^a_{co} > a^a_{cw}$ corresponds to alcohols whose number of carbon atoms is larger than 3 ("heavy" alcohols), while $a^a_{co} < a^a_{cw}$ corresponds to alcohols whose alkyl part contains less than 3 carbon atoms ("light" alcohols). Equation (5.9) shows clearly that R depends on the alcohol interfacial concentration through x_a and on the type of alcohol through a^a_{co}. Examples of these two effects are provided by Figs. 4.21 and 4.36. Obviously, x_a is related to the overall alcohol concentration and to its nature since these two factors govern its partitioning between the O, W, and C regions.

The presence of the alcohol in the C layer is responsible for the rotation of the phase map shown in Fig. 4.19. Because of its marked solubility in hexane, pentanol-1 exhibits a partition coefficient between O and C strongly favorable to O. At a given salinity and low surfactant plus alcohol concentration, pentanol-1 distributes mainly into the O region and x_a is small. The system exhibits type I behavior. When the surfactant and alcohol concentrations are increased together maintaining their ratio (S/A) constant, the interfacial alcohol concentration x_a increases. The balance of interactions of C with O and W becomes more favorable to O. Type III and then type II must follow, resulting in the rotation of the phase map. The partitioning of alcohol is described in quantitative terms in Section 5.VII.C based on the pseudophase model.

It must be emphasized that Eq. (5.11) embodies many of the important factors regarding the effect of alcohol on phase behavior, but its derivation makes use of arguable approximations. On the one hand, a_{cw}^s is probably affected by the presence of the neighboring alcohol molecules. It has been shown, for example, that the counterion binding to micelles of ionic surfactants is significantly decreased by the addition of alcohols [39,40] (see Section 2.IV). With nonionic surfactants, hydrogen bonding may occur between the surfactant and alcohol hydrophiles. On the other hand, a_{cw}^a is not constant but probably decreases due to a self-association mechanism through intermolecular hydrogen bonding, which reduces the ability of the OH group to form hydrogen bonds with water (see Section 1.II.B). Finally, the lateral interaction between the surfactant molecules is obviously decreased, as illustrated by the destruction of the G phase generally observed on increasing the alcohol concentration. However, the primary mechanisms underlying the influence of alcohol are taken into account by Eq. (5.11). Alcohol modifies the phase behavior primarily because of its presence in the C region, not because of any possible change in the solvent properties of the water and oil bulk phases. S_1-to-S_2 transformations have been observed at very low alcohol concentrations ($\approx 1\%$), which precludes any significant variation of these properties [24].

B. Temperature

The mechanism by which a change of temperature affects solubility in amphiphilic solutions is complex. In those amphiphilic solutions for which the simple miscibility of O and W, as well as the distribution of the amphiphile between the O, W, and C regions, are not expected to be changed much by variations in temperature, the following are likely to apply:

1. The lipophilic interaction A_{Lco} between the nonpolar moiety of the surfactant molecules and a nonpolar hydrocarbon is of the London type, essentially unaffected by changes in temperature (see Section 1.II.A).

2. Increasing temperature is expected to increase the average area occupied per surfactant molecule at the water/oil interface primarily because that force driving the adsorption, the hydrophobic effect, is decreased. This is an important issue in two respects. First, the solvency or solubilization tends to decrease with increased temperature (see Section 7.V.D) and because of decreased lateral interactions, the tendency to form liquid crystalline mesophases is decreased (Section 5.VI.B).

3. Increasing temperature yields a decrease in $A_{\ell\ell}$ and thus contributes to increase the R-ratio. When the polar group of the amphiphilic constituent is derived from ethylene oxide, A_{hh} is also expected to decrease and to compensate to some extent for the decrease in $A_{\ell\ell}$. The larger effect stems, however, with the interaction of the polyoxyethylene chain and water, which is commonly attributed to the formation of hydrogen bonds between the ether oxygen atoms and the water. A rise of temperature will tend to disrupt these hydrogen bonds and thus to diminish A_{cw}. The net result is that the R-ratio increases. An illustration is provided by Fig. 4.26, which shows that on increasing temperature, the I-III-II transition is obtained for intermediate water/cyclohexane ratios, while the I-IV-II transition is observed for both water-rich and oil-rich systems.

4. When the hydrophile is ionic, A_{hh} increases on increasing temperature (less negative) and thus contributes, together with the change in $A_{\ell\ell}$, to increase the R-ratio. However, an opposing effect arises, namely the facilitated dissociation of the counterions from the hydrophilic side of the C layer. This results in increased A_{cw} and thus in decreased R-ratio. The effects of temperature on the numerator of R and on the denominator compensate each other. The net result is that the effect of temperature on ionic surfactants is small and tends generally to produce the II-III-I transition [26,41,42]. In the case of alkyl p-ethylbenzene sulfonates, however, the effect of temperature has been found to depend on the point of attachment of the aromatic ring to the alkyl chain [43]. For some of them the R-ratio has been found to increase with temperature [43].

C. Pressure

Increasing pressure results in closer proximity of the molecules and thus in higher interactions, as shown by the distance-dependence expressions of the interaction potentials in Eqs. (1.1), (1.2), and (1.3). Indeed, the sensitivity of the effect is expected to be related to the compressibility of the various liquids. In systems essentially composed of hydrocarbon and water, the major effect should be produced in the oil phase since hydrocarbons are generally more compressible than water and should result mainly in increased cohesive energy A_{oo}, that is, in decreased R.

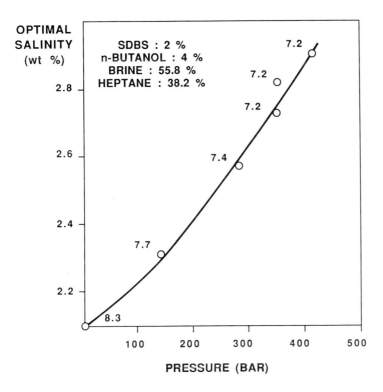

FIGURE 5.12 The effect of pressure on optimum behavior. SDBS is a commercial sodium dodecyl benzene sulfonate. The compositions are given in weight. The number next to each point indicates the solubilization parameter. (Adapted from Ref. 44.)

Figure 5.12 shows that the optimal salinity increases with the pressure [44]: when the pressure is raised on an optimized type III system, a type I is produced and increased salinity is required to return to optimum. The effect of pressure on optimal salinity is expected to become smaller if oils less compressible than heptane are to be used.

V. ENTROPIC EFFECT: WATER/OIL RATIO

The complete phase relationships of systems containing only three components (oil, water, and a single amphiphile) can at constant temperature and pressure be represented using a ternary diagram. For systems exhibiting type III behavior, the phase rule allows no degrees of freedom when three phases coexist. The composition of each one of the three phases is therefore invariant and is represented by a

single point (composition) on a ternary diagram. The triangle formed
by connecting the three points is called a tie triangle and any overall
composition corresponding to a point lying within this triangle will divide into three phases having the compositions represented by the
apexes of the triangle.

This is illustrated in Fig. 5.13, where ternary phase diagrams are
shown for a three-component system at four different temperatures [45].

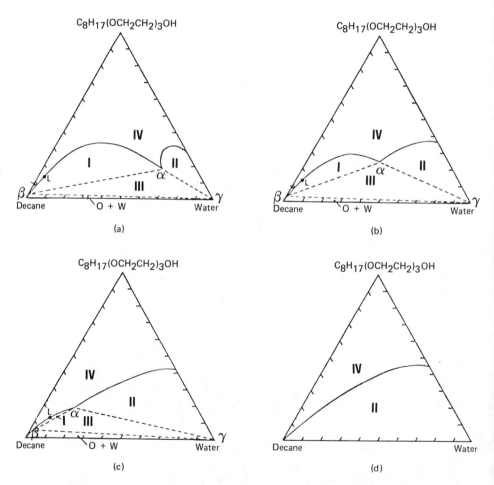

FIGURE 5.13 Phase diagrams for pure isomeric $C_8H_{17}(OCH_2CH_2)_3OH$-
water-decane system at various temperatures: (a) 15.8°C, (b) 21.5°C,
(c) 26°C, and (d) 30°C. α, β, γ are invariant points, L are plait
points. (Adapted from Ref. 45.)

Since the surfactant, $C_8H_{17}(OCH_2CH_2)_3OH$, is nonionic, a small rise in the temperature, about 10°C, sharply alters the phase diagrams, promoting a I-III-II transition. This progression of states is seen in Fig. 5.13. At the highest temperature (Fig. 5.13d) a two-phase region exists in which the micellar phase S_2 is in equilibrium with an excess aqueous phase. As the temperature is reduced, a three-phase region appears, characterized by a tie triangle in which the surfactant-rich vertex is positioned near the decane-surfactant leg of the ternary diagram, indicating that the micellar phase contains mostly decane and surfactant but little water. Further decreasing the temperature causes this surfactant-rich vertex to move from the decane-surfactant leg of the ternary diagram toward the water-surfactant leg. In Fig. 5.13b the micellar phase contains roughly equal proportions of decane and water. This, as has been discussed in Section 4.IV, corresponds to the optimum system (R = 1). Further decrease in the temperature shifts the micellar phase composition toward the water boundary. Ultimately, the excess aqueous phase will entirely disappear and a type I system (not shown) obtained.

For the present discussion the essential feature to be considered is the behavior of the system at optimum (T = 21.5°C, Fig. 5.13b). At a water/oil ratio of unity and within a range of surfactant concentrations, three phases exist. The micellar phase contains in this case equal proportions of oil and water and as noted above, R = 1. If the weight fraction surfactant is maintained constant and the water/oil ratio is increased, Fig. 5.13b shows that type II behavior will be obtained.

Similarly, if starting from equal proportions of oil and water, the water/oil ratio is decreased, a two-phase region exhibiting type I behavior is entered. Thus increasing the water/oil ratio promotes a I-III-II transition. This transition is not, however, attended by any change in the cohesive energies. The R-ratio is not affected by an alteration in the water/oil ratio except, perhaps, for small changes in the area per amphiphile within the C layer. Thus the influence of the water/oil ratio cannot be explained by consideration of the R-ratio.

The transitions exhibited in Fig. 5.13b resulting from a change in the water/oil ratio are believed to be in large measure a result of entropic contributions to the free energy of micellar solutions. Since the R-ratio includes only energetic terms, it cannot then be used to predict the influence of the water/oil ratio. Entropic contributions are considered in Chap. 8, where thermodynamic models of micellar systems are developed. One of these models treats the micellar solution as a collection of uniform spheres dispersed into a continuous phase. The optimum system, which is composed of equal volumes of oil and water, occurs when the free energy is the same regardless of which region, O or W, is the dispersed phase. Thus, for the optimum systems, the droplet model does not distinguish between water-continuous and oil-continuous systems. These two states then more or less coexist at the same time.

As the water/oil ratio is changed, the small entropy associated with a dispersion of drops (see Section 8.III.F) will favor that system having the largest number of drops. For large water/oil ratios, the number of oil drops, if water continuous, exceeds the number of water drops in an oil-continuous state, assuming a constant amphiphile concentration in the C layer.

The argument is simple if the volume occupied by the surfactant can be neglected. If N_O is the number of spherical oil drops per unit volume when water is continuous, then

$$N_O = \frac{3\phi_O}{4\pi r_O^3} \qquad (5.12)$$

and the surface area per unit volume is $3\phi_O/r_O$, where r_O is the radius of an oil drop and ϕ_O is the volume fraction of the oil region. Similarly, for an oil-continuous system, the number of water drops per unit volume is

$$N_W = \frac{3\phi_W}{4\pi r_W^3} \qquad (5.13)$$

The ratio of drop concentrations is therefore given by

$$\frac{N_O}{N_W} = \frac{\phi_O r_W^3}{\phi_W r_O^3} \qquad (5.14)$$

For a constant surface area this reduces to

$$\frac{N_O}{N_W} = \left(\frac{\phi_W}{\phi_O}\right)^2 \qquad (5.15)$$

Equation (5.15) indicates that large water/oil ratios, $\phi_W/\phi_O > 1$, yield values of N_O/N_W which are correspondingly large. Thus entropy favors oil drops in water, that is, type I systems. Conversely, $\phi_W/\phi_O < 1$ favors type II systems.

The concept applied throughout this book asserts that the Winsor R ratio dictates the structure of the micellar phase when the water/oil ratio is unity. This R ratio represents the curvature (see Section 8.III) that the C region would naturally assume all other factors being equal. Altering the water/oil ratio will, however, cause the actual curvature to differ from the natural one.

Thus the micellar phase at points just inside the type I and type II lobes surrounding the tie triangle in Fig. 5.13b is little different from that which exists within the micellar phase at the upper vertex of the tie triangle. Because of this evident similarity, these particular type I and type II systems have been designated as being in a "type III environment" [46]. This view is consistent with the concept expressed here; namely, $R = 1$ at all compositions of Fig. 5.13b.

VI. TRANSFORMATION FROM STRUCTURED MESOPHASES TO ISOTROPIC SOLUTIONS

A. General

The importance of thermal fluctuations that disrupt the interface for obtaining isotropic systems at optimum has been stressed in Chap. 1 (Sections 1.III.B and 1.V.B) and Section 4.I. Whenever these fluctuations are insufficient to prevent long-range structures from forming, anisotropic liquid-crystalline phases may appear or the amphiphile may be precipitated as a crystalline solid. In other circumstances, coarse emulsions, which may last for months, occur, making difficult the identification of the equilibrium phase behavior.

The physical properties and microstructure of various possible forms of surfactant precipitates are generally imprecisely characterized because of the difficulty caused by persistent emulsions. It appears, however, that two different types of microstructures can be considered, and these can best be described by considering their dependence on alkane carbon number. These two regimes are illustrated in Fig. 5.14, where for a given alkane, the I-III-II transition is produced for intermediate values of the ACN by varying the salinity [47]. At the lower ACN, liquid crystals tend to form. At higher ACN, which are generally associated with higher salinities, coarse emulsions that may be accompanied by surfactant precipitation are observed.

Figure 5.14 deals with an ionic surfactant. Similar behavior has, however, been observed with nonionic amphiphilic compounds [35]. This behavior can be explained by comparing the factors that promote the miscibility of the amphiphile with oil and water and those which promote its precipitation as a separate phase. At low ACN, those promoting miscibility are small compared to the lateral interactions between the surfactant molecules and the C region tends to crystallize. At high ACN, the interactions between the oil molecules dominate and the amphiphile is excluded from the O region, resulting in the formation of a precipitate. In both extremes the miscibility of the amphiphile is poor.

This discussion can be made more quantitative, as in Section 5.III.A, if one assumes that at optimum, miscibility is promoted by

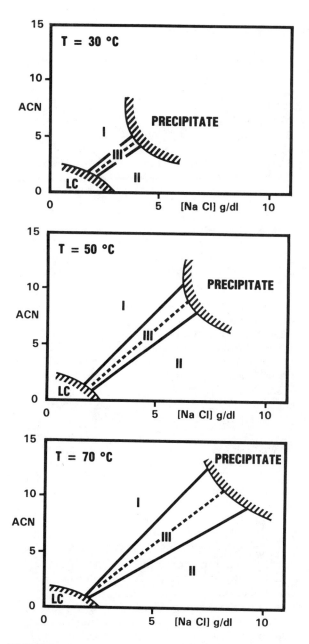

FIGURE 5.14 Phase maps in salinity/ACN space at 30°, 50°, and 70°C showing regions of liquid crystal in water (L.C.) behavior, the precipitation regime and the I, II and III regimes. The surfactant is the hexadecyl vinylidene sulfonate. Concentration: 1g/dl of aqueous phase. (After Ref. 47.)

$$M = A_{co} + A_{cw} \tag{5.16}$$

and resisted by

$$E = A_{oo} + A_{ww} + A_{\ell\ell} + A_{hh} \tag{5.17}$$

The difference

$$\Delta = E - M \tag{5.18}$$

is thus expected to determine which regime will be observed. The system tends to be isotropic when $\Delta < 0$. When $\Delta > 0$, the amphiphile is expected to precipitate as a separate phase in one of the two regimes described above.

At optimum, $R = 1$, and

$$A_{co} - A_{oo} - A_{\ell\ell} = A_{cw} - A_{ww} - A_{hh} \tag{5.19}$$

Combining Eqs. (5.16) to (5.19) yields

$$\Delta = 2(A_{oo} + A_{\ell\ell} - A_{co}) \tag{5.20}$$

From Eq. (5.5), Δ varies with ACN according to the relationship

$$\Delta = 2(a \cdot ACN^2 - c\sqrt{A_{ll}}\, ACN + A_{\ell\ell}) \tag{5.21}$$

and thus, as shown in Fig. 5.15, exhibits a minimum. Therefore, if $\Delta < 0$ is the condition for isotropic behavior, then only a window of ACN, bounded between ACN_1 and ACN_2, is accessible.[†] In fact, the experimental results show that anisotropic behavior is readily obtained when surfactant lateral interactions predominate over A_{co} (see Fig. 7.8); that is, the actual window for isotropic behavior is restricted to the range $ACN_2 - ACN_1'$ (Fig. 5.15).

Lateral interactions (see Section 5.III.B) and also the tendency for ionic amphiphiles to crystallize both promote crystal formation. The effect of the headgroup is evidenced, for example, by the solid-crystalline state of pure sodium dodecyl sulfate or sodium dodecanoate as compared to the liquid state of n-dodecane, dodecanol-1, or dodecanoic acid.

Various ways can be envisioned to modify the relative magnitude of the various interactions and thus to promote isotropic behavior:

[†]A similar argument would hold if one would consider, for a given oil, a change in surfactant alkyl chain length compensated by salinity for optimal behavior, for example. In that case, isotropic behavior is expected only for alkyl carbon numbers which range between two critical values.

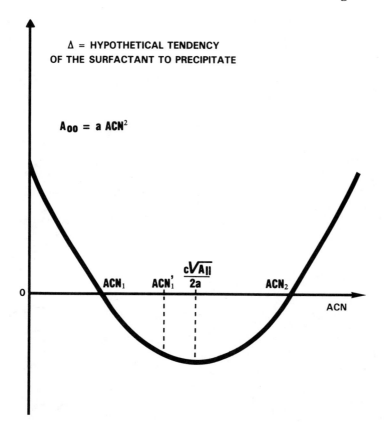

FIGURE 5.15 Hypothetical tendency of the surfactant to precipitate (at R = 1) when the alkane carbon number varies, assuming A_{oo} = a ACN^2, with c, $A_{\ell\ell}$, and a = constant.

Temperature can be raised primarily to reduce $A_{\ell\ell}$ in the C region.
Short amphiphilic compounds can be added to separate the surfactant
    molecules and to introduce disorder in the C region.
The surfactant lipophile can be branched or changed in some other
    way (i.e., additives on hydrocarbon chain or imposition of a
    double bond) that would tend to reduce the lateral interactions.

All of these methods effectively separate surfactant molecules in the C region; that is, they decrease Γ_s. According to Section 1.V.D, they are thus expected to be detrimental to the solubilizing power (see also Section 7.V).

B. Effect of Temperature

Increasing temperature results in decreased lateral interactions because of increased separation of the amphiphiles at the interface.
The impact of temperature on phase behavior is demonstrated experimentally in Fig. 5.14. The region at high ACN, where precipitation
is observed, tends toward higher ACN when temperature increases.
Similar results have been reported by Puerto and Reed [26] noting
the effect of temperature on the "very condensed phases" encountered
at high ACN.

In Fig. 5.14, the liquid-crystal regime (low ACN) is only slightly
affected by temperature. This is fortuitous, however, since in other
systems temperature has also been found to affect this regime profoundly. This is illustrated by Fig. 5.16 [48], which maps both the
optimal salinity for isotropic systems and the range where liquid crystals are observed as a function of the temperature for a series of
n-alkanes. The lower limit of the range of ACN for which isotropic
behavior is obtained is seen to decrease when temperature increases.
Above 60°C, all the n-alkanes investigated exhibit isotropic behavior.
Increasing the temperature thus appears to increase the range of ACN
for which isotropic behavior is obtained.

C. Effect of Alcohol (or Cosolvent)

Alcohols are short chain amphiphilic molecules (often called "cosolvents") which are used to promote isotropic behavior. Similar results can be obtained with other compounds, such as amines or ke-
tones.

The importance of alcohol for obtaining isotropic microemulsions
has been long recognized. Some authors even believed them to be
indispensable, especially when ionic surfactants are used.

Figure 5.17 [26] shows the effect of both alcohol and temperature
on optimal salinity for the series of n-alkanes. Tagged points represent very condensed phases, and dashed lines refer to regions where
a definite type III system was not observed. Clearly increasing the
alcohol concentration and/or temperature promotes isotropic behavior.
The trade-off between alcohol and temperature has also been extensively investigated by Wade and co-workers [47,49].

D. Effect of Surfactant Structure

Figure 5.15 shows that the ACN corresponding to the minimum value
of Δ depends on A_{ll}; that is, the range of conditions for which isotropic behavior is obtained depends strongly on the surfactant lateral
interactions. Various ways can be envisioned to modify the surfactant
lipophile to decrease lateral interactions: insertion of aromatic rings
[26], addition of side methyl groups to the main chain [25], insertion
of a double bond [47,50], or variation of the point of attachment of

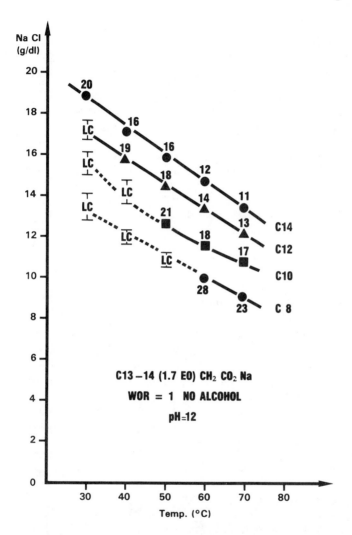

FIGURE 5.16   The effect of temperature on the phase behavior for four n-alkanes. The number next to each point indicates the solubilization parameter. C_{13-14} (1.7 EO) CH_2CO_2Na is a commercial alkylethoxylated carboxylate containing ethylene oxide on average.   The average point of attachment of the ethylene oxides is 3.9. (After Ref. 48.)

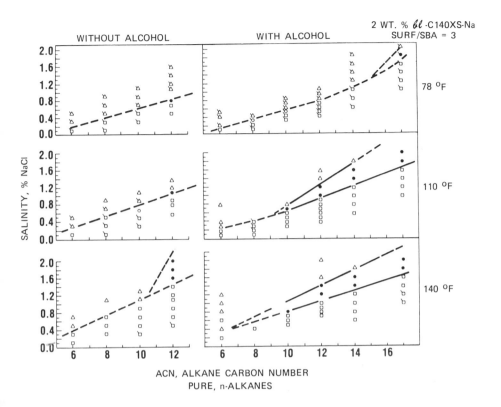

FIGURE 5.17 The effect of alcohol at three temperatures on the phase
behavior displayed by the series of n-alkanes. SBA = butanol-2;
bl-C14 OXS-Na is the sodium salt of tetradecyl orthoxylene sulfonate.
(After Ref. 27.)

the hydrophilic head to the main backbone [47,49], with or without
an intermediate link. These possible structures are schematically pic-
tured in Fig. 5.18. Alternatively, mixtures of straight-chain and
branched surfactants can be used for obtaining isotropic micellar solutions [51].
    The trade-off between surfactant branching and alcohol concentra-
tion is shown in Fig. 5.19 [52]. The series of isomers of octadecyl-
sulfonate has been used. For each isomer, optimal behavior has been
determined by varying the salinity at various butanol-2 concentrations.
Figure 5.19 shows the minimum alcohol concentration required with
each isomer to obtain isotropic behavior. When the point of attachment of the sulfonate group to the octadecyl chain recedes from a
near-terminal to a central position, this amount of alcohol decreases

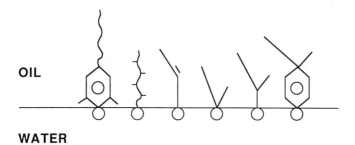

FIGURE 5.18   Various types of surfactant branching.

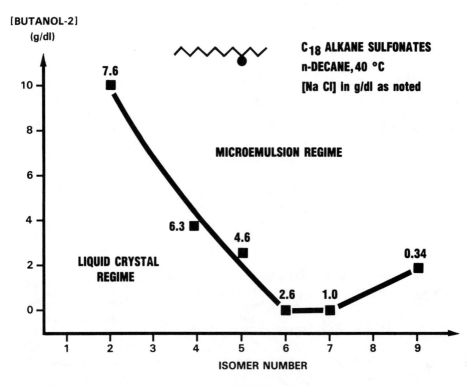

FIGURE 5.19   The effect of the point of attachment of the hydrophilic group along the octadecyl chain on the alcohol requirement for obtaining isotropic behavior.   Each point refers to a system at optimum. (After Ref. 52)

and vanishes for the -6 and -7 isomers. Interestingly, when the hydrophilic group is attached to the center of the alkyl chain, that is, when both secondary tails have about the same length, some alcohol is again required to produce isotropic behavior. Similar trends have been reported by Salter [43] using undecylparaethylbenzene-sulfonates, where the benzene ring is attached to various positions along the alkyl chain.

VII. EFFECT OF PARTITIONING ON PHASE BEHAVIOR: PSEUDOPHASE MODELING

A. Pseudoternary Diagrams

As noted in the preceding section, the phase rule requires, at a given temperature and pressure, the phase composition of a three-component three-phase system to be invariant. When three phases are present, increasing the surfactant concentration in a three-phase system such as that represented by Fig. 5.13 simply increases the volume of the micellar phase but does not alter the composition. Very few micellar systems contain only three components. Almost all commercial surfactant systems are a complex mixture of molecular types, as are the oleic phases of practical interest. Furthermore, the aqueous phase often contains electrolyte. The composition of the C layer of micellar systems composed of these complex mixtures will change with the water/oil ratio or the total surfactant concentration due to selective fractionation and partitioning between the various regions. Thus for many cases of practical interest the R-ratio will vary with water/oil ratio and surfactant concentration. These variations are called indirect effects. They stem from the complexity of the system.

Shown in Fig. 5.20 are ternary diagrams representing complex systems. In comparing these diagrams with the true three-component system shown by Fig. 5.13, it is seen that one notable difference is that the middle-phase composition, in fact, depends on the overall composition of all the components in the system. Since it is not possible to represent the phase behavior of systems having more than three components using ternary diagrams, those shown in Fig. 5.20 are known as pseudoternary diagrams. In Fig. 5.20, surfactant and alcohol taken in a certain proportion are treated as a single component. Since these two components certainly partition differently between the O, C, and W regions and they do not, therefore, appear in the various phases in fixed proportions, the use of a pseudoternary diagram is not strictly correct. The boundaries of the demixing curve are, however, accurately positioned, making the diagrams conceptually valuable. The similarity between the pseudoternary demixing curves shown by Fig. 5.20 and those of the true ternary system represented by Fig. 5.13 is evident.

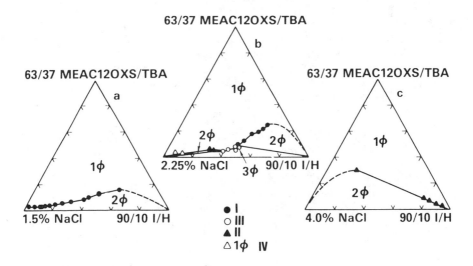

FIGURE 5.20 An example of pseudo ternary diagrams in which the surfactant to alcohol is kept constant and is essentially a pseudo-component. The system is the same as described in Fig. 5.28 with TBA (tertiary butyl alcohol) replacing TAA. (After Ref. 42.)

The representation that utilizes the surfactant plus alcohol (S + A) is used quite frequently in this book especially when focusing on the demixing curves. Other types of pseudoternary representations have been reported [53,54]. Vinatieri et al. [54] have defined an optimization procedure to select the "best" three pseudocomponents. The best has a well-defined mathematical definition which minimizes the length of tie lines that protrude out of the three-dimensional space of the ternary diagram.

B. Pseudophase Models: Fundamental Concepts

The phase separation model described in Section 2.VII treats micelles as a separate thermodynamic phase in equilibrium with an aqueous solution containing molecularly dispersed amphiphiles and in some cases electrolyte. The same approach can also be applied fruitfully to systems containing oil together with water and amphiphilic compounds. Thus the O, C, and W regions are, in this model, regarded as three separate and distinct phases in equilibrium with each other, and if excess oil and/or water phases do exist, these are similarly composed [55]. Since the C layer is not really a bulk phase, it must in fact be considered to be a pseudophase.

Figure 5.21 is intended to illustrate the essence of the pseudo-phase model as it is applied here. Whether excess phases are present

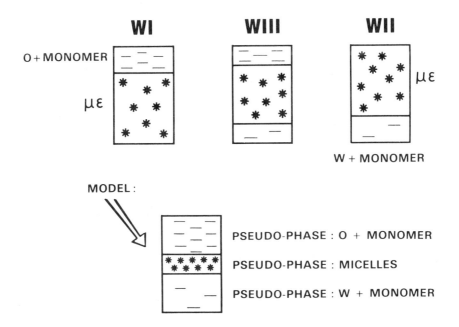

FIGURE 5.21 A schematic showing that any type of phase behavior is represented by a single pseudophase model composed of regions O and W together with the C layer. The symbol μϵ denotes micro-emulsion.

or not, all micellar systems can be regarded as composed of the O, C, and W regions. The model will be used here to determine the C-layer composition, which is, of course, the primary factor needed to evaluate the R-ratio. Thus the phase behavior of complex systems can be interpreted using this model.

C. Alcohol Partitioning

1. The Model

For a three-component system, the model increasing the water/oil ratio while maintaining all other factors constant yields a I-III-II transformation (Section 5.V) which can entirely be attributed to entropic effects. For four-component systems, such as those composed of surfactant, oil, water, and alcohol, varying the water/oil ratio will change the C-layer composition, thereby altering the R-ratio. Consequently, phase transitions promoted by increasing the water/oil ratio in the presence of alcohol will therefore be attributable not only to entropic contributions but also to changes in the R-ratio. The

C-layer composition can be calculated using the pseudophase model. This calculation starts with a formulation of the chemical potentials of alcohol in the O, C, and W regions [56].

The chemical potential of the alcohol in the O region must take into account the known tendency for association in nonpolar media [57]. At low concentrations, dimers tend to form, but as the alcohol concentration is raised, trimers, tetramers, and so on, also exist. The process is represented as follows:

$$S_{a1} + S_{a1} \; \overset{K}{\underset{\longleftarrow}{\longrightarrow}} \; S_{a2}$$
$$S_{a1} + S_{a2} \; \overset{K}{\underset{\longleftarrow}{\longrightarrow}} \; S_{a3} \tag{5.22}$$

where S_{ai} represents an association of i alcohol molecules. To a good approximation, the equilibrium constant K is the same for all reactions. Thus in terms of alcohol volume fraction ϕ_{ai}, the equilibrium conditions can be represented as follows:

$$K = \frac{\phi_{ai}}{\phi_{a1}\phi_{ai-1}} \qquad i = 2, \ldots, N \tag{5.23}$$

where

$$\phi_{ai} = \frac{V_{ai}}{V_{oil} + \sum V_{aj}} \tag{5.24}$$

V_{ai} is the volume of the alcohol i-mer in the oil phase and V_{oil} is the volume of oil.

The volumes are assumed additive. By manipulating Eqs. (5.23) and (5.24), it can be shown that[†]

$$\phi_{a1} = \frac{\gamma}{1 + \gamma(K + 1)} \tag{5.25}$$

where γ is the total volume fraction of alcohol in the oil phase. Equation (5.25) has a very simple interpretation. If γ is small, that is, if the oil phase contains little alcohol, then $\phi_{a1} \approx \gamma$. In this limit practically all of the alcohol is present as monomer. However, as γ increases, ϕ_{a1} approaches an upper limit which may, depending on the value of K, be substantially less than γ. Thus because of alcohol association the chemical potential at a given alcohol volume fraction is less than that of an ideal solution.

[†]The volume of water dissolved in the oil pseudophase is neglected.

The chemical potential of alcohol in the oil phase is

$$\mu_a^0 = \bar{\mu}_a^0 + RT \ln \phi_{a1} \tag{5.26}$$

and in the aqueous phase

$$\mu_a^w = \bar{\mu}_a^w + RT \ln \frac{\lambda}{1 + \lambda} \tag{5.27}$$

where λ is the volume of alcohol per volume of water. Furthermore, in the C layer the chemical potential is

$$\mu_a^c = \bar{\mu}_a^c + RT \ln \frac{S_a}{S_s + S_a} \tag{5.28}$$

provided that the C layer is treated as though it is a bulk phase where S_a and S_s are the interfacial areas of the alcohol and surfactant, respectively. To convert area fractions to a volume fraction, the quantity β, which is the ratio of the alcohol tail length to that of the surfactant, may be used. Assuming that the surfactant and alcohol have the same interfacial area per molecule, β is simply the molar volume ratio of alcohol to surfactant. Based on this definition,

$$\frac{S_a}{S_a + S_s} = \frac{\sigma}{\beta + \sigma} \tag{5.29}$$

where σ is the volume of alcohol per volume of surfactant in the C layer.

According to the pseudophase approximation,

$$\mu_a^w = \mu_a^0 = \mu_a^c \tag{5.30}$$

This leads to

$$k_w = \frac{\lambda[1 + \gamma(K + 1)]}{\gamma(1 + \lambda)} \tag{5.31}$$

where the partition coefficient, k_w, is a function of the standard chemical potentials and hence independent of the alcohol concentrations. Thus

$$k_w = \frac{\exp(\bar{\mu}_a^0 - \bar{\mu}_a^w)}{RT} \tag{5.32}$$

Similarly, equating the alcohol chemical potential in the oil phase to
that in the C layer yields

$$k_c = \frac{\sigma[1 + \gamma(K + 1)]}{\gamma(\beta + \sigma)} \qquad (5.33)$$

where k_c is a second partition coefficient, which is also concentration
independent. The parameters k_w, k_c, and K are to be determined
for each system.

The important quantity insofar as the R-ratio is concerned is
$\sigma/(\beta + \sigma)$, the fraction of alcohol within the C layer. To calculate
the value of σ, the following volume balance must be satisfied:

$$V_a = \lambda V_w + \gamma V_o + \sigma V_s \qquad (5.34)$$

Here V_i are the volumes of the alcohol, water, oil, and surfactant.
The system of Eqs. (5.31), (5.32), and (5.34) permits calculation of
σ as a function of the water/oil ratio, V_w/V_o, or as a function of the
surfactant concentration, V_s, while maintaining V_s/V_a constant.

2. Water/Oil Ratio

Parameters applicable to various systems are given in Table 5.3.
System A contains an alcohol, $\underline{n}$-butanol, for which $a^a_{co} > a^a_{cw}$ (see
Section 5.IV.A) and its increased presence within the C layer will

TABLE 5.3 Pseudophase Parameters for Systems which Include Alcohol

System	Alcohol	k_w	k_c	K
A	n Butanol	4.7	58	100
B	Isopropanol	120.7	727.3	600.7
C	n Pentanol	2.5	95	140
D	n Butanol	7.0	109	140
E	n Propanol	18.0	142	160

System	Surfactant	Oil	Water	Reference
A	Sodium dodecyl sulfate	Toluene	No added electrolyte	56
B	TRS-10-80	Paraffinic	2.5 wt% NaCl	58
C	Sodium dodecyl sulfate	Hexane	10.4 wt% NaCl	Unpublished
D	Sodium dodecyl sulfate	Hexane	3.8 wt% NaCl	Unpublished
E	Sodium dodecyl sulfate	Hexane	3.8 wt% NaCl	Unpublished

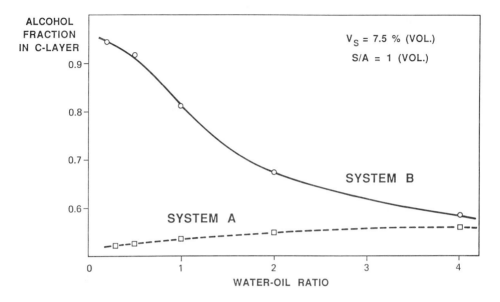

FIGURE 5.22 The fraction of alcohol in the C layer depends on the water to oil ratio.

promote a I-III-II transition. Based on the values of k_W, k_C, and K given in Table 5.3, the proportion of n-butanol in the C layer can be calculated. Figure 5.22 shows that it increases slightly with increasing water/oil ratio, and therefore the R-ratio will tend to increase. Thus increasing the water/oil ratio tends to yield a I-III-II transition because of increased alcohol partitioning into the C layer.

System B contains a hydrophilic alcohol, isopropanol. This alcohol partitions preferentially into the aqueous phase. Figure 5.22 shows the alcohol concentration in the C layer to decrease with increasing water/oil ratio. For this alcohol $a_{co}^a < a_{cw}^a$, and thus increasing the water/oil ratio tends to increase the R-ratio. This is precisely the trend found for the lipophilic alcohol, n-butanol. The reason that the R-ratio responds in a similar manner for both a hydrophilic and a lipophilic alcohol when the water/oil ratio is changed can be understood in terms of two compensating effects. The influence of the alcohols on the R-ratio differs and the change in C-layer composition of alcohol with changing water/oil ratios also differs. The net result is that the trend; namely, increasing R with increasing water/oil ratios, is the same. Baviere et al. [37] have observed that for some alcohols increasing the water/oil ratio produces trends op-

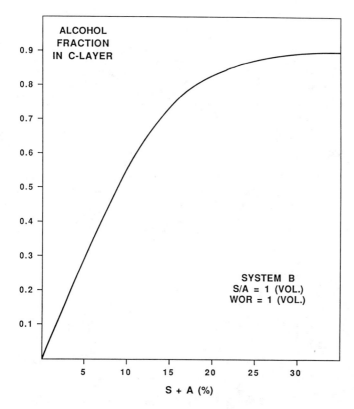

FIGURE 5.23 The fraction of alcohol which is contained in C layer increases as the amount of both alcohol plus surfactant increases even though their ratio is maintained constant. (The parameters are those of system B in Table 5.3.)

posite of that of the two systems considered here. This would occur if only one of the two factors differs from the trends found for system A and B.

3.   Alcohol Concentration

Increasing the surfactant plus alcohol concentration holding the ratio of the two constant also produces a change in the composition of the C layer. Figure 5.23 shows the isopropanol content in the C layer to increase as S + A is increased. Since for isopropanol $a_{co}^{a} < a_{cw}^{a}$, the trend shown will tend to decrease the R-ratio. A plot of the optimum electrolyte concentration as a function of (S + A) such as shown by

Fig. 4.19 will indicate a rotated phase diagram with the electrolyte concentration being lower at low values of (S + A) than at higher values. The partitioning of alcohol into the C layer is one of the primary causes for the demixing curve to be rotated in plots of (S + A) versus salt concentration (see Section 5.VII.H).

4.   Pseudophase Model and Pseudoternary Diagrams

The composition of a phase composed of four components can be represented as a single point positioned within an equilateral tetrahedron. Point D in Fig. 5.24 is an example. When conjugate phases coexist, their compositions can be connected by a tie line which lies entirely within the tetrahedral volume shown in Fig. 5.24; however,

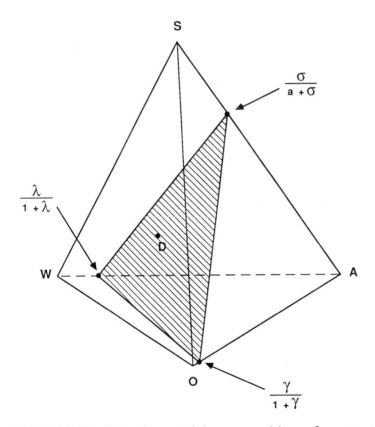

FIGURE 5.24   Triangle containing compositions of constant pseudophases (shaded triangle). The S, A, W, and O represents surfactant, alcohol, water, and oil, respectively.

the geometry of an entire demixing curve may be difficult to visualize in three dimensions. Of even greater complexity are the cases where three phases coexist and tie lines are replaced by tie triangles. A simpler, easier-to-visualize ternary representation is possible whenever the assumptions underlying the pseudophase model are accurate.

The composition D can be considered to be composed of a mixture of three pseudophases and given the parameters k_w, k_m, and K, the three compositions of these pseudophases can be determined [59]. These pseudophase compositions are shown in Fig. 5.24 as the vertices of the shaded triangle containing D. All compositions lying on this triangle are composed of identical pseudophases blended together in different proportions. If two phases form, the pseudophase model considers the conjugate excess phase to have the composition of either the O or the W region (see Section 5.VII.B), which is represented by one of the vertices of the shaded triangle and the micellar solution to be a blend of the three pseudophase compositions. Thus a tie line connecting mutually saturated phases will start at either O or W and terminate at the point representing the micellar phase. Tie lines connecting two phases having an overall composition lying on the shaded triangle will therefore lie entirely within the triangle. Similar arguments can be made for three-phase systems since the excess phases will be O and W and the tie triangles will lie in the shaded triangle.

Conjugate phases are therefore represented by points on the shaded triangle whenever the overall system composition is a point on the triangle. If these conditions are met, a ternary diagram with the pseudophases treated as pure components will provide a precise representation of the phase behavior. This representation is much simpler to use and to visualize than the tetrahedral one.

Ternary phase diagrams consisting of pseudophase planes for systems containing alkyl betaines [59], alkylsulfates [60], or petroleum sulfonates [58] have been found to contain tie lines and tie triangles. The pseudophase model therefore appears to be a useful tool for selecting appropriate pseudocomponents so that simple ternary diagrams can be constructed.

Difficulties may, however, arise when the system includes an inorganic electrolyte because partitioning of the salt may occur between the excess water and the micellar phase in type II and III systems [61]. This will result in complex phase behavior because brine cannot be treated as a single component, as is normally done. Electrolytes, as discussed in Section 5.VII.F, tend to be excluded from the micellar phase when a conjugate excess aqueous phase is present.

In the vicinity of critical endpoints where two micellar solutions having slightly different compositions may exist in equilibrium, the pseudophase model assumptions are invalid and representations based on pseudophase models may lead to serious errors and erroneous conclusions. Thus in a narrow range of compositions surrounding critical endpoints (see Chap. 8), pseudophase modeling should not be applied.

D. Ideal Mixtures of Nonionic Surfactants

1. The Model

The formation of mixed micelles has been treated in Section 2.VII. It is found that the mixed micelles do not generally have the same composition as the monomer, and furthermore, the compositions of both are functions of the total surfactant concentration. Thus the composition of the C layer is a function of the total surfactant concentration even in the absence of oil. When oil is present, the C-layer composition still depends on the total surfactant composition, with the additional complexity that the various surfactant species may now partition into the oil phase. This partitioning is particularly important when the surfactant is an ethoxylated alcohol because many members of such a series have substantial oil solubility.

Commercial polyoxyethylene surfactants, although characterized by an average ethylene oxide number (EON) m, actually represent a Poisson distribution of ethylene oxide units [see Eq. (2.96)]. Thus even though m may be large, the surfactant mixture will contain a significant fraction of molecules that have a small number of ethylene oxide units, and these have substantial oil solubility. This will markedly influence the phase behavior when the surfactant concentration or the water/oil ratio is changed.

To develop a pseudophase model, the chemical potential of each type of surfactant molecule must be written for each of the three regions. It is accurate to assume that all of the regions, including the C layer, are ideal mixtures. This assumption yields the following relationship between the oil and water concentrations of a surfactant molecule having i ethylene oxide units:

$$K_i = \frac{C_i^W}{C_i^0} \tag{5.35}$$

where K_i is the partition function and C_i^W and C_i^0 are the amphiphile concentrations in water and oil, respectively. Given K_i and CMC_i, the pure component CMC, the C-layer compositions can then be shown to depend on the water/oil ratio (WOR) and the total surfactant concentration.

The basic equation [62,63] is obtained by equating the chemical potentials of each of the surfactants in the pseudophases. This procedure is similar to the approach used in the preceding section to describe the partitioning of alcohol. The final equation is

$$N_i^T = y_i \left[V_w CMC_i \left(1 + \frac{1}{WOR \cdot K_i} \right) + \sum N_j^c \right] \tag{5.36}$$

where N_i^T is the total number of moles of component i, y_i its mole fraction in the C region, and the $\Sigma \ N_j^c$ is the total number of moles in the C region. V_W is the volume of water N_i^T can be calculated for a given overall surfactant concentration using the Poisson distribution.

If N_i^T, V_W, and WOR are specified, there are, for a surfactant system composed of N different components, N − 1 unknown mole fractions for the interfacial pseudophase, and there are N − 1 independent equations (5.36). These can be solved for y_i. Given y_i, then C_i^0 and C_i^W can be calculated. The results show that those species with small i (small EON) are found primarily in the hydrocarbon phase. Also, much more surfactant is in oil than in water. These trends are described in the following sections using the system octylphenol ethoxylates/isooctane/water as an illustration. Values of K_i and CMC_i given in Table 5.4 have been used in Eq. (5.36) [64]. The initial distribution N_i^T has been assumed be Poisson type.

2. Surfactant Concentration

Both the selective adsorption of nonionic surfactant molecules having small EO chain lengths and the substantial oil solubility of these components are illustrated by the results presented in Fig. 5.25. The concentration of each surfactant component in isooctane is shown. Even though the surfactant used to formulate the micellar system has an average of 5 EON units (m = 5), the large proportion of the molecules in the isooctane phase have less than 4 EON units with the average being slightly less than 3. The distribution of surfactant EON units in the oil differs markedly from that of the original surfactant. This severe fractionation can be anticipated by noting that K_5 is two orders of magnitude larger than K_2 (see Table 5.4).

The concentration of surfactant in the oil phase is significant. Figure 5.25 shows that for a total surfactant concentration of 7.7 g/liter, the isooctane phase concentration of that component having 3 EON units is very nearly 1 g/liter. The Poisson distribution defining the numbers of each is given by Eq. (2.90).

If Mw_i is the molecular weight of a surfactant component having i EON units, then the number of grams, G, corresponding to a feed consisting of $\Sigma \ N_i^T$ moles is

$$G = \sum_i Mw_i N_i^T \tag{5.37}$$

Using Eqs. (2.90) and (5.37) together with the values of the molecular weight given in Table 5.4, an octylphenol ethoxylate feed of 15.4 g (2 liters x 7.7 g/liter) will contain 0.937 g of molecules with 3 EON units. Thus almost all of these molecules partition into the isooctane phase as shown by Fig. 5.25. It can be shown that the amount of surfactant in the aqueous phase is quite small so that the

TABLE 5.4 Physical Properties of Ethoxylate Octyl Phenols

EO chain length	M_w	CMC in water 25°C (μmol/liter)	Partition coefficient water/isoctane at 25°C	HLB
1	250	49.5	1.84×10^{-4}	3.52
2	294	76.5	7.17×10^{-4}	5.99
3	338	103	3.13×10^{-3}	7.81
4	382	129	9.83×10^{-3}	9.21
5	426	172	2.46×10^{-2}	10.32
6	470	250	5.92×10^{-2}	11.23
7	514	268	1.02×10^{-1}	11.98
8	558	283	5.00×10^{-1}	12.62
9	602	304	1.39	13.12
10	646	323	3.84	13.62
11	690	340	10.6	14.02
12	734	361	29.4	14.39
13	778	382	81.3	14.70
14	822	401	225	14.99
15	866	418	622	15.24

Source: Ref. 64.

surfactant not in the isoctane phase will be in the C layer. Since almost all of the small EON molecules fractionate into the isoctane, the average number of ethylene oxide units in C layer must be greater than the average of the original surfactant. This is shown by Fig. 5.26. The average EON in the C layer approaches 5 as the surfactant concentration increases but is significantly larger at small surfactant concentrations. Therefore, the R-ratio must decrease as the total surfactant concentration increases. A consequence of this behavior is evidenced by the shape of phase maps as discussed in Section 5.VII.H.

The partitioning of the surfactant raises a question as to the proper definition of the solubilization parameter. As defined previously, the solubilization parameter is the ratio of the volume of oil (or water) solubilized within the micellar phase to the volume of

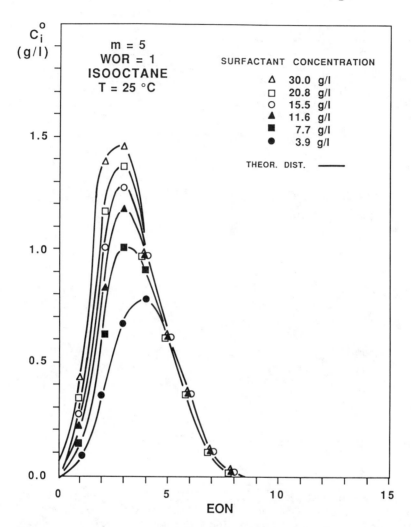

FIGURE 5.25 Concentration of oxyethylated octylphenol in the excess
oil phase shown as a function of the number of ethoxylates (EON).
The figure applies for m = 5, at water to oil ratio = 1, isooctane, and
25°C. (After Ref. 63.)

surfactant contained in the system (see Section 4.II.A). This definition makes physical sense only when the totality of the surfactant is
concentrated at the water/oil interface. When surfactant partitioning
occurs, this definition yields an apparent solubilization parameter.
Its value is lower than the actual solubilization parameter which would

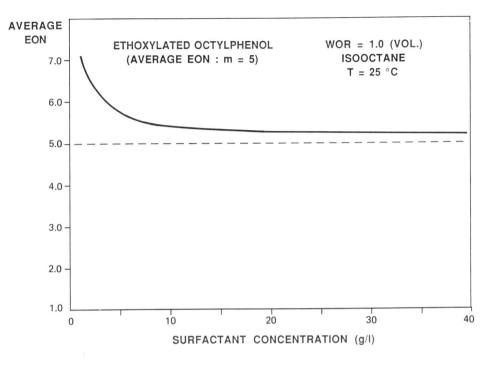

FIGURE 5.26 The average number of ethoxylates in the C layer shown as a function of the overall surfactant concentration. (After Ref. 62.)

be obtained by taking into consideration only those surfactant molecules adsorbed at the water/oil interface. Careful attention must
therefore be paid when comparing the solubilizing power of systems
involving surfactant partitioning [63]. Indeed, the same question
arises with regard to the definition of the solubilization parameter
for alcohol-containing systems. As defined in Section 4.II.A, the
alcohol is neglected, resulting in an overestimation of the solubilization parameter.

3. Mixing Rules

Mixing two (or more) surfactants to obtain intermediate properties is
a widely used technique and yields generally satisfactory results. In
the case of nonionics, it is especially convenient to blend compounds
of the same lipophilic moiety but differing by their ethylene oxide content in order to mimic the behavior of amphiphilic species of intermediate EON (see Fig. 5.1). This technique is especially applied for
the formulation of emulsions, where the HLB of the surfactant (see
Section 4.II.F) must match the required HLB of the oil (see Section

4.II.F) must match the required HLB of the oil (see Section 6.II.F). To achieve the proper HLB value, surfactants are often blended according to the following mixing rule:

$$HLB_m = \sum_i m_j HLB_j \tag{5.38}$$

where m_j represents the mass fraction of surfactant j in the mixture. As calculated, HLB_m refers to the HLB of the overall mixture, not to the HLB of the interfacial C region. Depending on the magnitude of the partitioning and fractionation, Eq. (5.38) may yield predictions grossly in error.

This is illustrated by Fig. 5.27, which shows the type of phase behavior observed for mixtures of ethoxylated nonylphenol, alcohol, hydrocarbon, and water. The hydrocarbon is varied by mixing members of the n-alkane series [65]. With lgepal CO 530, a commercial ethoxylated nonylphenol containing an average of 6 ethylene oxide units, type III is observed between heptane and tridecane. This behavior is then compared with mixtures of products EON = 5 and 7.5, 4 and 9, and 1.5 and 10.5, the proportions selected so that the average EON of the mixture is 6, assuming a linear mixing rule based

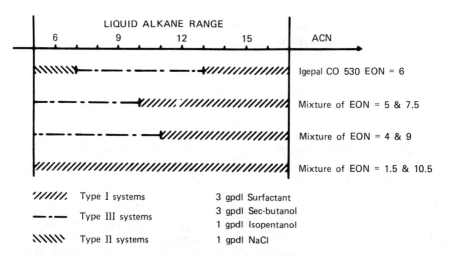

FIGURE 5.27 Comparison of range of type III behavior of lgepal CO530 (Commercial ethoxylated nonylphenol supplied by GAF) having EON = 6 with mixtures of EON = 5 and 7.5, 4 and 9, and 1.5 and 10.5 adjusted so that the average number for EON is 6. The concentrations refer to the aqueous phase. Water to oil ratio is 4 (vol). (After Ref. 65.)

on weight fraction. The results are seen to be different, and in some cases quite different. The mixture of EON = 5 and 7.5 shows type III behavior for ACN < 10. If the proportions are changed slightly so that the average EON is 5.9 instead of 6, the three-phase range is about the same as that for CO 530. This is interpreted to imply that the mixture does approximate the phase behavior of the original. Mixtures of 4 and 9 gave type III systems for ACN < 11, and the mixtures of 1.5 and 10.5 exhibited only type I behavior over the entire liquid alkane range. This profound difference in behavior of surfactant systems having the same average ethylene oxide chain length is due largely to the partitioning into the oil of the components with shorter EO chain lengths.

Since the phase behavior is governed by the species forming the C layer and their interaction with oil and water, the mixing rules, to be quantitative, must refer to those amphiphilic compounds present in the C layer rather than those added to the system. When partitioning occurs, calculating the C-layer composition is complex. In some cases mixing rules based on the pseudophase model have been applied successfully for predicting the phase behavior of nonionic amphiphiles [63]. Simple linear mixing rules are acceptable only in those cases where the EON of the blended surfactants are not too different, perhaps less than 2 units.

E. Nonideal Mixtures of Nonionic and Anionic Surfactants

Mixtures of nonionic and anionic surfactants show substantial negative deviation in the C region from ideal solution theory. For example, the mixture CMC for these two different amphiphiles is less than expected based on ideal mixing theory. This is accounted for in the pseudophase model by introducing an activity coefficient to account for the interaction between nonionic and anionic amphiphiles in the interfacial pseudophase [66]. Regular solution theory has been used as a model. The activity coefficient of the nonionic molecule having EON = i is therefore given by [66]

$$\ln \gamma_i = (\beta_i - \sum y_j \beta_j) y_a \qquad (5.39)$$

where β_i is an interaction parameter between i and the anionic surfactant (the nonionic/nonionic and anionic/anionic interactions are ideal). y_i and y_a are the mole fraction of i and of the anionic surfactant, respectively. In the interfacial pseudophase (C layer). Similarly, the activity coefficient of the anionic component is given by

$$\ln \gamma_a = (1 - y_a) \sum y_i \beta_i \qquad (5.40)$$

The β_i coefficients are obtained from CMC measurements of the nonionic-ionic mixture, using pure monoisomeric nonionics.

The basic equations are then [66]

$$N_i^T = y_i \left[V_w \gamma_i CMC_i \left(1 + \frac{1}{WOR \cdot K_i} \right) + N_a^c + \sum N_j^c \right] \qquad (5.41)$$

$$N_a^T = y_a \left[V_w \gamma_a CMC_a \left(1 + \frac{1}{WOR \cdot K_a} \right) + N_a^c + \sum N_j^c \right] \qquad (5.42)$$

where a refers to the anionic surfactant and the symbols are defined as in Eq. (5.36). Given N_i^T, N_a^T (from the overall composition of the mixture), β_i, K_i, K_a, WOR, and V_w, Eqs. (5.39) to (5.42) can be solved for y_i and y_a.

The consequences of the presence of the anionic surfactant on the partitioning of the nonionic are as follows [66]:

1. The total amount of surfactant partitioning into the oil phase is reduced. Fig. 5.28 shows the effect of increased anionic concentration on the nonionic partitioning. The decrease in the amount of nonionic partitioning into the oil is due primarily to the non-ideal nature of the mixed micelles.

2. The nonionic partitioning is reduced for polydisperse systems having an increasing average number of ethylene oxide units. This is attributable primarily to the reduction in the proportion of low EON species in the system.

3. Increasing the attractive interaction between the nonionic and anionic surfactants by changing the molecular structure favors the formation of mixed micelles and decreases the tendency of the nonionic to partition into the oil phase.

4. Because of the smaller degree of fractionation, the composition of the interfacial C layer exhibits a small variation on increasing the overall surfactant concentration. This is illustrated in Fig. 5.29, which shows a comparison of the change in the average EON of the nonionic component in the C region in the presence and in the absence of anionic surfactant.

F. Electrolyte Partitioning in Type II or Type III Systems

The basic assumption imposed in the development of the pseudophase model is that excess phases contain the same composition of molecularly dispersed components as O or W regions within the micellar phase (see Section 5.VII.B). This assumption has generally been found to be accurate; however, in type III or type II systems where an excess aqueous phase exists, it has been found that for sodium chloride, the electrolyte concentration is different in the micellar solution from that

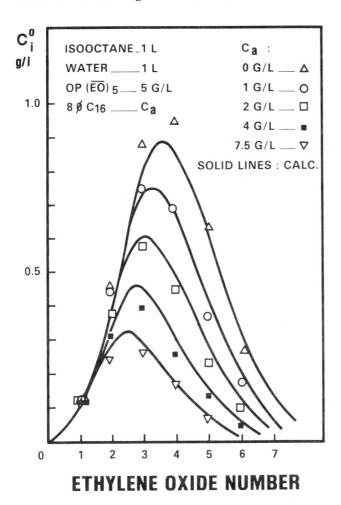

ETHYLENE OXIDE NUMBER

FIGURE 5.28 The concentration of oxyethylated octylphenols C_i^o in the isooctane phase shown as a function of the number of ethylene oxide units per surfactant. The parameter held constant on each curve is the overall concentration of anionic, dioctyl benzene sulfonate C_a. (After Ref. 66.)

in the excess phase [31,61,67,68]. This is illustrated by Fig. 5.30, which shows the difference in sodium chloride concentration in the excess aqueous phase (labeled S_w'') as the overall surfactant concentration varies while maintaining a constant sodium chloride concentration in the system.

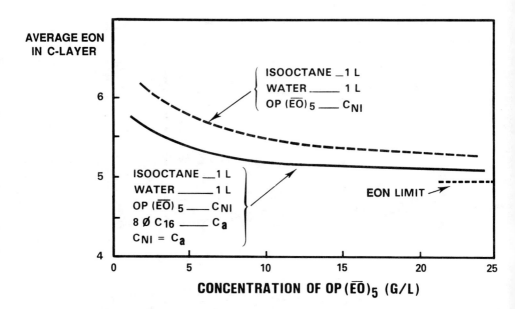

FIGURE 5.29   The average number of ethylene oxide units of the surfactant in the C layer shown as a function of the overall surfactant concentration.   The system is the same as that represented in Fig. 5.28 and the mole ratio of nonionic to anionic is maintained constant (solid line). $OP(\overline{EO})_5$, is a commercial ethoxylated octylphenol with an average EO number of 5.   (Overall concentration is C_{NI}).   (Adapted from Ref. 66.)

As seen in Fig. 5.30, the S''_w differs from the overall sodium chloride concentration for both nonionic and anionic surfactants.   For nonionic surfactants the concentration of salt in the aqueous excess phase as compared to the concentration in the micellar phase can be attributed to the salt-free water of hydration of the hydrophile in the micellar phase.   A similar mechanism is believed to account for the effect observed in the presence of anionic surfactants except that to account fully for the magnitude, ionization of the ionic hydrophile must be considered [61,68].

The partial exclusion of electrolyte from the micellar phase is most often small and has generally been ignored even though models have been proposed that can take it into consideration [61,68].

Multivalent cations, on the other hand, have a strong tendency to associate with anionic amphiphiles.   This leads to their concentration in the micellar phase of a type III or II system.   This is a trend opposite of that observed for NaCl as illustrated by Fig. 5.31.

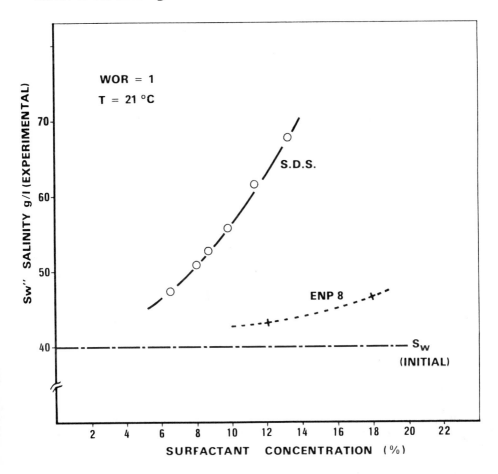

FIGURE 5.30 The excess aqueous phase electrolyte concentration (NaCl) S_w'' in type II systems as a function of the total surfactant concentration for the sodium dodecyl sulfate (SDS), butanol and toluene (surfactant/butanol = 0.5 wt.) and for the ethoxylated non-ylphenol (8 mole of ethyleneoxide, ENP8), pentanol and octane (surfactant/pentanol = 1.5 wt) systems. Equal weights of water and hydrocarbon are used. (After Ref. 61.)

G. Oil Partitioning in Type I or Type III Systems

When mixtures of hydrocarbons are used in micellar systems exhibiting type I or type III behavior, the question arises as to whether or not the composition of the excess oil phase is identical to that of the oil solubilized in the micellar phase. A difference in composition would

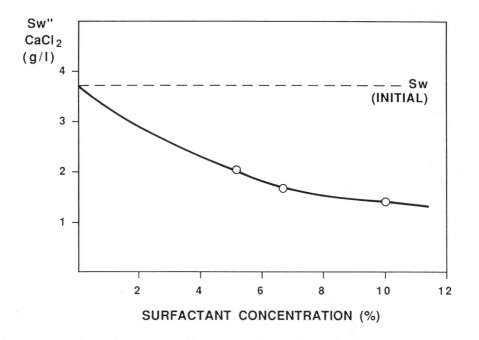

FIGURE 5.31   The excess aqueous phase calcium chloride concentration S_w'' in type II systems as a function of the total surfactant concentration.   The system contains sodium dodecylsulfate, butanol-1 (surfactant/alcohol = 0.5 wt), toluene, and brine (WOR = 1).   The electrolyte is a mixture of NaCl and CaCl$_2$ (NaCl/CaCl$_2$ = 9/1 wt).   (After M. Bourrel and J. Biais, unpublished results).

demonstrate the ability of micellar solutions to separate mixtures of hydrocarbons, which could have practical importance [69].

Recently, Graciaa et al. [70] have found oil fractionation.   Table 5.5 compares the composition determined by gas chromatography of the mixture of oils solubilized in the micellar phase of type III systems with the composition of the excess oil phase.   Because the solubilization parameter of oil is low (on the order of 2), the relative concentrations of the oils in the excess oil phase are essentially identical to those in the original mixture.   The results presented in Table 5.5 are averages among 20 measurements.   Although the effect is small, there is a consistent trend.   The aromatic hydrocarbon exhibits a higher concentration in the oil of the micellar phase than in the excess phase.   Similarly, a slight enrichment in octane versus the fluorinated oil is observed in the micellar phase.   The small magnitude of the fractionation suggests that it is due primarily to an interfacial

TABLE 5.5 Fractionation of Mixtures of Oils in Type III Systems[a]

Oil mixture	$C_{14}/\emptyset$	$C_{14}/C_2\emptyset$	$C_{14}/C_4\emptyset$	$C_{14}/C_6\emptyset$	$C_{14}/C_8\emptyset$	FC/C_8
Initial ratio (wt)	0.717	0.651	0.596	0.55	0.51	0.069
	0.283	0.349	0.404	0.45	0.49	0.931
Ratio in micellar phase (wt)	0.702	0.632	0.589	0.543	0.503	0.059
	0.298	0.368	0.411	0.457	0.497	0.941
EON	6.2	6.1	6.0	5.6	5.0	5.0

[a]Overall water/oil ratio: 1 (vol). Temperature: 25°C. 5% commercial ethoxylated (EON) octylphenol in water. $\emptyset$, $C_x\emptyset$, and C_{14} stand, respectively, for benzene, n-alkylbenzene (ethyl- to octyl-), and n-tetradecane. FC is a fluorinated oil $C_8F_{17}CH=CH_2$ and C_8 stands for n-octane.
Source: Ref. 70.

effect.   Some oil is present in the C layer (see Section 5.III.A) and
its composition there is probably different from the composition of
the oil in the O region (the core of a S_1 micelle, for example).

Two mechanisms may be invoked to interpret the "extractant'
property of the C region.   A specific interaction may exist between
the surfactant and one of the oils composing the mixture, and/or
the proximity of the water molecules may induce some attractive interaction with the most polar oil molecules.

In any case, the fractionation of hydrocarbon mixtures in micellar
systems is expected to be evidenced only in those systems rich in C
layer, that is, those exhibiting very low solubilization of oil.   This
may explain why attempts previously reported in the literature on
systems exhibiting reasonably high solubilization have failed to demonstrate the occurrence of oil fractionation [26,71].

H. Shape of Phase Maps

It has been seen that in complex systems, the composition of the C
region changes when the overall concentration of the amphiphilic compounds is varied.   This effect has direct consequences on the shape
of the phase maps such as those shown in Figs. 4.19 and 4.20.

1.   Alcohol Concentration Effect

As seen in Section 5.VII.C, increasing the surfactant plus alcohol
concentration in the system while maintaining the surfactant/alcohol
ratio constant results in increased alcohol concentration in the C
layer.   Depending on the relative magnitudes of a_{co}^a and a_{cw}^a (Section 5.IV.A), different behavior may be expected when S + A increas-
ses: low-molecular-weight alcohols, for which $a_{co}^a < a_{cw}^a$, should induce a II-III-I transition, while high-molecular-weight alcohols corresponding to $a_{co}^a > a_{cw}^a$ should yield a I-III-II transition.   Consequently, phase maps rotated in opposite directions are expected.   An
illustration is provided in Fig. 5.32, where the effects of isopropanol
and isopentanol are compared.

2.   Surfactant Concentration Effect

The fractionation of the various species generally contained in commercial surfactants is expected to produce changes in phase behavior
when the surfactant concentration varies, and thus to influence the
shape of phase maps.   This trend is readily found and interpreted
in the case of nonionic surfactants.   For these amphiphiles, the partitioning is described quantitatively in Section 5.VII.D.

Figure 5.33 provides an illustration.   For system c, the phase
map is highly rotated, and an increase in surfactant concentration at
EON = 7, for example, yields a I-III-II transition.   At low surfactant
concentration, the C region is enriched in molecules of high EON,

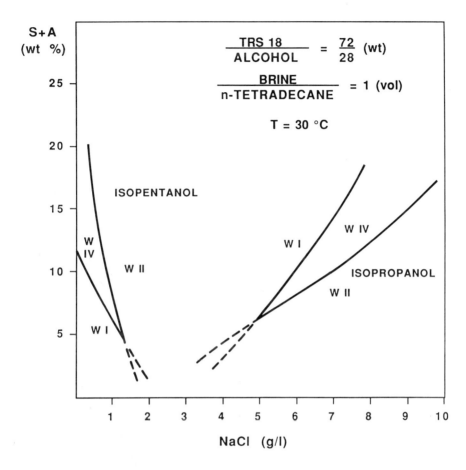

FIGURE 5.32 The effect of the type of alcohol on the shape of phase maps. TRS 18 is a sodium petroleum sulfonate (M_W = 490) supplied by Witco. (After A. Boix and M. Bourrel, unpublished results.)

which results in type I behavior. Increasing the overall surfactant concentration raises the concentration of a low EON species in the C region, promoting a decrease in A_{CW} and thus the I-III-II transition.

The upper part of the phase map is almost vertical, showing that for sufficiently high surfactant concentrations the interfacial layer attains a constant composition and phase transitions between types I, III, or IV and II do not occur [65]. Fractionation of the surfactant is therefore especially critical at low concentration, where it can greatly influence the composition of the C region.

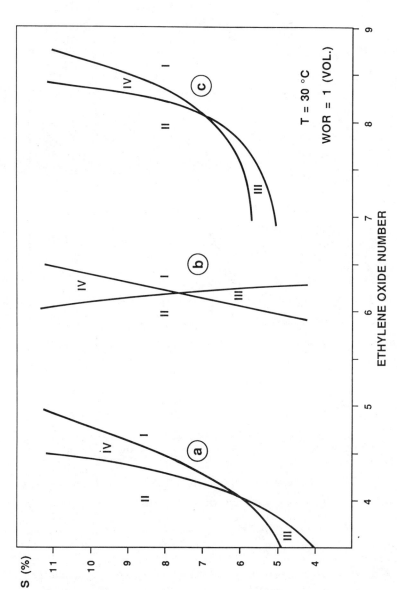

FIGURE 5.33 The effect of ethylene oxide number of commercial nonionics on phase behavior at various surfactant concentration. System a: ethoxylated isotridecanol, hexane, water. System b: ethoxylated nonylphenol, heptane, water. System c: ethoxylated nonylphenol, propylbenzene, water. (After M. Bourrel, P. Gard, J. L. Trouilly, unpublished results.)

Comparison of the overall feature of the phase maps obtained with systems a, b, and c in Fig. 5.33 provides an illustration of the effect of oil (c versus b) and surfactant structure (b versus a) on surfactant partitioning and fractionation. Fractionation is more pronounced at higher amphiphile concentration with propylbenzene than with heptane, perhaps because of the higher solvency of propylbenzene for the amphiphilic molecules. Fractionation also appears more important for ethoxylated isotridecanol than for ethoxylated nonylphenol, as shown by the comparison between curves a and b. Such a behavior is probably due to the lower ethylene oxide numbers used with ethoxylated isotridecanol.

The similarity of phase maps a and c in Fig. 5.33 with those shown in Fig. 4.19 is not coincidental. In Fig. 4.19, the amphiphilic mixture is composed of surfactant and alcohol in a fixed ratio, and as discussed in the preceding section, lipophilic alcohol molecules play the same role as the (lipophilic) species of low ethylene oxide numbers in the nonionic surfactant mixture. As discussed above, increasing the overall surfactant plus alcohol concentration results in an increase in the alcohol concentration in the C layer and thus in the I-III-II transition.

3. Effect of Mixtures of Electrolytes

As discussed in Section 5.VII.F, multivalent cations associate more strongly to the micelles of anionic surfactants than monovalent cations. The association, and thus the reduction in $A_{cw} - A_{hh}$, is, however, highly dependent on the surfactant (generally, a salt of monovalent cation) concentration, as one would expect simply from the law of mass action.

Figure 5.31 shows further that the concentration of calcium in the excess aqueous phase reaches a plateau as the surfactant concentration is increased, indicating, therefore, that the calcium/surfactant ratio in the micellar phase decreases. As a consequence, at a given calcium concentration, the interaction of the C layer with water increases on increasing the surfactant concentration, resulting in the II-III-I transition [7,10,72].

Figure 5.34 illustrates the effect of calcium on the shape of the phase map. With the calcium-free brine, the phase map exhibits the usual rotation due to the partitioning of isobutanol, as discussed above. This effect is nullified when a 90:10 NaCl/CaCl$_2$ brine is used. The presence of calcium induces a rotation of the phase map in the opposite direction. It may be conjectured that the same overall pattern should be observed with cationic surfactants in the presence of multivalent anions.

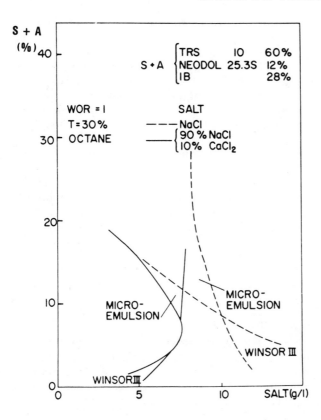

FIGURE 5.34 Influence of divalent cations on the phase boundaries at various amphiphile concentrations. TRS 10 is a sodium petroleum sulfonate supplied by Witco; Neodol 25.3S is a sodium alkyl ether sulfate supplied by Shell. IB denotes isobutanol and the percentages are in weight. (After Ref. 72.)

REFERENCES

1. P. A. Winsor, Solvent Properties of Amphiphilic Compounds, Butterworth, London (1954).

2. K. Shinoda, in Colloidal Surfactants (K. Shinoda, B. Tamamushi, T. Nakagawa, and T. Isemura, eds.), Academic Press, New York (1963).

3. D. Stigter, J. Phys. Chem., 79:1015 (1975).

4. G. Gunnarsson, B. Jönsson, and H. Wennerström, J. Phys. Chem., 84:3114 (1980).

5. B. Lindman and H. Wennerström, Topics in Current Chemistry: Micelles, Springer-Verlag, Berlin, p. 1 (1980).

6. B. Jönsson and H. Wennerström, J. Colloid Interface Sci., 2:482 (1981).

7. G. J. Hirasaki and J. B. Lawson, paper SPE 10921 presented at the 57th SPE Annual Technical Conference and Exhibition, New Orleans (Sept. 1982).

8. M. Celik, E. Manev, and P. Somasundaran, preprint 37a presented at the AIChE 90th National Meeting, Houston (Apr. 1981).

9. J. M. Peacock and E. Matijevic, J. Colloid Interface Sci., 77:548 (1980).

10. G. J. Glover, M. C. Puerto, J. M. Maerker, and E. L. Sandvick, Soc. Pet. Eng. J., 19:183 (1979).

11. M. Bourrel, C. Chambu, and F. Verzaro, Proceedings of the Second European Symposium on EOR, Technip, Paris, p. 39 (Nov. 1982).

12. W. D. Harkins, The Physical Chemistry of Surface Films, Reinhold, New York (1952).

13. P. A. Winsor, Trans.Faraday Soc., 44:390 (1948).

14. R. C. Nelson, D. R. Thigpen, and G. L. Stegemeier, paper SPE/DOE 12672 presented at the 5th SPE/DOE Symposium on Enhanced Oil Recovery, Tulsa (Apr. 1984).

15. S. Qutubuddin, C. A. Miller, and T. Fort, Jr., J. Colloid Interface Sci., 101:46 (1984).

16. K. Shinoda and H. Kunieda, J. Colloid Interface Sci., 42:381 (1973).

17. H. B. Klevens, Chem. Rev., 47:1 (1950).

18. A. M. Lipow, "The Effect of Surfactant Structure on the Phase Behavior of Surfactant-Brine-Oil Systems," M.S. thesis, The University of Texas at Austin (1979).

19. F. Tokiwa and K. Oki, J. Phys. Chem., 71:1343 (1967).

20. V. K. Bansal and D. O. Shah, Soc. Pet. Eng. J., 18:167 (1978).

21. I. Carmona, R. S. Schechter, W. H. Wade, and U. Weerasooriya, paper SPE 11771 presented at the SPE Symposium on Oilfield and Geothermal Chemistry, Denver (June 1983).

22. I. Carmona, "Ethoxylated Oleyl Sulfonates as Model Compounds for Enhanced Oil Recovery," M.S. thesis, The University of Texas at Austin (1983).

23. P. A. Winsor, Trans. Faraday Soc., 44:451 (1948).

24. F. Verzaro, M. Bourrel, and C. Chambu, paper presented at the 5th Symposium on Surfactants in Solution, Bordeaux (July 1984).

25. J. L. Salager, M. Bourrel, R. S. Schechter, and W. H. Wade, Soc. Pet. Eng. J., 19:271 (1979).

26. M. C. Puerto and R. L. Reed, Soc. Pet. Eng. J., 23:669 (1983).

27. W. H. Wade, E. Vasquez, J. L. Salager, M. El-Emary, C. Koukounis, and R. S. Schechter, in Solution Chemistry of Surfactants, Vol. 2 (K. L. Mittal, ed.), Plenum Press, New York, p. 801 (1979).

28. P. A. Spegt, A. E. Skoulios, and V. Luzzati, Acta Crystallogr., 14:866 (1961).

29. M. Bourrel, P. Gard, F. Verzaro, and J. Biais, in Interactions Solide-Liquide dans les Milieux Poreux, Collection Colloques et Séminaires No. 42, Technip, Paris, p. 303 (1985).

30. M. C. Puerto and W. W. Gale, Soc. Pet. Eng. J., 17:193 (1977).

31. S. J. Salter, paper SPE 12036 presented at the 58th SPE Annual Technical Conference and Exhibition, San Francisco (Oct. 1983).

32. Y. Barakat, L. N. Fortney, R. S. Schechter, W. H. Wade, S. Yiv, and A. Graciaa, J. Colloid Interface Sci., 92:561 (1983).

33. L. N. Fortney, "Criteria for Structuring Surfactants to Maximize Solubilization of Oil and Water," M.A. thesis, The University of Texas at Austin (1981).

34. P. A. Winsor, Trans. Faraday Soc., 44:463 (1948).

35. M. Bourrel and C. Chambu, Soc. Pet. Eng. J., 23:327 (1983).

36. S. J. Salter, paper SPE 6843 presented at the 52nd SPE Annual Technical Conference and Exhibition, Denver (Oct. 1977).

37. M. Baviere, R. S. Schechter, and W. H. Wade, J. Colloid Interface Sci., 81:266 (1981).

38. M. Bourrel, J. L. Salager, R. S. Schechter, and W. H. Wade, J. Colloid Interface Sci., 75:451 (1980).

39. R. Zana, S. Yiv, C. Strazielle, and P. Lianos, J. Colloid Interface Sci., 80:208 (1981).

40. A. K. Jain and R. P. B. Singh, J. Colloid Interface Sci., 81:536 (1981).

41. J. L. Salager, J. C. Morgan, R. S. Schechter, W. H. Wade, and E. Vasquez, Soc. Pet. Eng. J., 19:107 (1979).

42. R. N. Healy, R. L. Reed, and D. G. Stenmark, Soc. Pet. Eng. J., 16:147 (1976).

43. S. J. Salter, paper SPE 14106 presented at the SPE Meeting in People's Republic of China (1986).

44. A. Skauge and P. Fotland, paper SPE/DOE 14932 presented at the 7th SPE/DOE Symposium on Enhanced Oil Recovery, Tulsa (Apr. 1986).

45. H. Kunieda and S. E. Friberg, Bull. Chem. Soc. Jpn., 54:1010 (1981).

46. R. C. Nelson and G. A. Pope, Soc. Pet. Eng. J., 18:325 (1978).

47. C. Lalanne-Cassou, R. S. Schechter, and W. H. Wade, J. Disp. Sci. Tech., 7(A):479 (1986).

48. M. Abe, to be published.

49. Y. Barakat, L. N. Fortney, C. Lalanne-Cassou, R. S. Schechter, W. H. Wade, U. Weerasooriya, and S. Yiv, Soc. Pet. Eng. J., 23:913 (1983).

50. I. Carmona, R. S. Schechter, W. H. Wade, and U. Weerasooriya, Soc. Pet. Eng. J., 25:351 (1985).

51. C. Lalanne-Cassou, I. Carmona, L. Fortney, A. Samii, R. S. Schechter, W. H. Wade, U. Weerasooriya, V. Weerasooriya, and S. Yiv, paper SPE 12035 presented at the 58th SPE Annual Technical Conference and Exhibition, San Francisco (Oct. 1983).

52. M. Abe, D. Schechter, R. S. Schechter, W. H. Wade, U. Weerasooriya, and S. Yiv, J. Colloid Interface Sci., 114:342 (1986).

53. M. Bourrel, A. M. Lipow, W. H. Wade, R. S. Schechter, and J. L. Salager, paper SPE 7450 presented at the 53rd SPE Annual Technical Conference and Exhibition, Houston (Oct. 1978).

54. J. E. Vinatieri and P. D. Fleming III, Soc. Pet. Eng. J., 19:28 (1979).

55. J. Biais, B. Clin, and P. Lalanne, C. R. Acad. Sci., B294:457 (1982).

56. J. Biais, P. Bothorel, B. Clin, and P. Lalanne, J. Disp. Sci. Tech., 1:67 (1981).

57. I. Prigogine and R. Defay, Thermodynamique Chimique, (Desoer, ed.), Liege (1950).

58. L. P. Prouvost, T. Satoh, K. Sepehrnoori, and G. A. Pope, paper SPE 13031 presented at the 59th SPE Annual Technical Conference and Exhibition, Houston (Sept. 1984).

59. J. Biais, M. Bourrel, M. Barthe, B. Clin, and P. Lalanne, Proceedings of the Third European Symposium on EOR, Vol. 2, AGIP, editor, Rome, p. 229 (1985).

60. J. Biais, M. Barthe, B. Clin, and P. Lalanne, J. Colloid Interface Sci., 102:361 (1984).

61. J. Biais, M. Barthe, M. Bourrel, B. Clin, and P. Lalanne, J. Colloid Interface Sci., 109:576 (1986).

62. A. Graciaa, J. Lachaise, J. G. Sayous, P. Grenier, S. Yiv, R. S. Schechter, and W. H. Wade, J. Colloid Interface Sci., 93:474 (1983).

63. A. Graciaa, J. Lachaise, J. G. Sayous, M. Bourrel, R. S. Schechter, and W. H. Wade, Proceedings of the Second European Symposium on EOR, Technip, Paris, p. 61 (1982).

64. E. H. Crook, D. B. Fordyce, and G. F. Trebbi, J. Colloid Sci., 20:191 (1965).

65. M. Bourrel, C. Koukounis, R. S. Schechter, and W. H. Wade, J. Disp. Sci. Tech., 1:13 (1980).

66. A. Graciaa, J. Lachaise, M. Bourrel, I. Osborne-Lee, R. S. Schechter, and W. H. Wade, SPE Reservoir Eng., in press.

67. S. D. Robertson, paper SPE 14909 presented at the 7th SPE/DOE Symposium on Enhanced Oil Recovery, Tulsa (Apr. 1986).

68. A. W. Adamson, J. Colloid Interface Sci., 29:261 (1969).

69. M. Robbins, and D. W. Brownawell, "Microemulsion Separation of Organic Compounds in the Liquid State," U.S. Patent 3,641,181 (1972).

70. A. Graciaa, J. Lachaise, G. Marion, and M. Bourrel, to be published.

71. S. J. Salter, paper SPE/AIME 7056 presented at the 5th SPE/AIME Symposium on Improved Methods for Oil Recovery, Tulsa (Apr. 1978).

72. M. Bourrel, C. Chambu, R. S. Schechter, and W. H. Wade, Soc. Pet. Eng. J., 22:28 (1982).

6

Compensating Changes Between Formulation Variables

I. GENERAL

The influence of formulation variables on the R-ratio has been considered in Chap. 5 by examining those transitions that take place
when one of the variables is changed, holding all the others fixed.
This approach provides qualitative verification of the R-ratio concept. It is possible to obtain a quantitative comparison of the relative influence of formulation variables by employing the following
strategy. Start with an optimized system (R = 1). Change two
variables at the same time so that the final system is also optimized.
Thus the change resulting from varying one of the formulation variables tending to produce a III-II transition is compensated by the
change of a second variable tending to yield a III-I transition, with
the net result being that the type III behavior persists. By comparing the magnitude of the two compensating changes, a quantitative relationship defining the sensitivity of the R-ratio to various
formulation variables can be developed.

Compensating change in the cohesive energies which leaves the
R-ratio invariant obviously requires that at least two of the cohesive
energies appearing in Eq. (1.18) be varied. These changes can be
visualized as being essentially of three different types. Both of the
changes can be in cohesive energies appearing in the numerator of
the R-ratio so that the magnitude of the variation is the same but
the signs are opposite. These are called compensating lipophilic
changes. The individual terms in the denominator of the R-ratio
are practically unchanged.

Compensating hydrophilic changes are those that affect the cohesive energies in the denominator of the R-ratio. In this mode the

numerator is unchanged. Similarly, <u>compensating lipophilic-hydro-</u>
<u>philic</u> changes are those which cause both the numerator and the de-
nominator of the R-ratio to change simultaneously with the final value
of the R-ratio remaining unity.

These three distinctions are made carefully in structuring the discussion presented in this chapter; however, the great value of this
classification scheme will become most evident when considering the
solvency of amphiphilic compounds, a matter taken up in Chap. 7.

Indeed, because of the large number of parameters defining the
systems, many pairwise combinations can be considered; however, it
is not possible to discuss all of them, but rather to provide illustra-
tive examples.

In this chapter ionic amphiphiles are considered separately from
nonionic compounds. This division is convenient since the sensitivity
of the two types of amphiphiles to changes in certain formulation variables is often quite different. Thus changing the electrolyte concentration tends to promote the same transformation irrespective of the
type of hydrophile, but to compensate for a given change in electrolyte concentration quite different changes in the second variable are
required, depending on whether or not the hydrophile exhibits ionic
character.

II.   IONIC SURFACTANTS

A.   Compensating Hydrophilic/Lipophilic Changes

1.   Sodium Chloride Concentration Versus Alkane Carbon Number

When the sodium chloride concentration and the alkane carbon number (ACN) are varied simultaneously, results such as those shown
in Fig. 4.30 are typically obtained. The solid line drawn inside the
three-phase region denotes a locus of optimal systems[†] so that, for
example, one composed of heptane and water containing about 1 g/dl
sodium chloride is optimum (R = 1). If the heptane is replaced by
octane (ACN is increased), thereby decreasing R (see Section 5.III.A),
the system is no longer optimum. To restore optimum conditions, the
electrolyte concentration can be increased, thus decreasing the denominator (see Section 5.II.A) of the R-ratio. Figure 4.30 shows the re-

---

[†]The definition of optimality used to construct the line differs from
that usually applied. In this case it was defined as the salinity at
the center of the three-phase region. It is believed that for the
system studied this definition leads to optima which are very nearly
the same as those found by equating the solubilization of the oil and
the water.

lationship between the ACN and the salinity for a particular surfactant system. The slope of the curve increases with increasing ACN. Thus to maintain optimal conditions as the ACN is increased, larger changes in salinity will be required at higher values of ACN than at lower values. This is predictable since A_{hh} depends on $I^{-1/2}$, where I is the ionic strength of the solution [Eq. (5.2)].

A number of phase maps similar to Fig. 4.30 have been constructed by Salager [1,2], who found that for a large number of systems, plotting the logarithm of optimal salinity s* against ACN results in a straight line. An example is given in Fig. 6.1. The slope of these straight lines and those for many other alkyl aryl sulfonates were found to be essentially constant and equal to 0.16 when s* is expressed in grams of NaCl per 100 cm^3 of brine denoted as (g/dl). This result is restricted to the range of conditions studied by Salager: namely, relatively low salinities and high alcohol/surfactant ratios. With systems yielding higher optimal salinities, a linear relationship has been reported between s* and ACN [3] (see also Figs. 5.10 and 5.6). In addition, different slopes (around 0.10) have been found for other types of surfactants, such as alkylcarboxylates, alkylsulfates, or alkylsulfonates [1,4].

2. Sodium Chloride Concentration Versus Hydrocarbons
    Other Than Alkanes

Using similar experimental conditions corresponding to those of Fig. 6.1, Wade et al. have studied the homologous series of n-alkylcyclohexanes (heptyl to dodecyl) and n-alkylbenzenes (heptyl, nonyl, decyl, and undecyl) [5]. Again, the logarithm of optimal salinity was found to vary linearly with the number of carbon atoms of the alkyl part of the hydrocarbon molecule. With n-alkylbenzenes, however, the slope is larger than the value found for n-alkanes and n-alkylcyclohexanes.

Figure 6.2 shows optimal salinities obtained in the absence of alcohol as a function of the oil molar volume. For a given oil molar volume, the optimal salinity is seen to increase from n-alkanes, to alkylcyclohexanes, to alkylbenzenes. This trend indicates a larger interaction of the surfactant lipophile with alkylbenzene. This difference can be attributed either to a greater interaction between one oil molecule and the lipophile and/or because more oil molecules (better packing) interact with a single lipophile. This trend can be anticipated because of the high solvency of aromatics as compared with alkanes.

Another interesting feature of Fig. 6.2 is the apparently erratic behavior of the alkylbenzenes. The optimal salinity is not a monotonically increasing function of the oil molar volume. This is believed to be related to the difficulty of packing aromatic oils within the lipo-

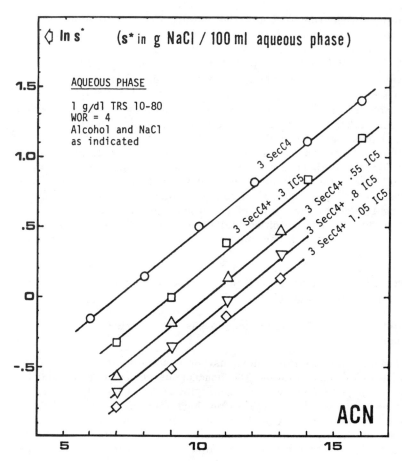

FIGURE 6.1   Correlation ln s* versus ACN at different alcohol concentrations.   SecC4 and IC5 stand for butanol-2 and isopentanol,
respectively.   The concentrations are expressed in grams per deciliter (g/dl) of water.   (After Ref. 2.)

philes in the C layer.   Further evidence to support this viewpoint is
provided by Fig. 6.3.   In the presence of alcohols a monotonical increasing optimal salinity with increasing oil molar volumes is found
even for alkylbenzenes.   The short-chain alcohols penetrate the C
layer (see Section 5.VI.C), separating the longer-chain lipophiles and
facilitating the packing of the oil molecules.   Typically at the alcohol/
surfactant ratios used to obtain the results shown in Fig. 6.3, at least
three or four alcohol molecules per surfactant molecule are expected
in the C region [6].

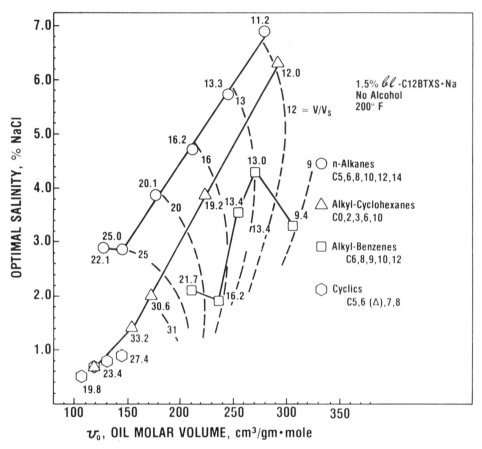

FIGURE 6.2   Optimal salinity measured from solubilization parameters versus oil molar volume for various homologous series of hydrocarbons. bl-C12BTXS·Na is a mixture of sodium alkylarylsulfonates. The number next to each experimental point is the solubilization parameter value which is discussed in Chap. 7.   (After Ref. 3.)

3.   Sodium Chloride Versus Length of Lipophile

Increasing the length of the surfactant's alkyl chain generally increases the numerator of the R-ratio (Section 5.III.B) and to maintain R = 1, the sodium chloride concentration can be decreased correspondingly.   This trend is shown by Fig. 6.4, where for constant ACN it is seen that increasing the lipophilic chain length is compensated by a decrease in the logarithm of the sodium chloride concen-

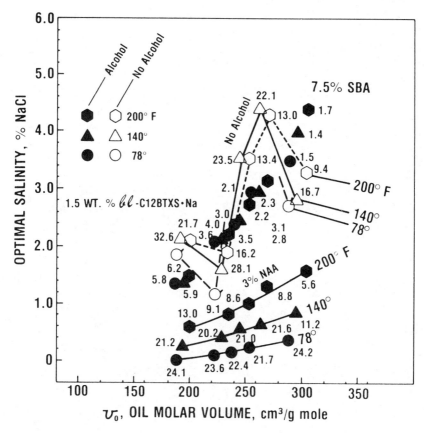

FIGURE 6.3   Effect of alcohol and temperature on the optimal salinity of the n-alkylbenzenes versus oil molar volume (see Fig. 6.2).   SBA and NAA stand for butanol-2 and pentanol-1, respectively.   (After Ref. 3.)

tration, which appears to be practically independent of the ACN. This observation will be used to help develop the empirical correlation discussed in Section 6.II.E.

B.   Compensating Lipophilic/Lipophilic Changes

1.   ACN Versus Length of Lipophile

The increase in the numerator of R resulting from an increase of the surfactant's alkyl chain can be compensated by increasing ACN. This can be visualized as an increase in A_{co} matched by an increase in A_{oo}.   It is interesting to note that for the results shown in Fig.

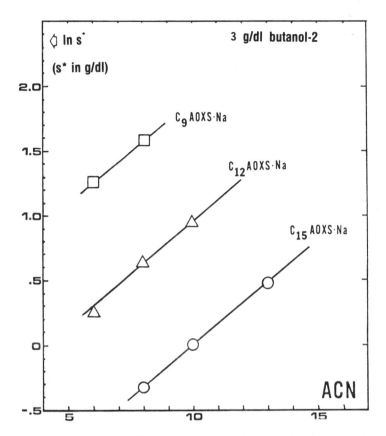

FIGURE 6.4 The effect of alkyl chain length on optimal salinity for various alkane carbon numbers. AOXSNa stands for sodium alkyl orthoxylene sulfonate (commercial products, supplied by Exxon Chem.). C_9, C_{12}, C_{15} = nonyl, dodecyl, pentadecyl. Experimental conditions are similar to Fig. 6.1. The proportion of surfactant is 1g/dl of aqueous phase. (After Ref. 2.)

6.4, adding three carbon atoms to the surfactant lipophile is compensated by increasing the ACN by 6 units. This linear relationship is further supported by results presented in subsequent sections; however, the ratio of 2 ACN units per added methyl group in the lipophilic chain is not universal. It depends to some extent on the alcohol concentration.

Figure 6.4 shows that at a given salinity and alcohol concentration, the system is optimum for an alkane carbon number (which can thus be noted as ACN*) that depends on the length of the surfactant alkyl

chain. ACN* can thus be seen as a characteristic of the surfactant
provided that all other conditions are fixed. The alcohol-free system at unit salinity (1 g/dl NaCl) has been taken as a standard state
[1,7,8], and the corresponding ACN* value, which may be extrapolated using a procedure described below, has been called both the
EPACNUS [for Extrapolated Preferred Alkane Carbon Number at Unit
Salinity and no alcohol [1]] and N_{min} [7,8] by others.

Typically, the EPACNUS has been determined based on phase behavior studies [2], whereas N_{min} was originally obtained from interfacial tension measurements using the spinning drop technique [9].

A typical plot of interfacial tension is shown in Fig. 6.5. For a
given surfactant, a deep minimum in interfacial tension which defines
an optimal alkane carbon number is observed. This ACN* has been
termed N_{min}. Presumably the minimum of interfacial tension corresponds to the same ACN which optimizes the phase behavior. Thus
for systems that are alcohol-free and are at 1 wt % NaCl, EPACNUS
should be identical to N_{min} [2,10,11]. Because of differences in the
method of measurement [e.g., many of the interfacial tension experiments were carried out at low surfactant concentrations in non-pre-
equilibrated systems [7–9,12,13], some difference in the two values
may be found, especially when the surfactant is a mixture of amphiphiles (see Section 6.II.G). These are not, however, important.
The minimum in the interfacial tension does correspond quite closely
to the same formulation, which yields optimized systems defined, for
example, by equal volumes of oil and water solubilized into a type
III micellar phase.

N_{min} is seen from Fig. 6.5 to depend on the surfactant tail length.
Moreover, N_{min} has been determined for mixtures of surfactant and
has been found to vary linearly with the average equivalent weight
of the mixture [calculated according to a linear mixing rule based on
mole fraction (see Section 5.VII.D)]. Figure 6.6 shows a typical example of the variation in N_{min} for a mixture of C_{12} and C_{15} orthoxylene sulfonates. The slope of the straight line indicates that roughly 3 ACN units are required to compensate for adding 1 carbon atom
to the surfactant lipophile. This is compared to the value of 2 found
from Fig. 6.5, which compares systems with alcohol added. Figure
6.6 indicates that a mixture of C_{12} and C_{15} orthoxylene sulfonate can
be characterized equivalently either by its average molecular weight
or by its N_{min}.

For the mixture, N_{min} can thus be calculated by

$$(N_{min})_{mix} = \sum_i x_i (N_{min})_i \qquad (6.1)$$

where x_i is the mole fraction of the ith component of the surfactant.

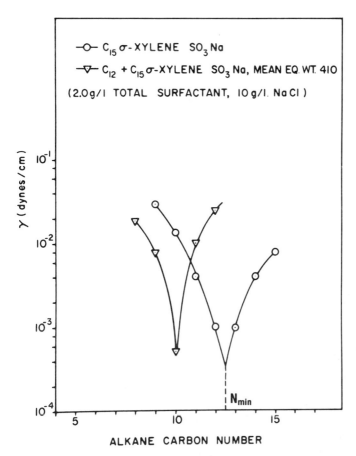

FIGURE 6.5 Typical interfacial tension variation along an ACN scan. The surfactants are C_{12}-orthoxylene sulfonate and a mixture of C_{12} and C_{15} orthoxylene sulfonates of mean equivalent weight 410. N_{min} for $C_{15}OXSNa$ is 12.5. (After Ref. 8.)

Alternatively, Eq. (6.1) can be used to estimate the N_{min} values characterizing "off-scale" surfactants, that is, those which do not yield minimal tensions with any of the liquid alkanes at 1 wt % NaCl. This requires experimentally determined values of $(N_{min})_{mix}$ of a mixture including the surfactant to be characterized and a surfactant of known N_{min}. Values of N_{min} less than 5 or greater than 16 can thus be determined even though these values lie outside the range of ACN corresponding to the liquid n-alkanes at room temperature. For some amphiphiles, a negative value of N_{min} may be found by this procedure.

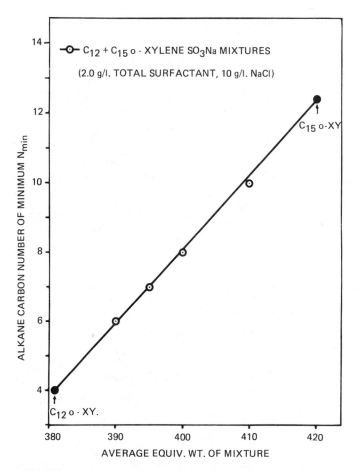

FIGURE 6.6   Dependence of N_{min} (obtained from interfacial tension measurements) on the average equivalent weight of mixtures of C_{12}- and C_{15}-orthoxylene sulfonates.   (After Ref. 8.)

Figure 6.7 provides an illustration of the similarity between EPACNUS and N_{min} [14].   In this plot the EPACNUS is seen to vary linearly with the surfactant molecular weight for the homologous alkylbenzene sulfonate series, and to increase by 2.2 units when one carbon atom is added to the alkyl chain [2].   This is very similar to the result shown by Fig. 6.6 for N_{min} based on interfacial tension measurements. Thus the EPACNUS of a surfactant mixture can be shown to be correctly calculated with a linear mixing rule [11] identical in form with that shown by Eq. (6.1) for N_{min}.

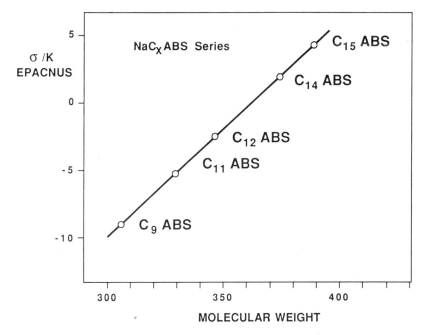

FIGURE 6.7 EPACNUS versus molecular weight for alkylbenzene sulfonate series. (Adapted from Ref. 2.)

The slope of N_{min} (or EPACNUS) versus the number n of carbon atoms of the surfactant alkyl chain is an indication of the interaction energies between the C and O regions. These interactions depend to some extent on the surfactant packing at the interface, as well as on the structure of the surfactant lipophile.

Figure 6.8 shows that in the absence of alcohol, the slope is 4.3 ACN units per methylene carbon atom added to the surfactant's alkyl chain. This value found at 92°C applies at two different salt concentrations. For comparison, a much smaller slope of 1.4 ACN units per methylene carbon atom in the presence of alcohol is shown. In this case the surfactant/isopentanol ratio is 60:40. It is believed that the smaller slope is due primarily to a decrease in surfactant interfacial density; thus increasing the length of the surfactant chain would be expected to have a proportionately smaller effect.

2. ACN Versus the Structure of the Lipophile

The structure of lipophiles of given molecular weight may differ in a number of ways. First, for a linear lipophilic alkyl chain, different points of attachment of the hydrophilic group confer different

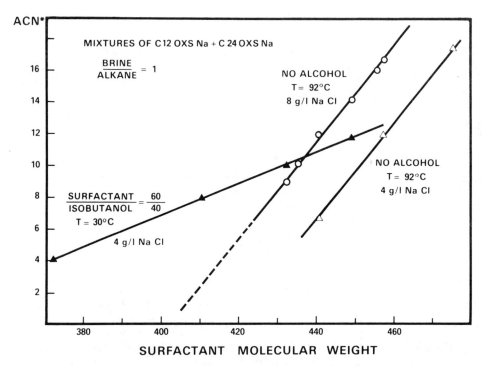

FIGURE 6.8 Optimal ACN against lipophile variation achieved by
mixing C_{12}- and C_{24} - orthoxylene sulfonates. Isobutanol decreases
the slope. (After Ref. 14.)

amphiphilic character. As noted in Section 5.III.B, the effect of
stepwise removal of the polar head from terminal to medial positions
in the alkyl chain is primarily to increase the area per surfactant molecule, and thus to decrease the number of surfactant molecules per
unit area of interface of the C layer. However, because of the decrease in the lateral interactions, and thus in $A_{\ell\ell}$, and the increase
in A_{hh}, R is increased. Whenever compensation is achieved by changing A_{oo}, ACN* is expected to increase. Figure 6.9 shows this prediction to be correct and that ACN* increases linearly when the benzene ring carrying the sulfonate group is moved along the hexadecyl
chain from the first to the eighth carbon atom [15].

Other possible modifications of the surfactant tail, generally termed
"branching," consist of different arrangements of the carbon atoms
of the lipophile; for example, short alkyl groups attached along the
main backbone is considered here to be branching. Very few systematic studies that examine the influence of lipophilic structure have

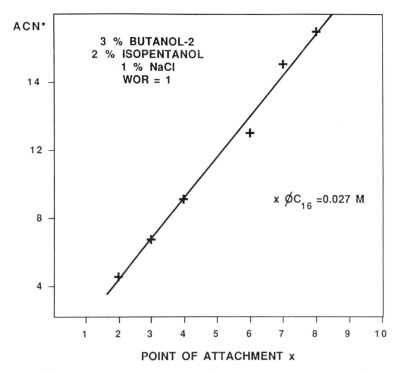

FIGURE 6.9 The effect of the lipophile structure on optimal alkane carbon number. $x\phi C_{16}$ refers to the isomer of hexadecylbenzene sulfonate whose benzene ring is attached to the xth carbon atom of the alkyl chain. The concentrations refer to the aqueous phase. (Adapted from Ref. 15.)

been reported, probably because of the difficulty of obtaining pure compounds. The trend is, however, that branching the surfactant tail results in a balance of interaction energies more favorable to oil, as compared to the behavior of the corresponding straight tail species [4,16] (see Fig. 6.12 and Section 7.V.B).

C. Compensating Hydrophilic Changes

1. Sodium Chloride Versus Hydrophile

The interaction of the C layer with water has been seen in Section 5.II.D to depend on the nature of the ionic hydrophile. One method of systematically increasing the hydrophilicity is the addition of ethylene oxide units. Figure 5.6 shows the optimum values for oleyl

ethylene oxide sulfonates. The presence of the ethylene oxide within the hydrophile has been found to increase the optimal salinity.
This is also true for other ionic groups, such as sulfates [17] or
carboxylates [18].

The hydrophilicity can also be increased by increasing the number of polar atoms in the hydrophile. Examples are the sodium salts
of n-acetyl- α-amino alcanoic acids, where the hydrophile is a combi-
nation of the −COONa and −NHCOCH$_3$ groups attached to the same
carbon of the alkyl chain. Hydroxy sulfonates, which are found in
large proportions in commercial α-olefin sulfonates (AOS), provide
a further example of surfactants with increasing hydrophilicity.
The sulfonation of α-olefins produces a mixture of alkene sulfonate,
$RCH=CHCH_2SO_3Na$; hydroxysulfonate, $RCHCH_2CH_2SO_3Na$; and di-
                                                           |
                                                          OH
sulfonate, $RCHCH_2CH_2SO_3Na$, roughly in the proportions 50:40:10.
         |
        SO$_3$

Figure 6.10 shows the results obtained with an octadecyl alkane
sulfonate, an octadecyl-3 hydroxysulfonate, and an octadecyl α-olefin sulfonate consisting of a mixture of sulfonate and hydroxysulfonate species but no disulfonate. The optimum salinity is, as expected, higher for the more hydrophilic sulfonate for given alkane than
for the alkene sulfonate. The α-olefin sulfonate exhibits intermediate behavior. Similar trends have been obtained in the presence of
calcium chloride [19]. Mixtures of these α-olefin derivatives may
not follow linear mixing rules because of selective partitioning (Section 5.VII.D).

2.   Sodium Chloride Compensated by pH
     (Carboxylated Surfactants)

An interesting experiment relates to compensation of a change in
electrolyte concentration by altering the degree of ionization of a
carboxylic acid, which depends, of course, on the solution pH (Section 5.II.C). Mixtures composed of various ratios of the carboxylic
acid and its alkaline salt yield solutions of varying pH. Figure 6.11
shows how the salinity is to be adjusted in response to variations
in the ratio of octanoic acid to sodium octanoate [20] to maintain optimum conditions.

When an acid molecule at the interface is replaced by its salt,
($A_{cw} - A_{hh}$) increases, as discussed in Section 5.II.C, because of
the increase in the overall degree of ionization. To compensate for
increased ($A_{cw} - A_{hh}$), the salinity must be increased. The evolution of the optimal salinity, as seen in Fig. 6.11, parallels the degree of ionization, which is simply calculated as the function of pH
for solutions of weak electrolytes [20] (see also Section 7.II.B).

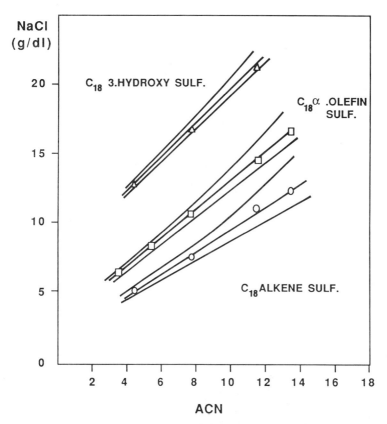

FIGURE 6.10 Phase behavior for systems containing 0.6 g/dl surfactant, 3 g/dl butanol-2,2 g/dl isopentanol with respect to the aqueous phase. Water—oil ratio is 1. Lines with data points are optimum
lines and the other lines are the extreme of the three-phase regions.
(After Ref. 19.)

D. Adding Alcohol to Compensate Hydrophilic
    or Lipophilic Changes

Figure 6.1 shows that upon changing the alcohol concentration, the
straight line ln s* shown as a function of ACN is displaced from the
original line but remains parallel to it. These curves show that upon
adding isopentanol, s* must be reduced to maintain an optimum system. Furthermore, if s* is fixed, ACN must be increased to compensate for increased isopentanol concentrations.

Since the curves in Fig. 6.1 are parallel, the change in ln s* resulting from the addition of a certain quantity of alcohol is independ-

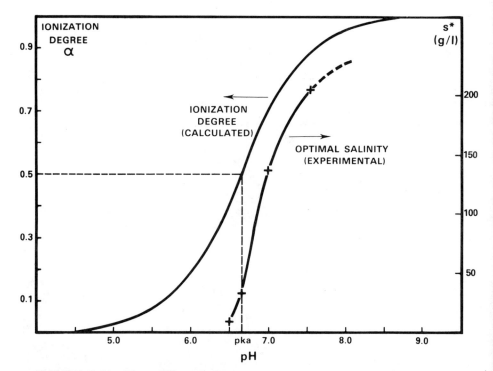

FIGURE 6.11   The effect of pH on optimal salinity for sodium octanoate.   The ionization degree α is calculated for dilute solutions
of electrolytes ($\alpha = 10^{(pH - pK_a)}/[1 + 10^{(pH - pK_a)}]$).   Brine/
decane = 1 (wt.).   Surfactant/isobutanol = 70/30 (wt.).   T = 30°C.
(After Ref. 20.)

ent of ACN.   This change (or shift as it is sometimes called), denoted as f(A), is recorded in Fig. 6.12.   It is noted that to achieve
a given shift, smaller quantities of hexanol-1 than isopentanol are required.   The experimental points shown on the isopentanol curve denote results obtained using different sulfonate surfactants.   That the
points fall on a single curve is interpreted to mean that f(A) does
not depend on the structure of the lipophile.   This conclusion is restricted to sulfonates and to relatively high alcohol/surfactant ratios.
For very small alcohol concentrations, f(A) will in general depend on
the surfactant structure [1].

The greater sensitivity to hexanol-1 compared with pentanol and
isopentanol shown in Fig. 6.12 agrees with the qualitative interpretation of the influence of alcohol presented in Section 5.IV.A.   It was
shown [see Eq. (5.11)] that if $(a_{co}^s - a_{co}^a) < (a_{cw}^s - a_{cw}^a)$, then in-

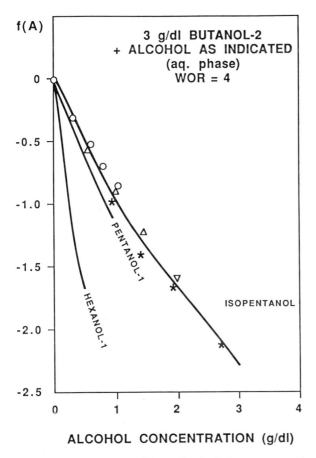

FIGURE 6.12   The effect of alcohol concentration and type on optimal line $\ln s^*$ versus ACN (see Fig. 6.1).   The surfactants are commercial sulfonates.   (After Ref. 2.)

creasing the alcohol concentration will promote I-III-II transformations and increasing the alcohol chain length further reinforces the inequality.   Thus a_{co}^a for hexanol-1 is larger than for isopentanol and smaller quantities of hexanol-1 are required to affect a given transformation. There is, however, some danger in extending this reasoning to longer and longer chain alcohols, for even though a_{co}^a most certainly increases with increasing chain length, the quantity of alcohol that partitions into the C layer decreases with increasing chain length.   The higher-molecular-weight alcohols will tend to partition preferentially into the oil phase.   Increasing the chain length of the higher-molec-

ular-weight alcohols therefore results in competing mechanisms—
higher cohesive energies per molecule but smaller C-layer concentrations.

Figure 6.12 shows that hexanol-1 decreases the optimum salinity to
a greater extent than an equivalent weight of pentanol-1, the greater
cohesive energy associated with the interaction of the hexanol-1 with
the oil is the more important factor. Continued increase in the length
of the carbon chain will ultimately reverse the trend shown in Fig.
6.12. The optimum alcohol chain length is discussed further in Section 7.V.C.

Thus for high-molecular-weight alcohols, $a_{co}^a > a_{cw}^a$ and the optimum salinity decreases with the addition of alcohol. For low-molecular-weight alcohols $a_{co}^a < a_{cw}^a$ and s* increases with alcohol concentration. This trend has been observed in the case of isopropanol,
propanol, and tertiary butanol [21-23]. The sign of f(A) (see Fig.
6.12) will, therefore, depend on the molecular structure of the alcohol.

Figure 6.12 shows that for the alcohols studied, the magnitude
of f(A) is practically independent of the surfactant structure or the
molecular weight of the alkane. This is true because the partitioning of a higher-molecular-weight alcohol between the interface and
the oil phase is relatively insensitive to these molecular structural
factors over the range of compounds studied. Only small quantities
of the higher-molecular-weight alcohols are found in the excess aqueous phase.

As the alcohol molecular weight is reduced, substantial partitioning between the oleic and the aqueous phase is observed and the degree to which an alcohol partitions into the oleic phase becomes a
strong function of the chemical nature of the oleic phase. For these
alcohols, f(A) depends on both surfactant and oil molecular structures
[1]. Thus the construction of graphs such as Fig. 6.12 has merit
only for that narrow range of alcohols which partition almost entirely
into the oil phase and yet still adsorb to some degree into the C layer.

E. Correlation Between Formulation Variables

Based on a systematic study similar to those reported in Figs. 6.1,
6.4, and 6.12, Salager et al. [2] have obtained an empirical correlation relating the salinity, ACN, and alcohol concentration which
yields optimum behavior. As is the case with most empirical equations, its value resides primarily in displaying the relationship that
exists between variables over the range studied. Extrapolation of
an empirical equation can often lead to difficulties. Recognizing the
shortcomings of empiricism, it has been found that the following
equation has considerable merit over a range of conditions limited

primarily to modest concentrations of monovalent electrolyte (NaCl generally) and to systems containing alcohols as well as other amphiphiles.

$$\ln s^* = K(ACN) + f(A) - \sigma \qquad (6.2)$$

This equation applies to systems for which the alcohol partitions primarily into the oleic phase and the C layer but not into the aqueous phase. As noted above, for this case $f(A)$ is essentially independent of ACN and Eq. (6.2) is valid. σ is a parameter that depends on the surfactant's lipophile structure. When $s^* = 1$ (unit salinity) and in the absence of alcohol [$f(A) = 0$], σ/K is the EPACNUS or N_{min}.

Equation (6.2) is satisfied by optimum systems and therefore, one can calculate, for example, the change in electrolyte concentration needed to compensate a change in hydrocarbon molecular weight (ACN) or the change in surfactant structure required to balance an increase in alcohol concentration. The EPACNUS (or N_{min}) can therefore be measured based on studies of optimum systems and Eq. (6.2). Figure 6.13 provides a plot of σ/K for a variety of amphiphilic compounds.

For a given number of methylene groups in the lipophile, the EPACNUS is seen to vary with the hydrophilicity in a sequence consistent with the trends described in Section 5.II.D. Also, as seen in Fig. 6.13, the addition of a benzene ring significantly increases σ/K.

Equation (6.2) shows that definitions other than σ/K (EPACNUS or N_{min}) could have been selected to characterize surfactant structure. For example, $\ln s^*$ for some (arbitrary) ACN or $f(A)$ would have served equally well. Furthermore, the HLB and N_{min} must, in some sense, be related to the other since both are related to the stability of macroemulsions. A precise relationship between the HLB requirement of a system and N_{min} has not been established. The underlying basis for asserting that a relationship does exist is described in Section 4.II.F.

The effects of temperature and water/oil ratio have also been investigated by Salager [11], who found that temperature (slightly) increases linearly the logarithm of optimal salinity (see Section 5.IV.B). This effect can be included in the correlation, Eq. (6.2), by the following form:

$$\ln s^* = K(ACN) + f(A) - \sigma + a_T(T - 25) \qquad (6.3)$$

where the temperature coefficient a_T is, for sulfonates, 10^{-2} ($\ln s^*$ units/°C). Salager [11] found little shift in optimum position when the water/oil ratio was varied between modest limits of 4 to 1.

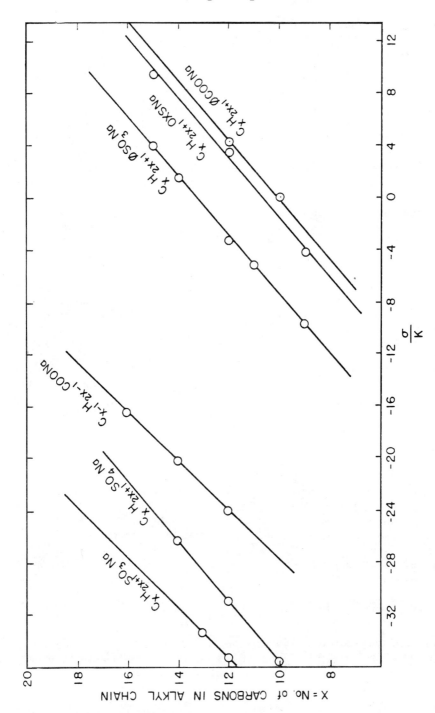

FIGURE 6.13 The effect of surfactant structure on σ/K (EPACNUS). ϕ represents the benzene ring. OXS stands for orthoxylene sulfonate. Water to oil ratio is 4. T = 25°C. (After Ref. 4.)

F. Mixing Rules: EACN Concept

Mixtures of surfactants (or alcohols) have been seen effective in promoting phase changes between types I, III, or IV and II (Figs. 5.10
and 4.36, for example) and it has been found that such mixtures can
be represented by an effective N_{min} or EPACNUS obtained by averaging on mole fraction basis according to Eq. (6.1). Surfactant blends
can also be characterized equivalently by an effective s^*. In this
case linear mixing rules assume that $\ln s^*$ for individual amphiphiles
are additive as follows:

$$\ln s^*_{mix} = \sum_i x_i \ln s^*_i \qquad (6.4)$$

where x_i is the mole fraction of component i. Experimental results
have been obtained by Salager et al. [11] showing the validity of
the equation. For different conditions (surfactant/alcohol ratio = 3),
Puerto and Gale [24] used a linear mixing rule on s^*_i to fit their data.
In fact, Eq. (6.4) has been found to slightly underestimate s^*,
whereas averaging s^*_i was found to overestimate it slightly [16]. In
general, for high-electrolyte concentrations a linear rather than log-
arithmic average of s^* to obtain s^*_{mix} is best, whereas at smaller electrolyte concentrations Eq. (6.4) should be better.

Equation (6.2) can also be used to determine the EPACNUS, σ/K,
of an unknown surfactant provided that the other terms in the equation are known. Alternatively, an oil can be assigned an effective
ACN based on phase behavior studies of optimized systems where the
salinity, f(A), and surfactant EPACNUS are known. In practice the
optimum is obtained by varying the salt concentration or the surfactant EPACNUS (using mixtures, for example). This characteristic
ACN, which corresponds to the ACN term in Eq. (6.2), has been
termed the equivalent alkane carbon number (EACN) of the oil under
consideration [13]. In other words, if a given system is, for example, optimized with an oil and if precisely the same system (electrolyte concentration, surfactant etc.) remains optimal when the oil is
replaced by octane, the EACN of the oil is 8.

Originally, EACN values were assigned to oils other than alkanes
by matching the alkane giving a minimum of interfacial tension under
the same conditions as the oil. A number of papers [7,9,10,12,13]
have appeared showing the EACN to be independent of the surfactant
structure (sulfonates were mainly used) and that relatively simple
equivalence rules allow correlation of the behavior of various hydrocarbons [9]. For example, an alkylbenzene was found to give minimal interfacial tension under the same conditions as an alkane having
the same number of carbon atoms as the alkyl side chain of the alkylbenzene. The contribution of the benzene ring to EACN was there-

fore considered to be zero.   Similarly, it was found that a cyclohexyl
ring can be considered to be equal to approximately four carbons in
an alkane.   The EACN concept has also been applied to complex mixtures of hydrocarbons such as crude oils [7].

Although it is possible to assign an equivalent alkane carbon number to almost any liquid hydrocarbon, it should be emphasized that
the alkane and its equivalent are not necessarily equivalent in all respects [25,26].   While the two may exhibit a minimum interfacial tension under the same conditions, the depth of the minimum often differs.   At optimum, for example, the interfacial tension of a crude
oil is often greater than that of the equivalent alkane.   An example
is seen in Fig. 6.14.   The crude oil at minimum exhibits an order of
magnitude larger interfacial tension than the alkane.

Phase behavior has also been used to assign values of EACN to
oils [2,27,28].   Care should be used in making such assignments
since phase behavior studies often use systems that include alcohols
and difficulties may arise when the alcohol partitioning depends markedly on the structure of the oil.   Tham and Lorenz [29], have, for
example, reported variations of the EACN assigned to a given crude
oil depending on the alcohol type or the water/oil ratio.   Furthermore,
discrepancies in the equivalence rules between alkylbenzenes and alkanes cited above were found.   Thus the EACN concept should be
used with care, especially in the presence of alcohols.

Results on solubilization obtained at higher surfactant concentrations (see Section 7.V.F) have confirmed that, in general, hydrocarbons cannot be completely represented by a single $\underline{n}$-alkane [3,14,29].
Puerto and Reed [3] have proposed that equivalent oils must have
the same molar volumes and optimal salinities to have the same solubilization.   Further discussion of solubilization is given in Chapter 7.
Qualitatively, however, the equivalence of $\underline{n}$-alkanes with certain $\underline{n}$-
alkylbenzenes and $\underline{n}$-alkylcyclohexane is a useful concept since it
provides a vehicle for understanding the effect of, for example, making the oil more aromatic.   For example, it can be seen from Fig. 6.2
that the (interpolated) optimal salinity increases from nonylbenzene
to octylcyclohexane to $\underline{n}$-tridecane, as expected from their respective
EACN (9-12-13) calculated according to the equivalence rule stated
above.

A mixing rule for EACN has also been established [13], which has
then been used for determining the EACN of an unknown hydrocarbon [7].   It is written

$$(EACN)_{mix} = \sum_i x_i (EACN)_i \qquad\qquad (6.5)$$

where x_i is the mole fraction of hydrocarbon i.   This equation is in
agreement with other mixing rules expressed by Eqs. (6.1) and (6.4).

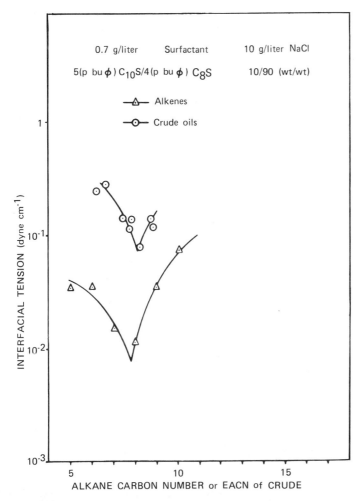

FIGURE 6.14 Comparison of alkane with crude oil interfacial tensions for sulfonated 4(diethylphenyl) nonane. Each crude oil is referred to on the abscissa by its EACN. (After Ref. 26.)

The procedure for assigning an EACN consists of blending the unknown oil in various proportions with a hydrocarbon of known EACN. Interfacial tensions are then measured against a given aqueous surfactant solution whose preferred alkane (N_{min}) is known. At optimum this gives the values of $(EACN)_{mix}$ and from the composition of the corresponding blend, the EACN of the unknown oil can be calculated based on Eq. (6.5).

Equation (6.5) shows that intermediate behavior representative of nonintegral number of carbon atoms can be obtained by blending hydrocarbons. The optimal salinity of mixtures of pentane and dodecane has been investigated and compared to the optimal salinity obtained with the homologous series of n-alkanes [3,11]. The agreement was good, although the optimal salinity of the mixtures was found to be slightly higher than that of pure alkanes. Further results are described in subsequent sections dealing with nonionic surfactants. Indeed, the similarity between the behavior of pure n-alkanes and their mixtures is not surprising in view of the "principle of congruence" put forward by Bronsted and Koefoed [30]. This principle states that the thermal properties of a multicomponent system of alkanes are determined not by the individual chain lengths but by the mean $\overline{\text{ACN}}$ = $\sum_i x_i \text{ACN}_i$. Thus the properties of congruent mixtures (mixtures with the same value ACN) differ only in the ideal entropy of mixing. As in the case of the optimal salinity of mixtures of n-alkanes mentioned above, some slight departures from the linear mixing rule have, however, been observed, such as negative excess volumes and positive excess enthalpies [31]. These (small) deviations are in fact consistent with the principle of corresponding states, as developed by Prigogine [32], which includes the empirical principle of congruence. Equation (6.5) has generally been found to give good qualitative predictions for mixtures of different types of oil, but some large quantitative discrepancies have been observed [3,14,29].

G. Compensation of Surfactant Concentration with ACN

It is noted in Section 5.VII.D that when a mixture of surfactants is used, the composition of the C layer depends on the total surfactant concentration. The conditions for which optimal behavior is observed are then expected to depend on the overall surfactant concentration. While general trends are difficult to predict, most often anionic surfactants have a very limited oil solubility and at low surfactant concentration the aqueous phase tends to be rich in the more hydrophilic amphiphiles, whereas the C layer is, on the other hand, rich in the more lipophilic compounds. When the surfactant concentration is increased, the composition of the C layer tends to approach that of the overall composition. Thus for many mixtures of anionic surfactants, the optimum salinity must be increased to compensate for the increasing hydrophilicity of the C layer, which attends an increase in surfactant concentration. Alternatively, the changes resulting from an increase in surfactant concentration can be compensated by a decrease in the ACN.

Figure 6.15 provides an illustration of the concentration effect for two commercial surfactants each of which is a complex mixture of amphiphiles. In both cases the optimum ACN is seen to decrease as the

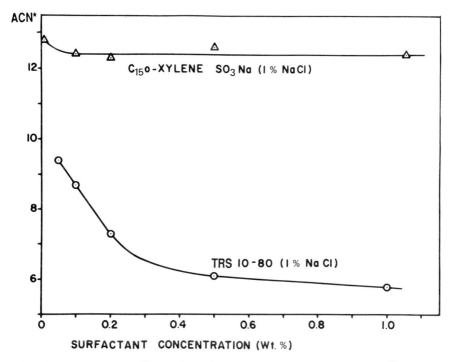

FIGURE 6.15 The effect of surfactant concentration on ACN* obtained from interfacial tension measurements. $C_{15}o$ - Xylene SO_3Na
is a synthetic sulfonate supplied by Exxon. TRS 10-80 is a petroleum sulfonate supplied by Witco. (After Ref. 33.)

surfactant concentration is increased [33]. There does exist a plateau value of ACN* which is reached as the overall concentration is
increased to a value sufficient so that the C-layer composition and
the overall composition coincide. This is an important effect to consider when dilute solutions of commercial surfactants are applied.
The total concentration is an extremely important variable.

III. NONIONIC SURFACTANTS

The amphiphiles considered in this section belong to the class of ethoxylated materials which represent most of the commercial nonionic surfactants that are available today. The section is organized in much
the same way as Section 6.II. Those compensating changes in which
both the numerator and the denominator of the R-ratio are changed
are classified as hydrophilic/lipophilic changes, and these are consid-

ered first.   Since the hydrophilic interactions of nonionic surfactants
are more profoundly influenced by temperature than are the lipophilic
interactions, temperature changes are considered here to alter primarily the denominator of the R-ratio.   This differs in the treatment of
temperature provided in the preceding section, where temperature was
recognized as only a minor contributor not classified as either lipophilic or hydrophilic.

A change in number of ethylene oxide units (EON) is also classified
as a hydrophilic change.   All other factors are treated in this section
similarly as in the preceding section dealing with anionic surfactants.
Again we stress that the organization of the material may appear to
be somewhat artificial, but the utility of the structure will become apparent in Chap. 7.

A. Compensating Hydrophilic/Lipophilic Changes

1. Temperature Versus ACN

Figure 6.16 shows the evolution of the phase boundaries when both
the temperature and the alkane carbon number are varied [34].   The
trends are predictable based on consideration of the R-ratio.   The numerator is decreased by increasing ACN [A_{oo} of Eq. (1.23)] and the
denominator is decreased by increasing $T(A_{cw})$.   Consequently, the
type III regime shifts toward higher temperatures on increasing ACN.

Not shown in Fig. 6.16 is the locus of optimal systems.   This curve
can be established by applying any one of a number of criteria, including minimum interfacial tension, equal solubilizations, or the phase
inversion temperature [35,36].   (See Section 4.IV for a further discussion.)

2. Temperature Versus Oil Mixtures

Arai and Shinoda [37] have investigated the phase inversion temperature (PIT) (Section 4.II.F) as a function of the composition of the oil
phase.   The PIT was found to vary linearly with the mole fraction of a
binary mixture as shown by Fig. 6.17.   These curves show that the
PIT for benzene is approximately 20°C, whereas the value for heptane
(ACN = 7) is about 90°C.   The PIT for various mixtures can evidently be calculated by a linear mixing rule based on the mole fraction of
each component in the oil phase.

3. Ethylene Oxide Number Versus ACN

Figure 6.18 shows the three-phase region obtained by varying EON
(by mixing the two surfactants having larger and smaller values than
the desired EON value) at constant temperature [34].   Increasing EON
is plotted in the negative direction along the ordinate to maintain the

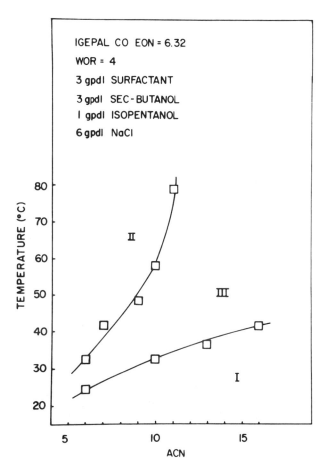

FIGURE 6.16 Phase behavior with temperature and alkane carbon number (ACN) as variables. Igepal CO is an ethoxylated nonylphenyl from GAF. EON = 6.32 is obtained by mixing the EON = 6 and EON = 7 species. The concentrations are expressed in g/dl of water. (After Ref. 34.)

same relative positions of types I, II, and III systems as shown in Fig. 6.16. The similarity between the two figures is evident.

In the present case, the decrease in the numerator of R caused by an increase in A_{oo} is compensated by a decrease in EON which results in a decrease in the denominator through A_{cw} (see Section 5.II.D). For the representation shown in Fig. 6.18, the three-phase region is therefore shifted upward on increasing ACN. The type

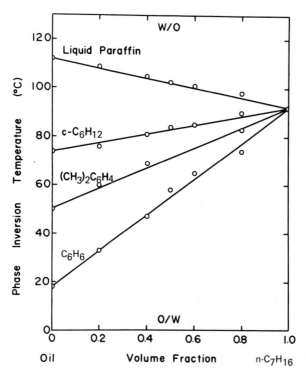

FIGURE 6.17 The effect of the mixture of n-heptane with various oils on the PITs of emulsions stabilized with polyoxyethylene (9.6) nonylphenylethers (3 wt % for water). (After Ref. 37.)

III domain in this figure has been extensively studied by a series of interfacial tension measurements varying both ACN and EON. The interfacial tensions between excess oil and the surfactant phase and excess water and the surfactant phase are equal at a point near the left boundary of the three-phase region when the ACN is varied at constant EON, but near the center of the three-phase region when EON is varied.

For this reason it is reasonable to characterize the three-phase region by defining an optimum EON (EON*) as being that in the center of the three-phase range. This is analogous to selecting the logarithm of the salinity at the center of the three-phase region as the optimum one when dealing with anionic surfactants. This definition of EON* has been shown experimentally to permit optimum systems to be represented by straight lines (within the range of EON investigated and for systems studied under similar conditions) such as shown

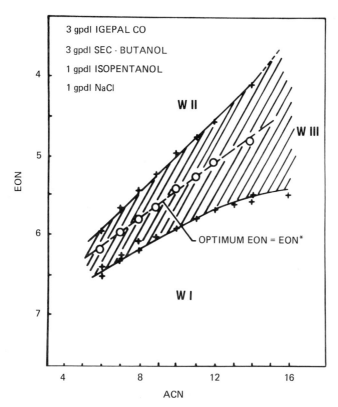

FIGURE 6.18 Phase behavior with ethylene oxide number and ACN as variables. Optimum EON (EON)* is defined in the text. The concentrations are expressed in g/dl of water. Igepal CO is an ethoxylated nonylphenol from GAF. The water to oil ratio is 4. (After Ref. 34.)

in Fig. 6.18. For the range of EON variations employed, it is also found that an alternative representation using HLB[†] is possible [38]. Thus HLB* also varies linearly with ACN just as EON* does as shown by Fig. 6.18.

Figure 6.18 shows the optimum line positioned in the center of the EON values which are required to effect the II-III-I transition. The line applies to a single temperature. Figure 6.19 shows a similar plot of the optimum line with temperature as a parameter. Near room tem-

---

[†]For ethoxylated amphiphiles the HLB is one-fifth of the weight fraction of the ethylene oxide content of the molecule.

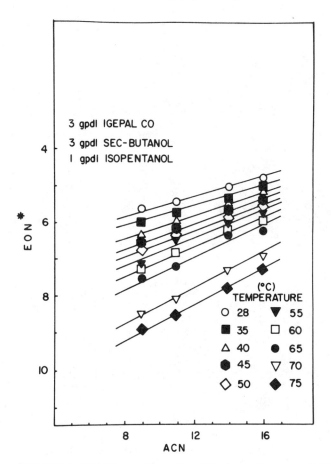

FIGURE 6.19 The influence of temperature on EON*/ACN. Same notations and conditions as in Fig. 6.16. (After Ref. 34.)

perature the plot of EON* versus ACN is a linear function of nearly constant slope. However, as the temperature is increased, there is a measurable increase in the slope and in some cases, curvature develops.

4. Ethylene Oxide Number Versus Lipophile Structure

Phase maps similar to that shown by Fig. 6.18 have been obtained using ethoxylated dodecylphenol [39], di-nonylphenol [39], octylphenol [34], and isotridecanol [34]. The change in surfactant lipophile structure is seen in Fig. 6.20 to shift the optimal EON*, but the slope is only slightly affected.

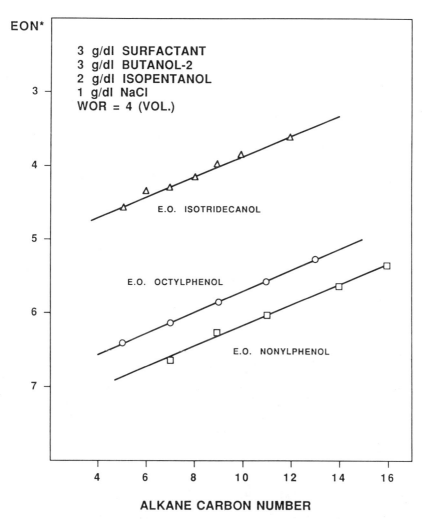

FIGURE 6.20 The effect of the surfactant lipophile structure optimal line EON*−ACN. The concentrations refer to the aqueous phase. (After M. Bourrel, unpublished results.)

As expected, increasing the weight of the surfactant lipophile raises the interaction with oil and increases ethylene oxide numbers required for compensation. It is worth noting, however, that the shift in EON* between ethoxylated isotridecanol and octylphenol is much larger than between octylphenol and nonylphenol, although the increment in number of carbon atoms is the same. This is a further illustration of the impact of the benzene ring on optimal phase behavior, as seen earlier in the case of anionic surfactants (Fig. 6.13). It has been observed that the optimal line, when plotted in HLB-ACN representation, is the same for ethoxylated octylphenol, nonylphenol, dodecylphenol, and di-nonylphenol [39]. The width of the three-phase region is, however, quite different, indicating different levels of solvency [39] (see Section 4.II.C).

5. Electrolyte Concentration Versus ACN

The primary reasons for discussing nonionic and ionic amphiphiles in separate sections of this chapter is that nonionic surfactant systems are quite sensitive to temperature (ionics are not) and are quite insensitive to electrolyte concentration (ionics are). Figure 6.21 shows that large changes in electrolyte concentration, say from 6 to 9 wt % NaCl, can be accommodated by a change of about 2 ACN units [40].

B. Compensating Lipophilic Changes

When compensating changes are made in the numerator of the R-ratio, the degree to which each variable is changed should not depend on the hydrophile except to the extent that the interfacial densities may vary. Thus the changes for nonionic surfactants may be expected to be similar to those for anionic surfactants. This similarity may be seen by comparing Eq. (6.2) with the correlation developed for nonionic surfactants expressed by Eq. (6.8).

C. Compensating Hydrophilic Changes

1. Temperature Versus Electrolyte Concentration

The effect of added salts, acid, and alkali on the phase inversion temperature is shown in Fig. 6.22. As discussed in Section 5.II.A, electrolyte is thought to disrupt the structure of water. The resulting decrease in interaction energy A_{cw} can be compensated by a decrease in temperature that tends to increase A_{cw} by restoring these bonds. This interpretation applies when salt and alkali are added to the aqueous phase as shown by Fig. 6.22. A different trend is observed when acid is added. Increasing acid concentration requires an increase in temperature to remain at optimum. This effect is probably related to the mechanism by which acids increase the solubility

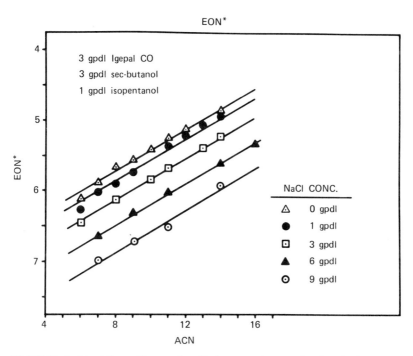

FIGURE 6.21   The effect of salinity on the optimal ethylene oxide number EON*.   Same conditions as in Fig. 6.18.   (After Ref. 40.)

of higher-molecular-weight alcohols in water.   The protonation of the hydroxyl group of the alcohol, conferring cationic character on the alcohol molecule, takes place, thereby increasing its solubility in water.   If the oxygen atoms of the ethylene oxides are similarly protonated, the interaction A_{cw} of the hydrophilic head of the surfactant with water is enhanced by the addition of acids.   Consequently, an increase in temperature is required to compensate for the addition of acid.

Results obtained on a different system containing an isomerically pure ethoxylated (EON = 6) tetradecanol, octane, water, and sodium chloride have recently been reported [41].   They show that the optimal temperature decreases linearly with the NaCl concentration.

2.   Ethylene Oxide Number Versus Electrolyte
      Type and Concentration

A variation in A_{cw} due to change in EON can also be compensated by an appropriate change in salinity.   Increasing the EON is expected to require higher salinities to keep the R-ratio equal to unity.

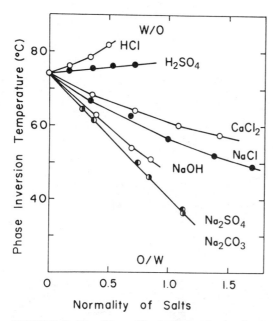

FIGURE 6.22   The effect of various electrolytes on the phase inversion temperature of a cyclohexane-water emulsion (1/1 volume) containing 3 wt % of i-$C_9H_{19}C_6H_4O(CH_2CH_2O)_{9.7}$ in water.   (After Ref. 41.)

Figure 6.21 provides an illustration. It is interesting to point out that for a given ACN a change of only one unit in EON is required to balance a 9 wt % change in NaCl. Thus small changes in EON compensate large variations in salinity. It is seen that the salt effect is independent of ACN, as one might expect since increasing electrolyte affects primarily the aqueous side of the C interfacial region.

The influence of the type of cation is shown in Fig. 6.23, which compares the shift in EON* due to sodium chloride, calcium chloride, and potassium chloride. For these three salts, the shift in EON* varies linearly with the electrolyte concentrations. The different symbols belonging to a given straight line refer to data obtained under a variety of conditions.

Figure 6.23 shows also that the effects are of the same order of magnitude for NaCl and $CaCl_2$. This is quite different from the behavior observed for sulfonates; for example, which are particularly sensitive to divalent cations (Section 5.II.A). When salinity is expressed in grams per deciliter, NaCl has a somewhat larger effect

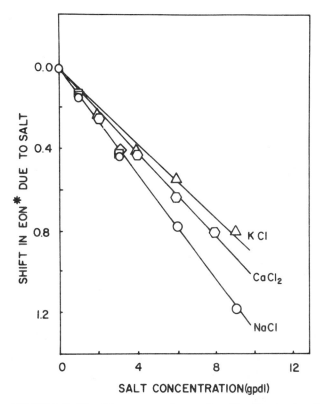

FIGURE 6.23 The comparison of salt effects for KCl, CaCl$_2$, and NaCl on optimal phase behavior. The ordinate axis is oriented downwards. (After Ref. 34.)

than CaCl$_2$, KCl having a smaller effect. When salinity is expressed in moles per liter, the NaCl/CaCl$_2$ order is reversed and the magnitude of the effect on EON* follows the order CaCl$_2$ > NaCl > KCl. These results are consistent with Shinoda and Takeda's observations on the effect of salt on PIT (Fig. 6.22).

The order of the effect of CaCl$_2$, NaCl, and KCl can be correlated with a classification of cations by their hydration numbers and thus to the degree of solvation by water. Some variation is found for these hydration numbers depending on the technique used for their evaluation, but there is a general agreement on their relative order: Ca^{2+} > Na$^+$ > K$^+$. The coincidence between this classification and the observed effects on the phase behavior provides support for considering the mechanism of divalent cation action on nonionic surfactants as being similar to that for univalent ions.

The sensitivity of the phase behavior to the salinity can be expressed in terms of a preferred ACN (ACN*) shift at constant EON. When the NaCl concentration goes from 1 to 2 g/dl (Fig. 6.21), the ACN is decreased by ≈ 0.9 unit. For anionic surfactants for which K = 0.16, the same change in salinity would produce a shift of 4.3 units, according to Eq. (6.2). This is a clear illustration of the reduced influence of the salinity on nonionic phase behavior as compared to anionics.

D. Adding Alcohol to Compensate Hydrophilic or Lipophilic Changes

Since alcohols influence both the numerator and the denominator of the R-ratio, it is not reasonable to classify its addition as either hydrophilic or lipophilic. The influence of alcohols is therefore treated separately. Figure 6.24 shows the effect of butanol-1 and pentanol-1 concentration on the optimal EON taking the alcohol-free system as a reference [20]. The alcohols are added at very low concentrations so as to magnify their influence.

For pentanol-1, EON* increases monotonically with increasing alcohol concentration over the range considered. This is due to the lipophilic character of pentanol-1, as discussed in Section 5.IV.A. The interaction energies of the alcohol and surfactant in the interfacial C region are higher on the oil side than on the water side, and R is then increased on adding pentanol-1. To return to R = 1, an increase in EON in the hydrophilic head of the ethoxylated nonylphenol is required.

The case of butanol-1 appears more complex. On increasing its concentration, EON* is seen first to remain nearly constant, and then to increase when the alcohol concentration is higher than about 1%. It is certain that butanol-1 is present in the C region even at small concentration. This is substantiated by the decrease in solubilization to be described in Section 7.V.C.

As illustrated by Eq. (5.11), the phase behavior depends on both the interfacial mole fraction x_a of the alcohol and on its balance of interaction. These factors can be more easily discussed with the aid of a simple model based on the R-ratio. If it is assumed, for simplicity, that the interaction of the surfactant head with water a_{cw}^s is proportional to EON, that is, $a_{cw}^s = \alpha \cdot EON$, where α is a constant, then Eq. (5.11) becomes

$$R \approx \frac{a_{co}^s - (a_{co}^s - a_{co}^a) x_a}{\alpha \cdot EON - (\alpha \cdot EON - a_{cw}^a) x_a} \qquad (6.6)$$

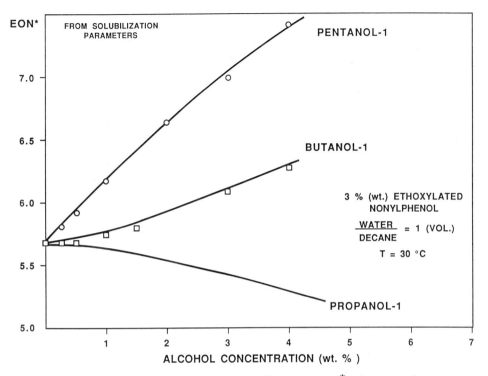

FIGURE 6.24   Alcohol concentration effect on EON^*.   The optimum
has been obtained from solubilization parameter measurements.   ○, □:
Experimental results; ——:   calculated from Eq. (6.7).   (After
Ref. 20.)

At optimum, $R = 1$ and EON^* is expressed as

$$EON^* = \frac{a_{co}^s - (a_{co}^s - \Delta^a)x_a}{\alpha(1 - x_a)} \tag{6.7}$$

where Δ^a is the difference $a_{co}^a - a_{cw}^a$ between the interaction energies of the alcohol with oil and water.

Based on Eq. (6.7), it can be argued that the slope of $EON^*(x_a)$
when $x_a \to 0$ (no alcohol) is $(a_{co}^a - a_{cw}^a)/\alpha$. For an alcohol, the hydrophile-lipophile interactions of which are equilibrated, $a_{co}^a = a_{cw}^a$
and the slope at the origin is zero; Eq. (6.7) shows, further, that
EON^* remains constant on adding this particular alcohol to the sys-

tem. For high-molecular-weight (lipophilic) alcohols, $a_{co}^a > a_{cw}^a$, the
slope is positive and EON* increases with x_a. For those low-molecular-weight (hydrophilic) alcohols for which $a_{co}^a < a_{cw}^a$, EON* decreases with x_a.

As discussed earlier in the case of anionic surfactants, it must
be pointed out that x_a is not the overall alcohol concentration but
its interfacial mole fraction. In particular, at a given overall alcohol concentration, x_a is known to depend on the type of alcohol [42,
43]. The pseudophase model [44] (see Section 5.VII.C), which treats
the interfacial C region as a phase in equilibrium with the water and
oil regions, provides a tool for evaluating the interfacial alcohol concentration. Calculations have been made using the values of the constants given in Table 5.3, and these are presented in Fig. 6.25.

At small overall alcohol concentrations, the interfacial concentrations x_a are in the order pentanol-1 > butanol-1 > propanol-1. At
high concentration the order may differ. EON* can be estimated as
a function of the overall alcohol concentration with the aid of Eq.
(6.7) and Fig. 6.25 provided that values of a_{co}^s, α, and Δ^a are known.
The values used for calculating the solid lines in Fig. 6.24 are reported in Table 6.1. The agreement between the calculated curves and
the experimental results for pentanol-1 and butanol-1 in Fig. 6.24
appears quite satisfactory. The change in EON* on increasing the
butanol-1 concentration is smaller than in the case of pentanol-1 simply because its x_a is smaller (Fig. 6.25). For propanol-1, $a_{co}^a < a_{cw}^a$
yields a decrease in EON* when the alcohol concentration increases.
As noted in Section 5.IV.A, many of the approximations used in the
simple calculations presented here are questionable. They do, however, serve to emphasize the main mechanisms that govern the action
of alcohol on phase behavior and which are related to its presence in
the C layer. The small alcohol concentrations used in the studies referred to above do not allow for any change in the bulk-phase solvent
properties of the O and W regions due to the presence of alcohols.
The effect of alcohol on phase behavior cannot be entirely or even
mainly attributed to its effect on the O and W regions as has sometimes
been proposed.

The effect of isopentanol concentration on EON* for a series of n-
alkanes is illustrated by Fig. 6.26. Additional data have been reported for various electrolytes, surfactants, and alcohols by Bourrel
et al. [34]. The results shown in Fig. 6.26 are typical.

EON* is seen to be shifted toward higher EON by the same distance independent of ACN, and the shift is proportional to the isopentanol concentration. This is also illustrated by Fig. 6.27. The
different points belonging to the same straight line refer to different
systems. The systems differed as to concentration of the surfactant
(1 to 3%), or its type, or the electrolyte concentration. The reason

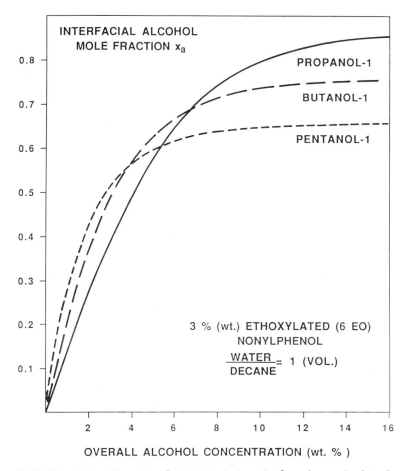

FIGURE 6.25   The interfacial alcohol mole fraction calculated as a function of the overall alcohol concentration according to the pseudo-phase model.   (After J. Biais, unpublished results.)

that the effect of isopentanol is the same for all of the systems test-ed is probably due to the fact that for high-molecular-weight alcohols, the partition coefficient greatly favors oil and is only slightly affected by the change in ACN or salinity over the range investigated.

In Fig. 6.27, one notes that pentanol-1 and hexanol-1 give larger positive shifts than isopentanol. The low-molecular-weight alcohols have an opposite effect on EON*, but the effect is small compared with that of high-molecular-weight alcohols.

TABLE 6.1   Values of the Parameters for Surfactant and Alcohol Used in Eq. (6.7) to Calculate EON* $(x_a)^a$

Alcohol	a_{co}^a	a_{cw}^a	Δ^a
Pentanol-1	5	3.5	1.5
Butanol-1	4	3.5	0.5
Propanol-1	3	3.5	−0.5

[a]Surfactant properties: $a_{co}^S = 6.8$; $\alpha = 1.2$.

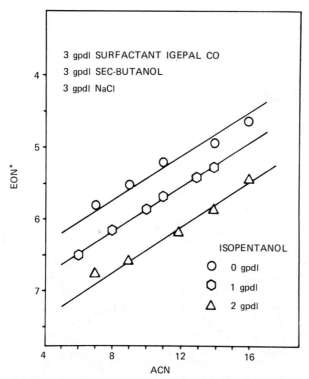

FIGURE 6.26   The influence of isopentanol on EON*/ACN.   Same notations and conditions as in Fig. 6.16.   (After Ref. 34.)

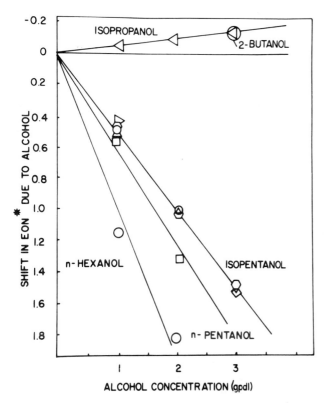

FIGURE 6.27   The comparison of the effect of a variety of alcohols
on optimal phase behavior.   The ordinate axis is oriented downwards.
The concentrations refer to the aqueous phase.   WOR = 4.   (After
Ref. 34.)

The classification of the various alcohols by the magnitude of their
effect on EON* as shown in Fig. 6.27 can readily be interpreted based
on Eq. (6.7).   The interaction energy a_{co}^{a} is larger for hexanol-1
than for pentanol-1 or isopentanol.   At the same interfacial concentration the increase in R is therefore larger for hexanol-1, and a larger
increase in EON* is required to compensate for its presence.   Again
it should be noted that the greater tendency for hexanol-1 to partition into the oil phase as compared to pentanol-1 or isopentanol will
cause its interfacial concentration to be less than that of pentanol-1
or isopentanol when all three alcohols are present in solution at the
same concentration.   The data show, however, that despite the smaller
interfacial concentration, hexanol-1 still provides a larger EON* shift
(Fig. 6.27), so that the smaller interfacial concentration is more than
offset by the value of a_{co}^{a}.

When the alcohol chain length is shortened, a_{co}^a may eventually become smaller than a_{cw}^a (see Fig. 6.24), and the addition of a short-chain alcohol is expected to decrease the R-ratio. This is evidently the case for isopropanol, which requires decreasing EON* for compensation, that is, a decrease in the interaction of C with W. The small magnitude of the effect indicates that for this alcohol a_{co}^a and a_{cw}^a are nearly equal. Butanol-2 also yields a small hydrophilic contribution to the C layer as shown by Fig. 6.27.

E. Correlation Between Formulation Variables

As in the case of anionic surfactants, an empirical correlation has been developed using results shown typically in Figs. 6.19 to 6.22, 6.26, and 6.27. The equation is a linear function of EON*, ACN electrolyte and alcohol concentrations and types, and surfactant lipophile which give optimal behavior as follows [34]:

$$a - EON^* = kACN - bs - mA - c(T - 28) \qquad (6.8)$$

where A and s are the alcohol and salt concentrations, respectively, m and b are constants characteristic of each type of alcohol and salt (their values are given by the slope of the straight lines in Figs. 6.23 and 6.27), a and k are constants for a given type of surfactant, and c is the coefficient of temperature, T, restricted to the low-temperature linear regime (28 to 60°C). The values of the constants pertinent to Eq. (6.8) are listed in Table 6.2. Equation (6.8) and Table 6.2 give some insight as to the relative impact of the formulation variables on the interaction energies.

Table 6.2 shows that k has the same value for ethoxylated nonyl- and octylphenols, is slightly different for ethoxylated tridecanol, and is independent of surfactant concentration within the range investigated (1 to 6% of the aqueous phase).

For a 28°C system without alcohol and without salt, the optimal ACN is given by (a − EON)/k, and it is of course a characteristic of the surfactant molecule. Therefore, since the hydrophilic part is represented by EON/k, a/k is a characteristic of the lipophile part, (a − EON)/k thus appears to have the same significance for nonionic surfactants as EPACNUS or N_{min} does for anionics and allows one to separate the hydrophilic and lipophilic contributions.

Table 6.2 shows that an increment of 1 unit in EON yields a decrease in the optimal alkane of about 6.6 units, while adding 1 methylene group to the lipophile increases the optimal alkane by only 2 units. This shows that the phase behavior of nonionic surfactants is much more sensitive to the ethylene oxide number of the molecule than to the number of carbons of its alkyl chain. Commercial nonionic sur-

TABLE 6.2   Correlation for Optimal Phase Behavior of Nonionic Surfactants

	Surfactant		
	Ethoxylated nonylphenol	Ethoxylated octylphenol	Ethoxylated isotridecanol
k (EON/ACN)	0.15	0.15	0.14
a (EON)	6.65	6.35	4.34
c (EON/°C)	0.057	—	—

	Alcohol			
	IsoC$_3$OH or 2-C$_4$OH	IsoC$_5$OH	n-C$_5$OH	n-C$_6$OH
m (EON/g/dl)	−0.05	0.5	0.58	0.98

	Salt		
	NaCl	CaCl$_2$	KCl
b (EON/g/dl)	0.13	0.10	0.09
b (EON/mol/liter)	0.77	1.11	0.69

Source: After Ref. 34.

factants which exhibit a broad distribution of EON display a wide variation in hydrophile-lipophile balance, and this contributes to the large degree partitioning of nonionic surfactants between oil and water.

The left-hand side of Eq. (6.8) clearly characterizes the surfactant molecule, while the right-hand side characterizes the system in which the amphiphile is dissolved and separates the constributions of the various components of the system. (Alternatively, the alcohol being considered as an amphiphile, mA could be transferred to the left side and collected with the characteristics of the surfactant.) When both characteristics are identical, the overall mixture is optimized. Equation (6.8) is the condition for optimal phase behavior and again can be compared to the condition for minimum emulsion stability discussed in Section 4.II.F.

HLB = RHLB   (6.9)

For the range of EON generally investigated, HLB can be approximated by a linear function of EON and Eq. (6.8) can be converted
into a form making use of HLB rather than EON. A systematic study
of phase behavior has been carried out under conditions that differ
only slightly from those used in Fig. 6.16; namely, the surfactant
concentration was maintained constant in mole instead of weight percent and HLB was used instead of EON [38,39]. It was found that
the upper (UPB) and lower (LPB) phase boundaries, as well as the
optimal phase behavior (OPB) line are straight lines, as illustrated
by Fig. 6.28. Intermediate HLBs were obtained by blending surfac-

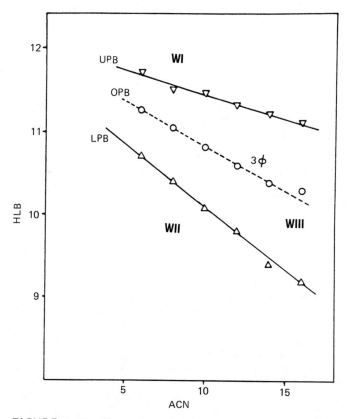

FIGURE 6.28 Phase map in the HLB-ACN representation. Ethoxylated nonylphenol: 0.0227 mole/l. Water-oil ratio = 4 (vol). Aqueous phase: 3% butanol-2 (vol); 2% isopentanol (vol). 1% NaCl (wt.).
(After Ref. 38.)

TABLE 6.3 Experimentally Determined Values for
the Constants of Eq. (6.10)[a]

	UPB	OPB	LPB
k'	0.080	0.115	0.160
m'	0.55	0.45	0.40
b'	0.150	0.138	0.062
c'	0.056	0.037	0.015
a'	11.0	11.0	10.8

[a] m is for Isopentanol.
Source: After Ref. 38.

tants on a mole fraction basis. The significance of these lines regarding the stability of emulsions has been discussed [38].

The positions of UPB, LPB, and OPB were examined for temperatures of 20, 30, 40, and 50°C; isopentanol concentrations of 0, 1,
2, and 3 % (vol) and NaCl concentrations of 1, 3, 6, and 9 % (wt).
The experimental observations have shown that the UPB, LPB, and
OPB are all linear functions of the form

$$HLB = a' - k'ACN + m'A + b's + c'(T - 28) \qquad (6.10)$$

where a', k', m', b', and c' are constants whose values differ for the
three lines and are given in Table 6.3. A and s are the alcohol and
salt concentrations, respectively.

F. Compensation of Surfactant Concentration by EON

In Section 5.VII.D it has been shown that increasing the surfactant
concentration generally tends to decrease the average EON of the C
layer when the surfactant system is a polydistributed nonionic. Thus
it is anticipated that to retain an optimum system an increase in the
overall surfactant concentration must be compensated by a decrease
in the average EON of the surfactant. Essentially, the average EON
of those nonionic molecules residing in the C layer must remain constant.

An illustration of this trend is provided by Fig. 6.29. The system studied is the same as that represented in Fig. 6.18. The shift
of EON* is plotted as a function of surfactant concentration taking
the optimum EON at 3 g/dl as a reference. It is seen that a large
increase in EON is required to maintain optimal conditions when the

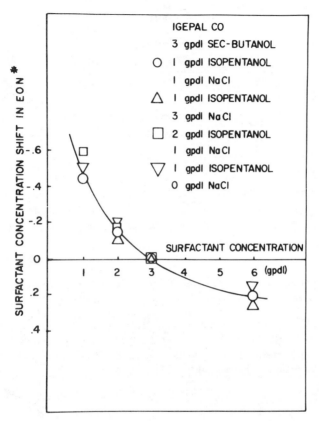

FIGURE 6.29 The influence of surfactant (ethoxylated nonylphenol) concentration on EON*. Water to oil volume ratio is 4. The concentrations refer to the aqueous phase. (After Ref. 34.)

surfactant concentration is changed at small concentrations. Smaller changes in EON are required to compensate changes at larger concentrations. These results are in accord with the trends found based on the pseudophase model described in Section 5.VII.D.

IV. MIXTURES OF ANIONIC AND NONIONIC SURFACTANTS

A. Optimal Salinity

The main purpose for blending nonionic surfactants (sometimes called "cosurfactants") with anionics is to take advantage of the lower sensitivity of the former to electrolytes to enhance the salinity tolerance

of the latter [11,46—52]. In other instances, mixtures have been
used to optimize the system by varying the proportions of the components [53,54]. The nonionic compound plays there a role similar
to that of alcohol.

Figure 6.30 shows the effect on optimal salinity of mixing an ethoxylated (EON = 7.5) nonylphenol with a dodecyl orthoxylene sulfonate, the overall surfactant concentration being kept constant. The
optimal salinity of these two compounds are 13 and 1.5 g/dl, respectively.

One of the main features displayed by Fig. 6.30 is that the optimal salinity does not vary linearly with the composition of the mixture. The optimal salinity remains almost constant for nonionic mole
fractions in the mixture lower than 0.2. Similar trends have been
obtained with different anionic surfactants [46,52]. In some cases,
a decrease in optimal salinity has even been observed for small fractions of nonionic compound in the mixture, although its optimal salinity is higher than that of the anionic compound [46,52]. When the
nonionic mole fraction is high enough, its effect becomes dominant
and the optimal salinity increases.

Because of the high overall surfactant concentration used in the
experiments reported in Fig. 6.28, the composition of the C layer is
probably identical to the overall composition. The nonlinear behavior of the optimal salinity displayed in Fig. 6.30 cannot therefore be
explained by a higher concentration of the anionic compound in the
C layer (see Chap. 2 for the discussion of the effect of the surfactant concentration on the composition of mixed micelles).

An interpretation of the deviation of the optimal salinity of the
mixtures from linear behavior may be proposed from the R-ratio.
Consider first the system containing the anionic surfactant at optimum, that is, for which R = 1. At the same salinity, the nonionic
surfactant exhibits higher interactions with water than with oil. When,
in these conditions, some anionic molecules are replaced by nonionic
ones, the average term A_{cw} of the mixture increases and tends to
decrease R. This is, however, compensated by an increase in A_{hh}
(less negative) due to the reduced electrostatic repulsion since the
anionic molecules are screened by the nonionic ones [55—57] (Section
5.VII.E). As a result of these opposing effects the R-ratio may even
increase with the addition of small quantities of nonionic compound and
then decrease.

The values of the interfacial tension at optimum are shown in Fig.
6.28. It is seen that they increase gradually from the pure anionic
to the pure nonionic system. Similar behavior has been reported for
the solubilization at optimum by Haque and Scamehorn [52]. It may
be conjectured that decreased rather than increased interfacial tensions would be obtained upon increasing the nonionic content in the
mixture if the nonionic system exhibited lower interfacial tension at
optimum than the anionic one.

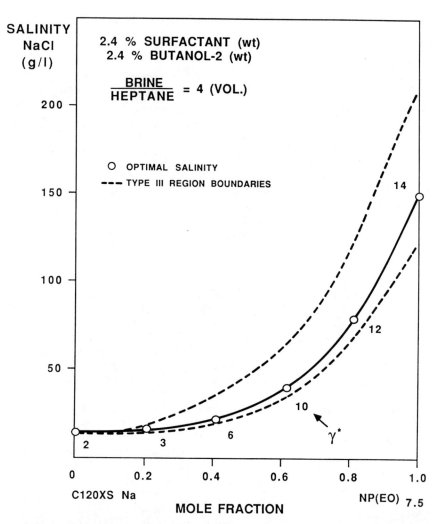

FIGURE 6.30   The optimal salinity and three-phase range of mixtures of anionic and nonionic surfactants. Anionic: dodecylorthoxylene sulfonate (C12)XSBa); nonionic: commercial ethoxylated nonylphenol with an average of 7.5 ethylene oxide units (NP (EO)$_{7.5}$). The number by each experimental data point indicates the values of the interfacial tension, in mdyne/cm. (After Ref. 46.)

B. Temperature Effect

It has been seen (Section 5.IV.B) that, generally, increasing temperature tends to promote type I behavior with anionic surfactants
and type II with nonionics. Anionic and nonionic amphiphiles have
thus been blended in order to obtain mixtures insensitive to temperature [11]. Figure 6.31 shows the effect of temperature on the optimal salinity of various mixtures of an ethoxylated (EON = 6) nonylphenol and a petroleum sulfonate. For a small mole fraction of the
nonionic surfactant, the effect of the anionic predominates, but at
higher mole fractions, the change in optimal salinity with respect to

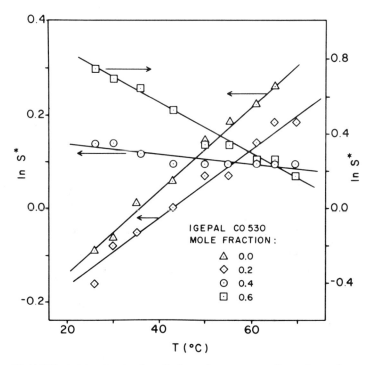

FIGURE 6.31 Optimal salinity shown as a function of temperature
for various mixtures of a petroleum sulfonate (TRS 10 - 80, supplied
by Witco) and a commercial ethoxylated nonylphenol with an average
of 6 ethylene oxide units (IGEPAL CO530 supplied by GAF). Water/
heptane = 4. 3 g/dl butanol-2 and 1.25×10^{-2} M surfactant (with respect to the aqueous phase). s* (NaCl) in g/dl. (After Ref. 11.)

changes in temperature is negative. The temperature coefficient vanishes when the surfactant mixture is about 40 mol % nonionic. This mixture exhibits a phase behavior independent of temperature over the range of temperatures studied.

V. MIXTURES OF ANIONIC AND CATIONIC SURFACTANTS

Among the large body of work devoted to the mixtures of various surfactants, the investigation of blends of cationics and anionics has been developed rather recently [58–60], even though the existence of anionic-cationic mixed micelles has been known for a long time [61]. Fundamental studies have been carried out on mixed monolayers of cationics and anionics at the aqueous solution/air interface [60,62], as well as on the micellization phenomenon [59,60]. From the standpoint of industrial applications, formulations involving mixtures of anionic and cationic surfactants have been claimed in a number of recent patents, mainly for detergency [63].

Bourrel et al. [64] have investigated the phase behavior of mixtures of tetradecyltrimethylammonium bromide (TTAB) and sodium n-octylsulfonate (C8SO$_3$). Phase maps such as that shown in Fig. 6.32 were determined. For a given mixture of TTAB and C8SO$_3$, the phase behavior can be determined for the n-alkane series and the optimum ACN* and (S + A)* found. The change of the optimal ACN is shown in Fig. 6.33 as a function of the TTAB mole fraction, x_T, in the C8SO$_3$-TTAB mixture. The corresponding change in (S + A)* is discussed in Chap. 7. Figure 6.33 shows that ACN* goes through a sharp maximum for a TTAB mole fraction of 0.55.

When starting from an optimized system, the cationic proportion of the surfactant mixture is increased up to 0.5, the following changes of the cohesive energies appearing in the R-ratio are to be expected:

1. A_{co}, which is the weighted sum of the interaction of TTAB (14 carbons) and of C8SO$_3$ (8 carbons) with oil, will increase.
2. $A_{cw} - A_{hh}$, is decreased sharply because of charge neutralization.
3. ACN* increases to compensate for changes 1 and 2.

When x_t varies between 0.5 and 1, A_{co} still increases monotonically while $A_{cw} - A_{hh}$ also increases sharply. Thus $A_{cw} - A_{hh}$ goes through a minimum when x_t varies and requires ACN* to go through a maximum to maintain R = 1. Clearly, as shown by Fig. 6.33, this is the most important factor. In other words, one can visualize the ion pair resulting from the anionic-cationic interaction (mole to mole) as a neutral compound exhibiting a very small interaction with water.

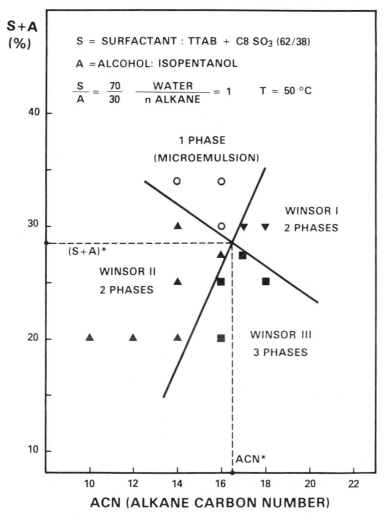

FIGURE 6.32 Phase map for a 62/38 (molar) mixture of TTAB and $C8SO_3$ (see text). The surfactant/alcohol ratio is kept constant in weight. WOR = 1 (vol). (After Ref. 64.)

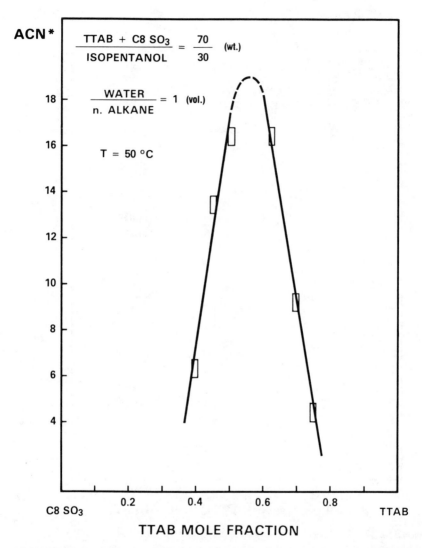

FIGURE 6.33 Effect of cationic–anionic molar ratio on optimal alkane. (After Ref. 64.)

REFERENCES

1. J. L. Salager, "Physico-chemical Properties of Surfactant Water-Oil Mixtures: Phase Behavior, Microemulsion Formation and Interfacial Tension," Ph.D. dissertation, The University of Texas at Austin (1977).

2. J. L. Salager, J. C. Morgan, R. S. Schechter, W. H. Wade, and E. Vasquez, Soc. Pet. Eng. J., 19:107 (1979).

3. M. C. Puerto and R. L. Reed, Soc. Pet. Eng. J., 23:669 (1983).

4. A. M. Lipow, The Effect of Surfactant Structure on the Phase Behavior of Surfactant-Brine-Oil Systems, M.S. thesis, The University of Texas at Austin (1979).

5. W. H. Wade, E. Vasquez, J. L. Salager, M. El-Emary, C. Koukounis, and R. S. Schechter, in Solution Chemistry of Surfactants, Vol. 2, (K. L. Mittal, ed.), Plenum Press, New York, p. 801 (1979).

6. A. Graciaa, J. Lachaise, A. Martinez, M. Bourrel, and C. Chambu, C.R. Acad. Sci., Paris, B282:547 (1976).

7. R. L. Cash, J. L. Cayias, R. G. Fournier, J. K. Jacobson, T. Schares, R. S. Schechter, and W. H. Wade, paper SPE/AIME 5813 presented at the 3rd SPE/AIME Symposium on Improved Methods for Oil Recovery, Tulsa (Mar. 1976).

8. W. H. Wade, J. C. Morgan, J. K. Jacobson, and R. S. Schechter, Soc. Pet. Eng. J., 17: 122 (1977).

9. J. L. Cayias, R. S. Schechter, and W. H. Wade, in Adsorption at Interfaces, ACS Symposium Series No. 8, p. 234 (1975).

10. W. H. Wade, J. C. Morgan, J. K. Jacobson, J. L. Salager, and R. S. Schechter, Soc. Pet. Eng. J., 18:242 (1978).

11. J. L. Salager, M. Bourrel, R. S. Schechter, and W. H. Wade, Soc. Pet. Eng. J., 19:271 (1979).

12. R. L. Cash, J. L. Cayias, G. R. Fournier, J. K. Jacobson, C. A. Le Gear, T. Schares, R. S. Schechter, and W. H. Wade, in Detergents in the Changing Scene, American Oil Chemical Society, p. 1 (1977).

13. R. L. Cash, J. L. Cayias, G. R. Fournier, D. MacAllister, T. Schares, R. S. Schechter, and W. H. Wade, J. Colloid Interface Sci., 59:39 (1977).

14. M. Bourrel, F. Verzaro, and C. Chambu, SPE Reservoir Eng., 2:41 (1987).

15. Y. Barakat, L. N. Fortney, R. S. Schechter, W. H. Wade, S. Yiv, and G. Graciaa, J. Colloid Interface Sci., 92:561 (1983).

16. M. C. Puerto and R. L. Reed, paper SPE/DOE 10678 presented at the 3rd SPE/DOE Symposium on Enhanced Oil Recovery, Tulsa (Apr. 1982).

17. K. Shinoda and T. Hirai, J. Phys. Chem., 81:1842 (1977).

18. M. H. Akstinat, Proceedings of the First European Symposium on EOR, p. 43, Bournemouth (Sept. 1981).

19. Y. Barakat, L. N. Fortney, R. S. Schechter, W. H. Wade, and S. Yiv, Proceedings of the Second European Symposium on EOR, Technip, Paris, p. 11 (Nov. 1982).

20. F. Verzaro, M. Bourrel, and C. Chambu, in Surfactants in Solution, (K. L. Mittal and P. Bothorel, eds.) vol. 6, p. 1137, Plenum Press, New York (1986).

21. S. J. Salter, paper SPE 6843 presented at the 52nd SPE Annual Technical Conference and Exhibition, Denver (Oct. 1977).

22. M. Baviere, R. S. Schechter, and W. H. Wade, J. Colloid Interface Sci., 81:266 (1981).

23. M. Baviere, Compte Rendu de Recherche, No 26928, DGRST, Paris (1979).

24. M. C. Puerto and W. W. Gale, Soc. Pet. Eng. J., 17:193 (1977).

25. P. H. Doe, M. El-Emary, W. H. Wade, and R. S. Schechter, J. Am. Oil Chem. Soc., 54:570 (1977).

26. P. H. Doe, M. El-Emary, W. H. Wade, and R. S. Schechter, paper presented at the Symposium of Chemistry of Oil, ACS Division of Petroleum Chemistry, Anaheim (Mar. 1978).

27. G. R. Glinsmann, paper SPE 8326 presented at the 54th SPE Annual Technical Conference and Exhibition, Las Vegas (Sept. 1979).

28. A. K. Sharma, Y. K. Pithapurwala, and D. O. Shah, paper SPE 11772 presented at the SPE Symposium on Oilfield and Geothermal Chemistry, Denver (June 1983).

29. M. K. Tham and P. B. Lorenz, Proceedings of the First European Symposium on EOR, Bournemouth (Sept. 1981).

30. J. N. Bronsted, and I. Koefoed, K. danske Vidensk. Selsk. (Mat. Fys. Skr.), 22, No. 17 (1946). See also J. S. Rowlinson and F. L. Swinton, Liquids and Liquid Mixtures, 3rd Ed., p. 298, Butterworths, London (1982).

31. See for example K. N. Marsh, J. B. Ott, and M. J. Costigan, J. Chem. Thermodyn., 12:343 (1980); ibid 12, 897 (1980).

32. I. Prigogine, N. Trappeniers, and V. Mathot, Discuss. Faraday Soc., 15:93 (1953).

33. J. C. Morgan, R. S. Schechter, and W. H. Wade, in Improved Oil Recovery by Surfactant and Polymer Flooding, (D. O. Shah and R. S. Schechter, eds.), Academic Press, New York, p. 101 (1977).

34. M. Bourrel, J. L. Salager, R. S. Schechter, and W. H. Wade, J. Colloid Interface Sci., 75:451 (1980).

35. K. Shinoda and H. Arai, J. Phys. Chem., 68:3485 (1964).

36. H. Saito and K. Shinoda, J. Colloid Interface Sci., 32:647 (1970).

37. H. Arai and K. Shinoda, J. Colloid Interface Sci., 25:396 (1967).

38. A. Graciaa, Y. Barakat, R. S. Schechter, W. H. Wade, and S. Yiv, J. Colloid Interface Sci., 89:217 (1982).

39. A. Graciaa, L. N. Fortney, R. S. Schechter, W. H. Wade, and S. Yiv, Soc. Pet. Eng. J., 22:743 (1982).

40. M. Bourrel, C. Koukounis, R. S. Schechter, and W. H. Wade, J. Disp. Sci. Tech., 1:13 (1980).

41. K. Shinoda and H. Takeda, J. Colloid Interface Sci., 32:642 (1970).

42. L. A. Verkruyse and S. J. Salter, paper SPE 13574 presented at the SPE Symposium on Oilfield and Geothermal Chemistry, Phoenix (Apr. 1985).

43. M. Bourrel and C. Chambu, Soc. Pet. Eng. J., 23:327 (1983).

44. M. Bourrel, P. Gard, F. Verzaro, and J. Biais, in Interactions Solide-Liquide dans les Millieux Poreux, Collection Colloques et Seminaires No. 42, Technip, Paris, p. 303 (1985).

45. J. Biais, P. Bothorel, B. Clin, and P. Lalanne, J. Disp. Sci. Tech., 1:67 (1981).

46. M. Bourrel, Compte Rendu de Recherche, No. 7870740, DGRST, Paris (1979).

47. B. B. Maini and J. P. Baticki, paper SPE 9350 presented at the 55th SPE Annual Technical Conference and Exhibition, Houston (Sept. 1980).

48. D. R. McCoy, C. G. Naylor, U.S. Patent No. 4,288,334 (1981).

49. P. E. Hinkamp, D. C. Tomkins, N. J. Byth, and J. L. Thompson, European Patent 32070 (1981).

50. P. K. Shankar, J. H. Bae, and R. M. Enick, paper SPE 10600 presented at the SPE Symposium on Oilfield and Geothermal Chemistry, Dallas (Jan. 1982).

51. C. Koukounis, W. H. Wade, and R. S. Schechter, Soc. Pet. Eng. J., 23:301 (1983).

52. O. Haque and J. F. Scamehorn, J. Disp. Sci. Tech., 7:129 (1986).

53. H. Sagitani, T. Suzuki, M. Nagai, and K. Shinoda, J. Colloid Interface Sci., 87:11 (1982).

54. K. Shinoda, H. Kunieda, T. Arai, and H. Saito, J. Phys. Chem., 88:5126 (1984).

55. M. J. Schick and D. J. Manning, J. Am. Oil Chem. Soc., 43:133 (1966).

56. C. P. Kurzendorfer, M. J. Schwuger, and H. Lange, Ber. Bunsenges. Phys. Chem., 82:962 (1978).

57. J. F. Scamehorn, R. S. Schechter, and W. H. Wade, J. Colloid Interface Sci., 85:494 (1982).

58. J. M. Corkill, J. F. Goodman, S. P. Harrold, and J. R. Tate, Trans. Faraday Soc., 63:247 (1967).

59. H. Lange and M. J. Schwuger, Kolloid-Z. Z. Polym., 243:120 (1971).

60. E. H. Lucassen-Reynders, J. Lucassen, and D. Giles, J. Colloid Interface Sci., 81:150 (1981).

61. A. B. Scott, H. V. Tartar, and E. C. Lingafelter, J. Am. Chem. Soc., 65:698 (1943).

62. J. Rodakiewicz-Nowak, J. Colloid Interface Sci., 85:586 (1982).

63. See for example U.S. Patent 4,267,077 (1981), Kao Soap Co.; U.S. Patent 4,321,165 (1982), Procter and Gamble; European Patent 880,537 (1980), Colgate-Palmolive; British Patent 77/11, 685 (1977), Unilever.

64. M. Bourrel, D. Bernard, and A. Graciaa, Tenside Detergents, 21:311 (1984).

7
Solubilization

I. THE FUNDAMENTAL PREMISE

A. Measures of Solvency

In Chap. 5 it is seen that the variation in phase behavior resulting from a change in any one of the formulation variables is predictable based on a consideration of the influence of that variable on the R-ratio. This is one of its important applications. There is a second and perhaps even more important characteristic of the R-ratio, namely, its relationship to the solvency of amphiphilic compounds. This chapter is devoted to a study of this relationship.

Before stating the premise on which all discussion of solvency in this chapter is based, it is necessary to examine the various measures of this important property of amphiphilic compounds so that their limitations are clearly understood. First, it is of the greatest importance when comparing the solvency of different amphiphiles that they all be compared at corresponding or equivalent states. In the discussion presented here, only type III systems (R = 1) will be compared. One must understand that certain applications may require systems other than type III, but the selection of a surfactant is best achieved by comparing the performance at optimum.

There are several measures of solvency that have been introduced. The solubilization parameters, $SP_O = V_O/V_S$ and $SP_W = V_W/V_S$, where V_O, V_W, and V_S are the volumes of oil, water, and surfactant, respectively, contained within the micellar phase of a multiphase system represent one measure. At optimum, R = 1, where three conjugate phases coexist, the micellar phase contains equal volumes of oil and water [1]. Therefore, $SP_O = SP_W = SP^*$, where SP^* is the solubilization parameter at optimum.

At first thought it might appear that SP^* provides an unambiguous measure of the solvency of amphiphilic compounds. This, however, is unfortunately not the case. Compare, for example, the phase maps shown in Fig. 4.19 and 4.20. The surfactant system used to prepare Fig. 4.20, while not monoisomeric, is a narrow-molecular-weight mixture of orthoxylene sulfonates, and alcohols have not been used to help adjust the phase behavior. Thus SP^* measured at any one of the three points (numbers in parentheses) is practically the same and corresponds to the minimum volume of surfactant required to achieve a single phase, S^*. In this case SP^* is a relatively precise measure of the solvency. It does not strongly depend, for example, on the surfactant concentration or on the proportions of oil and water present in the overall system.

In many cases, SP^* is calculated assuming that all of the surfactant present in the system resides within the C layer and that none is contained in the O and W regions of the micellar phase nor in either of the excess conjugate phases. To the extent that this is true, SP^* is an unambiguous measure of solvency. However, the partitioning of the surfactant into either the O or W regions or into both may be substantial. In such cases it is necessary to obtain an accurate measure of the amount of amphiphile in the C layer to calculate SP^* or at least to recognize that if the partitioning is ignored, SP^* is an approximation and will depend to some extent on the overall composition.

Whenever the system includes a mixture of amphiphiles having disparate properties such as the blend of sodium dodecyl sulfate and pentanol used to construct the phase map shown by Fig. 4.19, then SP^* is both difficult to define and variable, depending on the proportions of amphiphile used to prepare the optimum system. Two extreme values can be considered. At a given total $(S + A)$, SP^* calculated including all of amphiphile added to the system, for example in the case considered both SDS and pentanol, is an underestimate of the solvency since all of the alcohol does not reside within the C layer (see Section 5.VII.C). On the other hand, ignoring the alcohol in calculating SP^* results in an overestimate since a portion of the alcohol does reside in the C layer and does contribute to the solvency.

Figure 7.1 illustrates the difficulty in using SP^* as the measure of solvency. The solubilization parameter has been calculated by neglecting the pentanol contribution and assuming that all of the SDS molecules reside within the C layer [2]. Note that the solubilization parameter based entirely on the SDS concentration is a function of the total concentration of amphiphile, indicating the difficulty of selecting any one SP^* as a measure of the solvency. Furthermore, by including that portion of the pentanol which partitions into the C layer as amphiphile, smaller and thus more variable SP^* would be

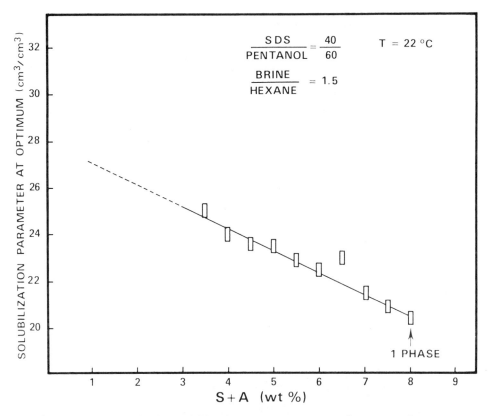

FIGURE 7.1 Optimal solubilization parameter as a function of the
surfactant and alcohol concentration, for systems indicated on the
R = 1 curve in Fig. 4.19. (After Ref. 2.)

found. The reason for the decrease in solubilization upon increasing
the surfactant plus alcohol concentration is discussed in Section 7.V.C.
The important point to be noted here is the variability of the solubilization.

It is difficult, therefore, to compare values of SP* when the concentration of alcohol and surfactant are increased simultaneously, maintaining their ratio constant.

In this chapter SP* will sometimes be used as a measure of solvency. When it is reported, it is calculated by neglecting the amount of
alcohol in the C layer, and in each case the alcohol concentration is
held constant rather than being maintained in a fixed proportion with
the concentration of the other amphiphiles as is the case in the preparation of Fig. 7.1. It is believed that by holding the alcohol con-

centration fixed, SP* will, in general, not be as sensitive to changes in amphiphile concentration as is the case for the results shown in Fig. 7.1.

Another measure of solvency which is used extensively in this chapter is (S + A)*, the amount of amphiphile (alcohol included) required to achieve a single phase (see Fig. 4.19 and the associated discussion). This quantity is easily measured, unique for a given proportion of surfactant to alcohol and of practical importance since it refers to the total amount of chemical required to solubilize equal volumes of oil and water. Thus in many of the graphs a number of different systems will be compared and only the value of (S + A)* for each system represented. The reader is reminded that these are determined by constructing phase maps such as shown by Fig. 4.19. (S + A)* is easily found from such a map.

Solubilization has been also characterized by the amount of oil (or water) which can be dissolved into a moderately concentrated aqueous (or organic) soap solution. The term "solubilization" was introduced by McBain [3] to describe the solubility in these solutions. On changing a formulation variable, such as the molecular weight of the hydrocarbon phase, for example, the amount of water solubilized by the organic solution containing amphiphiles has been found to go through a maximum. This type of experiment will be described later in this chapter, but care should be used in drawing conclusions relative to solvency of different surfactant systems based on these experiments since conditions may not be comparable; that is, one system may be more nearly optimum than another. Comparisons, as noted above, are meaningful when all systems are optimized.

B.   Correlation with Cohesive Energy Densities

The rule for improving solubilization is stated in Section 1.V.D. According to Winsor [4], the solvency of an amphiphilic compound will be higher the greater are the cohesive energies between the amphiphile in the C layer with both the O and W regions as compared to the cohesive energies that tend to favor phase separation. Thus simultaneously increasing $\Gamma_s(a_{co} - a_{oo} - a_{\ell\ell})$ and $\Gamma_s(a_{cw} - a_{ww} - a_{hh})$, which are, respectively, the numerator and denominator of the R-ratio as given by Eq. (1.19), will increase solubilization.

This hypothesis has prompted the organization of Chap. 6. A brief review of that chapter which treats compensating changes between formulation variables will reveal that compensation can be accomplished by either hydrophilic, lipophilic, or lipophilic/hydrophilic changes. These three distinctions are useful in understanding the variation in solubilization that takes place when the formulation variables are changed. For example, compensating lipophilic changes—one tending to increase and the other tending to decrease—the numerator of the R-ratio should not change the solubilization. This is true since lipophilic changes do not affect the denominator of the R-ratio and hence if the compensating changes leave R = 1, these changes do not, therefore, change either the numerator or the denominator, leaving solubilization invariant. Similarly, compensating hydrophilic changes should leave solubilization unchanged.

On the other hand, compensating lipophilic-hydrophilic changes will increase (decrease) the solubilization whenever these increase (decrease) the numerator and denominator simultaneously. The changes in solubilization that take place upon making compensating changes are considered first.

The factors influencing the packing density of the amphiphilic molecules in the C region, Γ_s, are also important. Discussion of this factor is needed to interpret a large body of solubilization data in the presence of alcohols, surfactant branching, temperature, nature of oil, and so on. It must, however, be pointed out that any separation of those variables that affect the interaction terms in the R-ratio from those which affect primarily the packing density is somewhat artificial and is done here primarily for pedogogical reasons. Clearly, both effects are interrelated. For example, increasing the ethylene oxide number of a surfactant or its alkyl chain length modifies not only its interaction with water or oil, but also its interfacial molecular area. Despite evident difficulties, the experimental evidence related to changes in solubilization because of changes in packing density is considered separately in Section 7.V. It will be seen that quite complex trends can be understood only by considering the role of surface density.

Finally, it is shown in Section 7.VI that solubilization in dilute micellar solutions is also subject to interpretation based on a consideration of the R-ratio. The fundamental premise that relates solubilization to cohesive energy densities is unequivocally demonstrated by the large body of experimental evidence presented in this chapter. In Chap. 8 a thermodynamic basis for the correlation is suggested.

II. INFLUENCE OF COMPENSATING CHANGES: IONIC SURFACTANTS

A. Lipophilic/Hydrophilic Changes

1. Sodium Chloride and Surfactant Lipophile

As discussed in Section 5.III.B, under most experimental conditions, an increase in the length of the surfactant lipophile results in an increase in the numerator of R. To compensate, the salinity can be decreased (see Section 6.II.A) to increase the denominator of R. Since the cohesive energy densities in both the numerator and denominator increase correspondingly, it is predicted that the solubilizing power of the C region increases, and smaller quantities of surfactant are required to reach the single-phase region. Figure 7.2 provides an illustration in which the lipophilic tail is varied by mixing sulfonates of different molecular weights [5]. Note that the predicted trends are realized. Longer lipophiles balanced by smaller salinities decrease S*.

Puerto and Gale [6] have reported that for a series of alkylorthoxylene sulfonates, the optimal salinity, s*, and solubilization parameter at optimum SP* are related to the carbon number n of the lipophile according to

$$s^* = a \ln \frac{b}{n} \qquad\qquad (7.1)$$

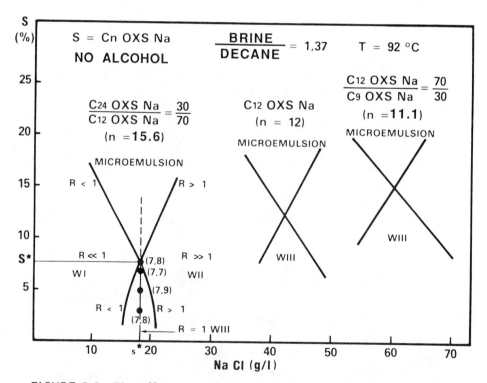

FIGURE 7.2 The effect on solubilization of compensating the length of surfactant lipophile by salinity. Same conditions as in Fig. 4.20. The lipophile variation is achieved by mixing C_9-, C_{12}- and C_{24}-orthoxylene sulfonates. n indicates the average carbon number. (After Ref. 5.)

and

$$SP^* = a \ln bn \qquad\qquad\qquad (7.2)$$

where a and b are positive constants that depend on the type of alcohol used in the system.   Combining Eqs. (7.1) and (7.2) shows that SP^* varies linearly with s^*, according to

$$SP^* = 2a \ln b - s^* \qquad\qquad\qquad (7.3)$$

This equation provides further verification of the trends predicted based on the fundamental premise.

Because increased solubilization corresponds to decreased interfacial tensions (Section 4.II.B), Eq. (7.3) indicates that the interfacial tension at optimum, γ^*, decreases with decreasing s^*.   In Fig. 7.3, γ^* is seen to decrease when the number of carbon atoms of the surfactant lipophile increases and salinity decreases to compensate the change in the lipophile [7].

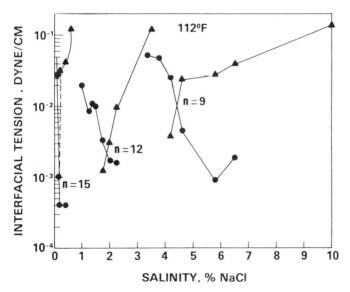

FIGURE 7.3   Dependence of γ on salinity for the system 3% 63/37 MEACnOXS/TAA, 48.5% 90/10 I/H, 48.5% X % NaCl, where n = 9, 12, and 15 at 112°F, showing improved interfacial tension as n increases. See Fig. 4.28 for acronyms.   (After Ref. 7.)

2. Salinity and Oil Type

As discussed in Section 5.III.A and shown by Fig. 5.8, under most
conditions, but not all, the numerator of R decreases on increasing
the alkane carbon number. If this is the case, then whenever an
increase in ACN is compensated by increasing the salinity, the solubilization is expected to decrease. This trend has been reported for
a wide variety of surfactants, such as sulfonates [8,9], α-olefin sulfonates [10], and ethoxylated sulfonates [11]. A typical example is
given by Fig. 7.4, which shows an increase in (S + A)* corresponding to the predicted decrease in solubilization upon increasing ACN.

   Figure 5.8 shows that for some systems at small ACN, an increase
in ACN yields an increase in the numerator of the R-ratio rather than
the decrease normally observed. In this region an increase in ACN
would be compensated by a decrease in the salinity and the solubilization would increase correspondingly. Data to support this prediction are limited, but Puerto and Reed [12] have observed, as shown
by Fig. 6.2, that increasing the ACN from pentane to hexane is com-

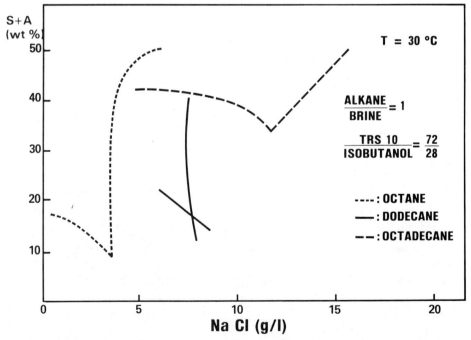

FIGURE 7.4 The effect on solubilization of compensating the alkane
carbon number by salinity. TRS 10 is a petroleum sulfonate supplied
by Witco. (After Ref. 9.)

pensated by a small decrease in salinity and an increase in solubilization is obtained.

3. ACN and Mixtures of Anionic and Cationic Surfactants

The relationship between the optimum ACN and the proportions of cationic-anionic surfactants is described in Section 6.V, where it is shown that the charges of opposite sign attract one another and the resulting neutral compound has a very small cohesive energy density A_{cw} and a large A_{hh} [13]. Thus in accordance with the fundamental premise, the solubilization using an amphiphilic system containing mixtures of anionic and cationic would be expected to decrease as more and more of the neutral species is formed whenever compensation is achieved by a lipophilic change. This is precisely the experimental finding shown by Fig. 7.5.

When the mole fraction of the anionic and cationic surfactants are both close to 0.5, the amount of amphiphile $(S + A)^*$ required to achieve a single phase is greatly increased. This value is reduced as the mole fraction of the cationic surfactant decreases or increases. In either direction the amount of the neutral species formed is reduced. This result shows that the solvency of complex systems such as mixtures of surfactants having opposite charge is predictable based on a consideration of the R-ratio.

B. Hydrophilic/Hydrophilic Changes

In the preceding section data are presented which confirm the premise that solubilization or solvency increase with a rise in the cohesive energy densities. In this section it is expected that compensating hydrophilic changes which maintain R = 1 will leave the solubilization unchanged since the numerator is not affected by variations that influence only hydrophilic interations. Extensive data are presented here to show this to be true.

1. Inorganic Electrolytes

Healy et al. [1] have reported that the same value of optimal interfacial tension is obtained for two different concentrations of the electrolytes NaCl and 10:1 NaCl/CaCl$_2$·2H$_2$O dissolved in the aqueous phase. The effect of adding Ca^{2+} to the water phase is to decrease the optimal salinity (see Section 5.II.A) but does not change the solubilization (interfacial tension), which is in agreement with the fundamental premise.

Table 7.1 shows the influence of the anion associated with sodium in the electrolyte used for achieving optimal behavior with three sulfonated surfactants. As expected, for a given surfactant, the type

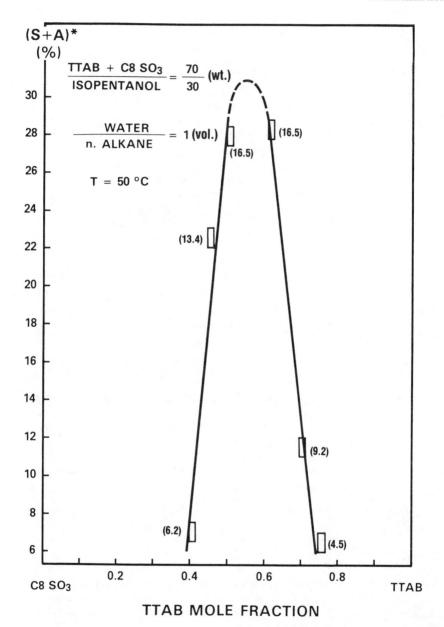

FIGURE 7.5   Effect of cationic—anionic ratio on minimal surfactant
and alcohol concentration to obtain a single micellar phase.   The
number next to each point refers to the optimal ACN (see Fig. 6.33).
TTAB and C8SO$_3$ stand for trimethyl-tetradecyl ammonium bromide
and sodium octylsulfonate, respectively.   (After Ref. 13.)

TABLE 7.1 Effect of the Anion on Optimal Phase Behavior of Anionic Surfactants[a]

Surf. Mol. Wt.	Electrolyte	s* (g/liter)	s* (mol Na^+/liter)	$(S + A)*$ (%)
420	NaCl	14.25	0.24	27.5
420	Na_2SO_4	24.75	0.353	28.5
495	NaCl	5	0.085	4
495	Na_2SO_4	7.75	0.110	4
485	NaCl	6.75	0.115	18.2
485	Na_2CO_3	8.25	0.2	19

[a]The surfactants S are petroleum sulfonate sodium salts (62% active material) supplied by Witco. Alcohol A: isobutanol; S/A = 72/28; oil: octadecane; WOR = 3.67; T = 30°C.
Source: After A. Boix, A. Saint-Marc and M. Bourrel, unpublished results.

of electrolyte has no effect on $(S + A)*$. Depending on the anion, however, the amount of salt required to obtain an optimum may vary. Na^+ decreases the interaction of the surfactant with water or the interaction between neighboring surfactants with each other more readily when present as sodium chloride than as sodium sulfate or in sodium carbonate. This phenomenon is related to the activity of Na^+. Recent results for a different system containing a synthetic sulfonate, tetradecane, and brine have been reported. Again, it has been found that replacing NaCl by Na_2CO_3 to achieve optimal salinity does not affect the solubilization parameter at optimum [14]. Similar conclusions have also been drawn based on a systematic investigation of the effect of the anion associated with the sodium ion when electrolyte is added. The following anions have been studied SO_4^{2-}, HSO_4^-, PO_4^{3-}, HPO_4^{2-}, NO_3^-, NO_2^-, CH_3COO^-, HCO_3^-, CO_3^{2-}, and Cl^- [15].

2. Salinity Versus pH

The effect of pH on the phase behavior of systems that include carboxylated surfactants is described in Sections 5.II.C and 6.II.C. It is seen that changes in pH lead to variations in the degree of ionization of the carboxylates. This variation then tends to change the R-ratio, which can in turn be compensated by changing another variable. One possible variable is the salinity.

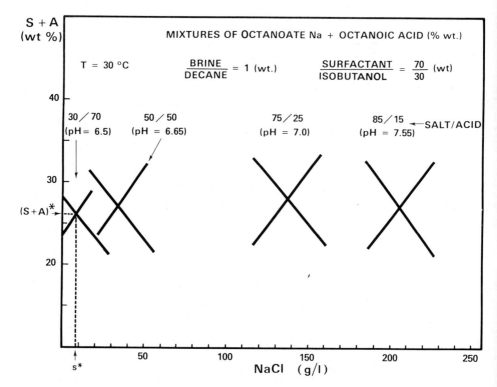

FIGURE 7.6 pH effect on optimal salinity and solubilization of carboxylates. (After Ref. 5.)

On compensating the change in pH by salinity, the interaction of
the C layer with O remains unaffected. Therefore, the numerator is
unchanged and constant solubilization at optimum is expected. Figure
7.6 shows the expected behavior.

Although the results shown in Fig. 7.6 are in good agreement with
the fundamental premise, it must be noted that the mechanisms attending the change in the proportions of the acid to the salt form are complex and the almost constant $(S + A)^*$ shown in Fig. 7.6 must, to
some extent, be fortuitous. When the overall surfactant plus alcohol
concentration is 20%, then composition is below the demixing curve.
For overall acid/salt ratios 50:50, 75:25, and 85:15, an analysis of the
excess oil phase which will exist for type I and type III systems reveals the average concentration of octanoic acid to be substantial in
some cases. These average concentrations and the fraction of the
total surfactant inventory present in the O region are listed in Table
7.2.

TABLE 7.2 Octanoic Acid Partitioning into the Excess Oil Phase of type I and III Systems for the Conditions of Fig. 7.6.

Salt/acid (overall)	50:50	75:25	85:15
Acid concentration in excess oil (wt %)	6.6	1.5	0.7
Fraction of surfactant dissolved in oil (%)	18.9	4.3	2

Source: After T. Outassourt and M. Bourrel, unpublished results.

Based on these results it is seen that the total amount of surfactant present in the C layer decreases with decreasing salt/acid ratio. This is true because more of the surfactant partitions into the O region. This result indicates that the effective solubilization has increased. At least two explanations may be put forward to explain the increased solubilization. First it might be thought that Γ_s increases as the surfactant becomes converted to the acid form or both a_{co} and a_{cw} may increase because alcohol (isobutanol, see Fig. 7.6) is displaced from the surface by the acid form of the surfactant. Thus apparently both the numerator and the denominator remain constant. The factors that influence Γ_s are complex and require further discussion. This is deferred until Section 7.V. The issue is, however, raised here to emphasize the real complexity of systems containing a mixture of disparate amphiphiles, such as the acid, salt, and alcohol used to construct Fig. 7.6.

C. Lipophilic/Lipophilic Interactions

1. Surfactant Lipophile Versus ACN

Changing the alkane carbon number of the oil has been shown to affect primarily the numerator of R. Compensation by modifying the surfactant lipophile to maintain the numerator constant should keep the solubilization constant provided that the number Γ_s of surfactant molecules per unit area of the C layer is not varied. Figure 7.7 shows a series of phase maps obtained using mixtures of two petroleum sulfonates having molecular weights 430 and 490 [16]. The composition of a mixture is referred to by its average molecular weight M_w calculated on a mole fraction basis. On increasing M_w, the optimal ACN (ACN*) is seen to increase while the minimal amount of amphiphile S* required to reach the single-phase regime remains essentially constant.

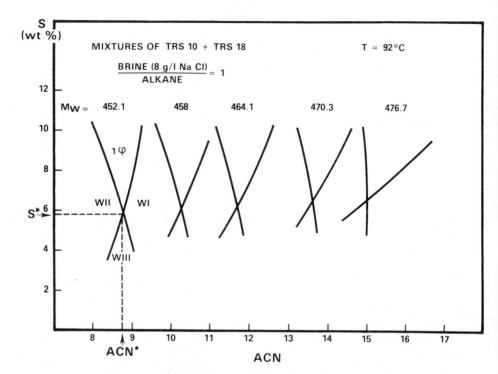

FIGURE 7.7 The effect on solubilization of compensating the alkane
carbon number by the surfactant lipophile. TRS 10 and TRS 18 are
petroleum sulfonates supplied by Witco, with equivalent molecular
weights of 430 and 490, respectively. (After Ref. 16.)

Similar behavior has been obtained over wide variety of conditions
and for many different systems [9,16,17] (see also Figs. 5.10 and
7.31). Again there is an overwhelming weight of evidence in support
of the fundamental premise.

III. INFLUENCE OF COMPENSATING CHANGES: NONIONIC SURFACTANTS

The results of this section dealing with nonionic surfactants are not
basically different from those found for anionic surfactants in Section
7.II. Indeed, the trends here are also in full accord with predictions
based on the fundamental premise. The rationale for separating the
discussion of nonionic from ionic surfactants was compelling when de-
veloping the arguments presented in Chap. 6. The same basic out-

line is retained in this chapter, so that the reader can easily cross-reference the compensating changes described in Chap. 6 with the discussion of solubilization presented in this chapter.

Nonionic surfactants are particularly interesting because they do provide an extra degree of freedom as compared with ionics. The hydrophilic interactions can be varied by changing the number of ethylene oxide units. By doing this, some striking results in support of the fundamental premise are obtained.

A. Lipophilic/Hydrophilic Interactions

1. Ethylene Oxide Number and Molecular Weight of Pure n-Alkanes

Increasing ACN is shown in Section 5.III.A to generally decrease the numerator of R. To compensate, the EON must be decreased to decrease the denominator of R. Simultaneous changes in which the ACN is increased and EON decreased to keep the system at optimum are thus expected to result in decreased solubilization [2]. Figure 7.8 provides a series of examples that confirm this hypothesis [18].

For a given oil, Fig. 7.8 also shows that the solubilization parameter at optimum increases, as expected, with the length of the surfactant lipophile as the ethylene oxide number is increased correspondingly. This increase shows that two different surfactants each similarly balanced (same HLB) can exhibit quite different solubilizations.

The solubilization is higher for the ethoxylated octadecylphenol than for the ethoxylated dinonylphenol even though their lipophiles contain the same number of methylene units. This is readily understood by considering the respective areas per molecule. Clearly, the former occupies a smaller area at the interface than the latter, yielding larger values of the numerator and denominator of R, and thus higher solubilization. This is discussed further in Section 7.V.B.

As seen in Fig. 5.8, decreasing ACN to very small values may ultimately result in a decrease in the numerator of R. When compensated by decreasing the EON, the solubilization in this range of ACN is then expected to decrease. This is clearly illustrated by Fig. 7.9, where the S^* is plotted against the optimal EON for the series of n-alkanes.

This trend is in agreement with Fig. 5.9, where the ACN corresponding to the maximum of the numerator of R ("preferred ACN") is seen to depend on the surfactant lateral interaction $A_{\ell\ell}$: the higher $A_{\ell\ell}$, that is, the longer the lipophile, the higher the preferred ACN. This trend has been further confirmed by investigating a series of nonionic surfactants having different length of lipophilic chain [16].

With the ethoxylated nonylphenol surfactant (labeled ETHOX.NP) it is seen in Fig. 7.9 that upon decreasing ACN, the S^* initially decreases, reaches a minimum at ACN = 6, which we call the "preferred"

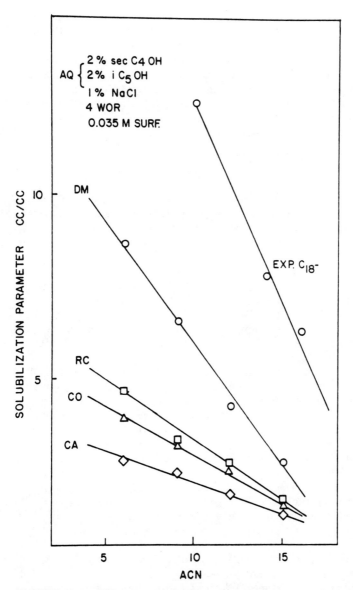

FIGURE 7.8 Effect on solubilization at optimum of simultaneous changes in EON and ACN. EON decreases on increasing ACN for each surfactant. CA, CO, RC, DM, and EXP C_{18} refer to ethoxylated octyl-, nonyl-, dodecyl-, dinonyl-, and octadecyl-phenols respectively. The concentrations refer to the aqueous phase. (After Ref. 18.)

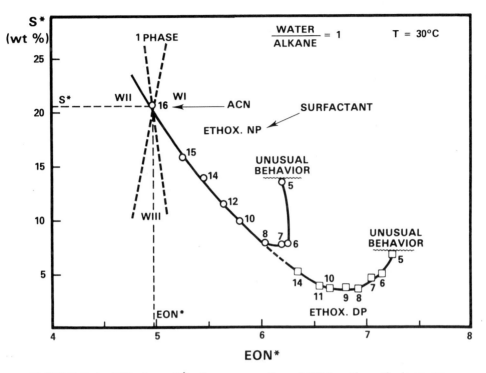

FIGURE 7.9 Effect on S^* of compensating ACN by the ethylene oxide number. Ethox. NP and ethox. DP stand for ethoxylated nonylphenol and dodecylphenol, respectively. The number next to each
point indicates ACN. S^* is obtained from phase map typically shown
for hexadecane. (After Ref. 16.)

alkane, and then increases. For the ethoxylated dodecylphenol, the
minimum occurs at ACN = 8, so that the "preferred" alkane increases
from 6 to 8 when the lipophile increases from 9 to 12 methylene groups.
It is seen that decreasing ACN to a value less than 6 yields a decrease
in solubilization (an increase in S^*).

The level of solubilizing power attained in each case depends on
the level of cohesive energy densities between the C layer with oil
and water. As expected, the values of SP^* shown in Fig. 7.8 are
generally less for ethoxylated nonylphenol than for ethoxylated dodecylphenol.

The "unusual phase behavior" noted in Fig. 7.9 and observed at
very low ACN (<5) refers to a behavior characterized by a small solubilization of oil in the micellar phase of type III systems [16]. For
butane there is no intersection of the solubilization parameter SP_0

and SP_w curves when EON is varied. This precludes determination
of an optimum. Furthermore, for small EON, viscous "gels" are observed. The interaction A_{co} between the oil and the surfactant tail
becomes so small compared to the interaction $A_{\ell\ell}$ between tails that it
is not sufficient to promote miscibility of surfactant with oil. Consequently, at low EON, the surfactant, which is no longer soluble in
water, forms a separate phase containing very little oil and less and
less water as the EON decreases. [It can be anticipated that the
phase behavior can be improved by using an amphiphile with a shorter lipophile. This conjecture has been demonstrated experimentally
[16]].

It is interesting to recall that difficulties in characterizing phase
behavior are also encountered with heavy alkanes or oils of large molar volume [19]. This corresponds to small EON*, which are equivalent to high optimal salinities. "Very condensed phases" appear in
this region, for now the interaction A_{oo} between the oil molecules is
large compared to A_{co} and prevents miscibility of the surfactant with
the oil.

As pointed out in Section 5.VI, there exists, for a given lipophile,
a window in ACN for which isotropic regular phase behavior is possible. When the ACN is too small, $A_{\ell\ell}$ dominates, and when the ACN
is too large, A_{oo} dominates. This window can be shifted toward the
smaller ACN by reducing the surfactant lipophile, but it should be
stressed that this is detrimental to solvency, as shown in Fig. 7.9,
because of the decrease in cohesive energy density.

2. Mixtures of n-Alkanes: Effect of Methane

Figure 7.10 shows that the behavior of mixtures of n-alkanes tends
to follow that observed for the series of pure n-alkanes. When pentane is mixed with dodecane or tetradecane or hexadecane, a minimum occurs in the S*-EON* curve at some intermediate oil composition,
just as in the case of the n-alkane series shown in Fig. 7.9. The
locus of optima as a function of the mole average ACN of a mixture
does not, however, coincide with that of the corresponding pure alkane. The difference is larger when the two alkanes mixed have
larger differences in ACN. Mixtures of octane and butane are interesting since they appear to model the behavior of alkanes blended
with methane at elevated pressures. "Gels" appear as the proportion
of butane is increased and the phase behavior becomes more and more
complex, especially at low surfactant concentration. Based on the
results shown in Fig. 7.9, it appears certain that for the particular
surfactant system, the EON* corresponding to butane is less than
5.8. Thus EON* decreases and S* increases rapidly as the ACN decreases below 5. It is also interesting to note that EON* increases
with the initial addition of butane and reaches a value that is larger
than the EON* of the pure components.

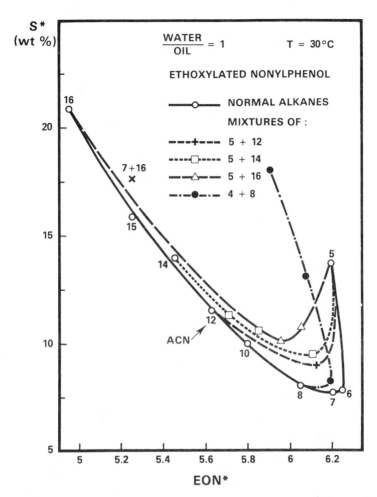

FIGURE 7.10 Mixing effect for n-alkanes in the S^*-EON^* representation. The number next to each point refers to ACN. (After Ref.
16.)

If the alkanes that are blended are all larger than the one yielding the maximum in the numerator of R shown in Fig. 5.8, a simple
mole fraction mixing rule is expected [19,20] to apply; that is, the
mixture will closely approximate the phase behavior of a pure alkane
having ACN equal to the mixture average. An example is the mixture of heptane and hexadecane shown in Fig. 7.9. For the pure
n-alkanes ranging between ACN = 6 and ACN = 16, a plot of EON^*
as a function of mixture average ACN is approximately a straight

line [16]. This agrees with the results shown in Figs. 6.18 and 6.20.
To the extent that the principle of congruence applies, that is, if
the properties of a multicomponent system of alkanes are determined
by the mean $\overline{ACN} = \sum_i x_i ACN_i$ (see Section 6.II.F), a linear mixing
rule for EON* applies

$$EON^*_{mix} = \sum_i x_i EON^*_i \qquad (7.4)$$

where x_i is the mole fraction of the alkane i, having an optimal ethylene oxide number EON^*_i [16]. Similarly, since the descending portion of the curve in Fig. 7.9 corresponding to the higher alkanes is
approximated by a straight line, then for alkane mixtures of molecular weight heavier than hexane, a linear mixing rule on S* can be
applied:

$$S^*_{mix} = \sum_i x_i S^*_i \qquad (7.5)$$

where S^*_i is the minimal amount of surfactant required to obtain a
single-phase micellar solution with ACN equal i. Similar equations
have been reported for optimal solubilization parameters of anionic
surfactants [6,17].

Equations (7.4) and (7.5) should be used with some care, for as
noted in Section 6.II.F, deviations are to be expected. Figure 7.10,
showing both S* and EON*, provides a clear illustration of the difficulty. If one is interested in the optimal EON and in the solubilization at optimum, it is readily seen that mixtures of n-alkanes do
not fully represent a single pure n-alkane at a point S*.

Figure 7.11 shows the effect on the optimal phase behavior of
dissolving methane under pressure in various alkanes. For the alkanes investigated, EON* decreases and S* increases when the methane pressure is raised. The unusual behavior described previously
is encountered at sufficiently high pressures. The limiting pressure
for the appearance of unusual behavior depends on the oil blended
with the methane. The lower the ACN of the oil, the lower is the
limiting pressure. The results shown in Fig. 7.11 can readily be interpreted if it is assumed that the point representing methane on this
diagram is located at very low EON* and very high S*.

Oil C120X indicated in Fig. 7.11 is a commercial tetrapropyl orthoxylene (molar volume: 245 cm^3/mol). For this oil a different trend
is observed. The EON* and S* are found to increase and to decrease
slightly, respectively, with a rise in the methane pressure (at 200
bar, however, the gas/oil ratio is only 56 standard cubic meters of
gas per cubic meter of oil). This trend can be understood by com-

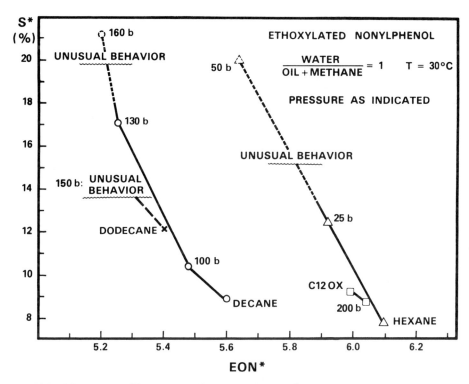

FIGURE 7.11 Effect on optimal behavior of pressurizing various oils with methane. The number next to each point indicates the pressure in bars. Points at 50 bars for hexane and 160 bars for decane are estimated. See text for discussion of C120X. (After Ref. 16.)

paring with mixtures of alkanes and butane shown in Fig. 7.10. These exhibit similar trends; namely, for some mixtures the EON* increases when butane is initially added, reaches a maximum, and then decreases toward the value corresponding to pure butane. At the same time, S* goes through a minimum. It is postulated, therefore, that the different trends shown in Fig. 7.11 simply correspond to different branches of the curve located on either side of the minimum S*.

Similar results on the effect of methane have been reported by Nelson [21], who observed that an optimized system was shifted toward type II or type I when methane was added to a stock tank oil or to a synthetic (isooctane-rich) oil, respectively.

3. Temperature and ACN

As seen in Section 6.III.A, the decrease in the numerator of R induced by an increase in ACN (increase in A_{oo}) can be compensated
by a decrease in temperature which decreases A_{cw} in the denominator. The decrease in both the numerator and the denominator of R
is thus expected to result in a decrease in solubilization at optimum.
Figure 7.12 provides an illustration. The proportion of the surfactant required to reach the single-phase regime increases when ACN
is compensated by temperature [22]. In addition, as ACN increases,
the coexistence curve between type III and type II shifts to higher
temperatures than the coexisting curve between type III and type I,
and the three-phase region becomes wider. This is also an indication of decreased solubilization (see Section 4.II.C).

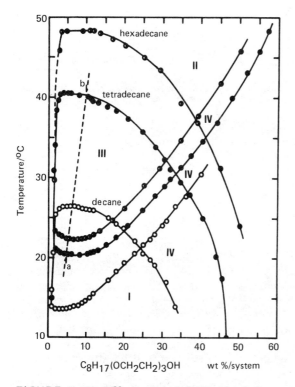

FIGURE 7.12 Effect on solubilization of compensating ACN by temperature. Surfactant: $C_8H_{17}(OCH_2CH_2)_3OH$. Water—oil ratio = 1
(wt). (After Ref. 22.)

B.   Hydrophilic/Hydrophilic Interactions

1.   Ethylene Oxide Number and Salinity

As seen in Section 5.II.D, increasing EON results in a larger interaction with water, and thus requires increased electrolyte concentrations to maintain optimum behavior (Section 6.III.C). Since both the
surfactant lipophile and the hydrocarbon are fixed, the numerator
of the R-ratio (and thus the denominator at optimum) is expected to
remain constant. As a consequence, the solubilization should remain
unaffected on compensating EON by salinity.

Figure 7.13 [23] shows the evolution of S^* and $(S + A)^*$ for various EON obtained from phase maps such as those shown by Fig. 5.1.
The corresponding optimal salinities are given next to each point. It
can be seen in Fig. 7.13 that S^* (for the alcohol-free system) increases slightly with EON. This effect is probably due to the increase in the molecular area of the surfactant with EON [24], yielding a corresponding decrease in Γ_s (see Section 7.V.A).

For systems containing alcohol, the slope of $(S + A)^*$ versus EON
is smaller than for the alcohol-free system. This is probably due to
the dilution of the surfactant by the alcohol in the C region, which
reduces the change in the surfactant molecular area with increasing
EON. Indeed, one striking feature shown in Fig. 7.13 is the sharp
decrease in solubilization in the presence of alcohol. This important
feature is discussed further in Section 7.V.C.

IV.   IMPROVING SOLVENCY

The amount of amphiphile required to create a stable blend of oil and
water is obviously a quantity of considerable practical importance. In
this section the steps required to reduce this quantity are reviewed.

Certainly, the first step is to ensure that the amphiphile is properly balanced. Is R unity? If not, the solvency can be improved,
perhaps substantially, by adjusting appropriately the structure of the
amphiphile so that R = 1. This assumes that the composition of both
the water and the oil are dictated by the application and not subject
to variation. This is a most important step. Often the search for
an amphiphile is terminated when a suitable but not optimum solubilization is attained and a complete phase diagram is not routinely investigated.

If the solvency at optimum is still not adequate, the cohesive energy density between the C layer and both the O and W regions must
be increased but in such a way that R remains unity. Figure 7.14
provides an example.

The optimal salinity, defined as the electrolyte concentration corresponding to $(S + A)^*$, is the same for both surfactants [sodium dodecylsulfate (SDS) and the sodium salt of the n-acetyl α-amino eico-

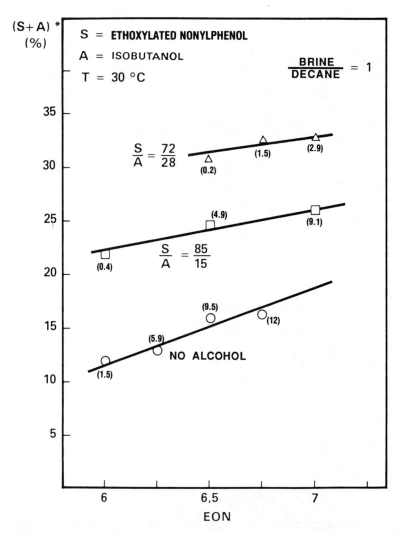

FIGURE 7.13 Effect on solubilization at optimum of simultaneous changes in ethylene oxide number and salinity for systems containing ethoxylated nonylphenol, brine, decane, and isobutanol. Surfactant/ alcohol ratio (wt) S/A as indicated. Brine/decane = 1 (wt). The optimal salinity in g/dl is indicated by each datum point. (After Ref. 23.)

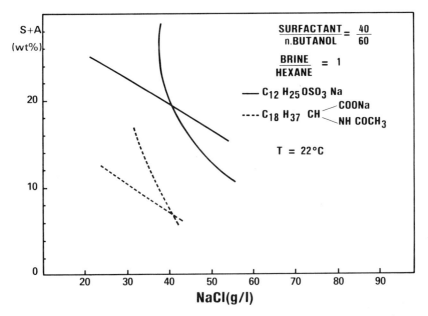

FIGURE 7.14 Example of the effect of simultaneous increase in surfactant hydrophile and lipophile. (After Ref. 2.)

sanoic acid (SAAE)]; that is, both of these surfactants are balanced under the conditions defined in Fig. 7.14. $(S + A)^*$ for SAAE is less than half the value of $(S + A)^*$ for SDS, because of the larger interactions of the former surfactant with both oil and water.

This process of simultaneously increasing the hydrophile and the lipophile to increase solvency has limits. First a decrease in Γ_S and hence in the numerator and denominator of the R-ratio may ultimately result. Therefore, an increase in solubilization upon lengthening the lipophile and strengthening the hydrophile may not be realized. Moreover, as discussed in Sections 5.VI and 7.III.A, another limitation stems from the lateral interactions between the surfactant molecules, which may be raised to the point where they become predominant and cause the separation of the amphiphile as either a solid, a gel, or a liquid crystal.

A convenient way of varying the hydrophilic and lipophilic moieties of the molecule simultaneously is to use ethoxylated nonionic amphiphiles. The alcohol can be considered as the first member of the series. Using monodisperse ethoxylated butanol, hexanol, and octanol, Kilpatrick et al. [25] have illustrated the continuity of the solvent properties of amphiphilic compounds from the lower to the higher

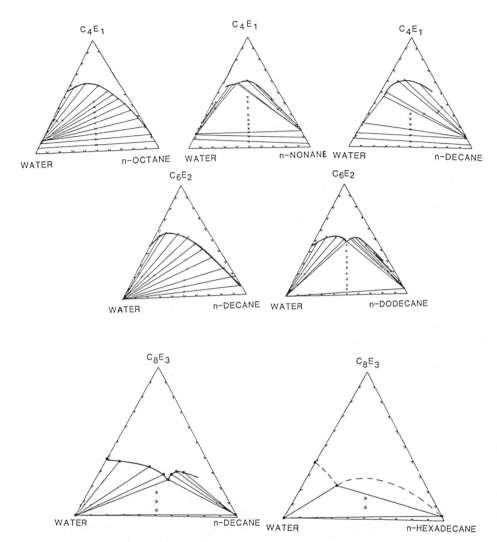

FIGURE 7.15 Ternary phase diagrams for water-n-alkane and ethoxylated monoisomeric alcohols C_nE_m; n and m indicate the number
of carbon atoms in the alkyl chain and the ethylene oxide number,
respectively. (After Ref. 25.)

members of the series. Figure 7.15 shows selected examples. Whatever the molecular weight of the amphiphile, increasing ACN promotes qualitatively the same type of phase transition. Some quantitative differences between the amphiphiles are, however, worth noting. First, the lower-molecular-weight homologs exhibit a larger partitioning into the excess oil and water phases in three-phase systems
than in the higher ones. Second, although optimum behavior is not
obtained for the same ACN with all three amphiphiles, the proportion
of amphiphilic compound required to dissolve equal amounts of water
and hydrocarbon clearly decreases on increasing the length of both
polar and nonpolar moieties. Finally, according to the authors [26],
the fluid microstructure of the amphiphilic phases seems to differ
qualitatively.

Figure 7.15 shows the pronounced decrease in the height of the
three-phase triangle that can be obtained by increasing the cohesive
energy densities. It is this gain that is available by careful optimization of the surfactant system.

V. FACTORS INFLUENCING THE INTERFACIAL
DENSITY OF AMPHIPHILE

Sections 7.II and 7.III are devoted to demonstrating the profound
relationship that exists between the cohesive energies per unit area
and solubilization. The primary variables studied in these sections
are A_{co}, A_{cw}, $A_{\ell\ell}$ and A_{hh}. These are, of course, the variables
that can most readily be adjusted; however, the adsorption density,
Γ_s, is shown by Eq. (1.19) to multiply both the numerator and the
denominator of the R-ratio and is therefore also an important factor
governing solubilization. The factors influencing Γ_s are often subtle
and complex. These are considered in this section. Very often unexpected trends in solubilization can be attributed to variations of Γ_s.

The adsorption density of amphiphile in the C layer is difficult to
measure and, as shown in Section 8.III.E, dependent on the curvature of the interface. It is perhaps not difficult to assess the total
number of amphiphilic molecules in the C layer. Determination of the
interfacial area is, however, a most difficult proposition. Because of
this difficulty, a number of the trends that in this section are assert-
ed to be related to changes in the surface density are inferred rather than actually established by measurement. Despite the difficulties
associated with the direct determination of surfactant adsorption in
the C layer, a chapter dealing with solvency would be incomplete if
a discussion of it were omitted.

A. Surfactant Hydrophile

1. Ionic Surfactants

The relationship between the structure of ionic hydrophiles and optimal salinity has been discussed in Section 5.II.D. Table 5.2 gives
the solubilization parameters at optimum corresponding to different
ions. For the octadecyl series, one might expect the solubilization
to be constant since neither the surfactant lipophile nor the hydrocarbon is changed. As seen in Table 5.2, the experimental results
show that the solubilization parameter SP* may vary widely, indicating that the area per surfactant molecule at the interface, changes
with the type of hydrophilic group. To interpret the results presented in Table 5.2, the molecular area must decrease in the order

$$N(CH_3)_3Cl \gg -CO_2Na > -SO_4Na > -SO_3Na \gg -PO_4Na_2$$

The carboxylate, sulfate, and sulfonate, which yield approximately the same solubilization, exhibit the same magnitude interfacial area
and thus their order may be interchanged in some experiments[†]. The
larger molecular area of the trimethyl ammonium suggested by the solubilization experiments is reasonable because of steric hindrance
caused by the presence of the three methyl groups attached to the
hydrophile. The classification of the phosphate is somewhat unexpected; however because this surfactant yields significantly higher solubilization, a smaller molecular area for the system studied is indicated.

2. Nonionic Surfactants

Increasing the ethylene oxide number of a nonionic surfactant is
known to increase its interfacial area [24]. This is shown by Fig. 7.16.
These results, which apply to an air/water interface, reveal that the
area per molecule increases proportionate to $(EON)^{1/2}$, which is consistent with a random chain configuration [27]. Thus one might anticipate a decrease in Γ_s as EON increases. The decrease in Γ_s must
account for some but certainly not all of the increase in $(S + A)^*$

[†]Depending on how we define the carboxylate alkyl chain length, its
ranking could be different (see Section 5.II.D) than that shown here.
$C_{18}CO_2Na$ yields lower optimal salinities than does $C_{17}CO_2Na$ and therefore higher solubilization parameters. Thus if the carbon in the car-
boxyl group is counted as a carbon in the lipophilic tail rather than
as a part of the hydrophile as is done above, $C_{17}CO_2Na$ would be
classified as a member of the octadecyl series leading to the following possible order of decreasing molecular areas: $-N(CH_3)_3Cl \gg$
$-SO_4Na > -SO_3Na > -CO_2Na \gg -PO_4Na_2$.

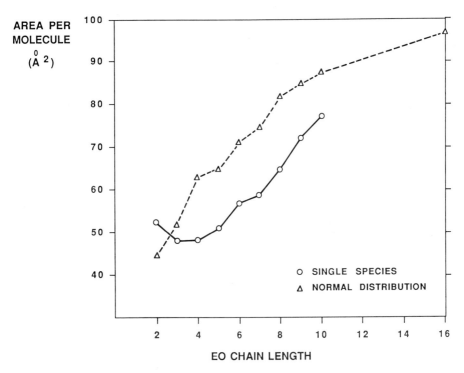

FIGURE 7.16 The effect of ethylene oxide number on the molecular area of monoisomeric ethoxylated octylphenol at the water-isooctane interface. (After Ref. 24.)

shown by Fig. 7.13. Another factor contributing to the increase in $(S + A)^*$ is surfactant partitioning into the O region, which perhaps increases with EON since the electrolyte concentration is increased correspondingly.

3. Ethoxylated Ionic Surfactants

Figure 7.17 shows the solubilization parameter at optimum for a series of oleyl ethylene oxide sulfonates (see Section 5.II.D). The optimal salinities associated with a given alkane is listed next to each datum point.

Figure 7.17 shows that for a given ACN^*, increasing the ethylene oxide number from 1 to 2 does not change the solubilization at optimum when the increase in the hydrophile is compensated by an increase in the salinity. This is the expected result since the compensating changes are both in the denominator of the R ratio.

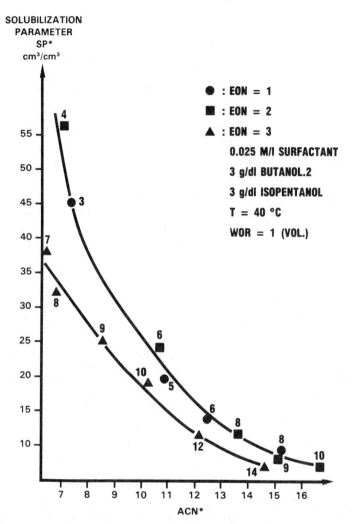

FIGURE 7.17 The effect of ethylene oxide number EON of ethoxylated anionics on solubilization for various n-alkanes. For a given alkane, EON is compensated by salinity. The optimal salinity in g/dl is indicated by each datum point. The concentrations refer to the aqueous phase. (After Ref. 11, replotted data.)

When the EON goes from 2 to 3, a decrease in SP* is observed. This is probably due to a decrease in Γ_s, a conclusion that is consistent with the trend and the magnitude shown in Fig. 7.16.

B. Surfactant Lipophile Structure

In general the stepwise removal of the polar head from a terminal to a more central position along the alkyl chain increases the interfacial area per amphiphile. This in turn decreases $a_{\ell\ell}$ and for ionic surfactants, increases a_{hh}. Both of these trends serve to increase the R ratio, and the general observation is that R increases with surfactant "branching." Here the use of the term "branching" is not intended to mean a certain restricted class of lipophilic structures, but instead is used simply to convey the concept of surfactant "bulkiness" (see, e.g., Fig. 5.18). Any change in the lipophilic structure that results in a decrease in Γ_s is deemed here to be branching. Thus adding a double bond at some point along the lipophilic chain, which tends to increase the rigidity of it, is, in this context, considered to be branching. This point is strongly emphasized here because the term "branching" may bring to mind a certain limited range of possible lipophilic structures which is more restrictive than the meaning considered here.

Even though branching tends to increase R, solvency at optimum will depend to some extent on the particular variable that is used for compensation. This complexity arises because while increased branching decreases Γ_s, it also leads to the following inequality:

$$(a_{co} - a_{oo} - a_{\ell\ell}) > (a_{cw} - a_{ww} - a_{hh}) \qquad (7.6)$$

which is certainly true because R has increased. Thus if the increase in amphiphile branching is compensated by reducing the larger of the terms in Eq. (7.6), that is, by reducing the cohesive energy on the oil side of the interface, then almost certainly the numerator of the R-ratio, $\Gamma_s(a_{co} - a_{oo} - a_{\ell\ell})$, will decrease leading to decreased solubilization. Thus lipophilic compensation for increased branching will probably lead to decreased solubilization.

If, on the other hand, compensation is accomplished by increasing the smaller of the terms in inequality (7.6), that is, a hydrophilic change, then the denominator of the R ratio may increase or decrease depending on the relative magnitudes of the variation of Γ_s and $(a_{cw} - a_{ww} - a_{hh})$. For changes in the molecular structure of the lipophile which greatly decrease the interfacial density, solubilization is expected to decrease.

Because these trends are complex, but certainly crucial in terms of evaluating the solvency of amphiphilic compounds, they are examined in some detail here.

1. Compensating Lipophilic Changes

If compensation for increased branching is achieved by decreasing the lipophile or increasing the ACN, then, as noted above, a smaller solubilization at optimum is expected. Figure 7.18 shows the solubilization parameter at optimum for isomers, noted as $x\phi C_n$, of sodium alkyl benzene sulfonates of various chain length [17]. n denotes the total number of carbon atoms in the alkyl chain and x refers to the carbon to which the benzene ring is attached.

For each surfactant, the ACN is varied and the particular alkane giving optimal behavior identified. The optimum is denoted as ACN*, which is shown in Fig. 7.18. Certain amphiphiles shown in Fig. 7.18 do not exhibit optimal behavior for an ACN within the range of liquid

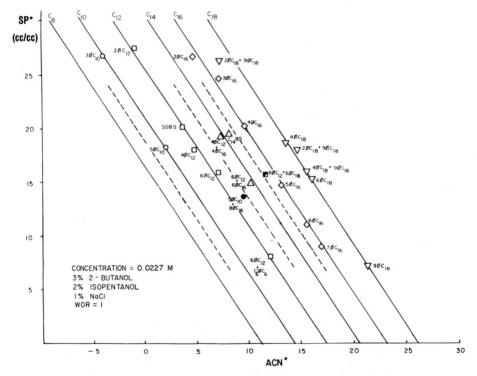

FIGURE 7.18 The effect of lipophile structure on solubilization for several families of pure alkyl benzene sodium sulfonates and some mixtures. SDBS and $C_{14}BS$ stand for sodium dodecyl and tetradecyl benzene sulfonate, respectively. The points refering to mixtures have been positioned assuming linear mixing rules on ACN*. The concentrations refer to the aqueous phase. (After Ref. 17.)

alkanes at the conditions under consideration. These have been
blended with other surfactants whose optimum ACN* and SP* are
known. The ACN* and SP* of the unknown surfactant were then
calculated using the linear mixing rules for ACN* and SP* as expressed by Eqs. (6.5) and (7.5).

For a given lipophile chain length, Fig. 7.18 shows a strong decrease in SP* when the point of attachment of the benzene ring recedes from a terminal to a medial position in the chain. This agrees
with the discussion above since the effect of branching is compensated by an increase in ACN.

For a given ACN*, the solubilization parameter also decreases when
the compensation for changes in point of attachment is achieved by
reducing the alkyl chain length (compare $3\phi C_{16}$ and $6\phi C_{12}$, for in-
stance). Thus when branched surfactants are used in place of longer chain, less branched surfactants, the solvency of the system is
decreased. Larger quantities of a more branched surfactant will be
required for a given application.

2.   Compensating Hydrophilic Changes

The increase in the R-ratio due to branching can be compensated by
increasing the hydrophilic interactions. This can be accomplished,
for example, by decreasing the electrolyte concentration or, in the
case of nonionic surfactants, increasing the ethylene oxide number.
When this method of compensation is used, the solubilization at optimum may increase, decrease, or remain the same depending on whether $\Gamma_s(a_{cw} - a_{ww} - a_{hh})$ has increased, decreased, or remained constant.

Figure 7.19 shows the solubilization parameter when $2\phi C_{16}$ and
$8\phi C_{16}$ are blended in varying proportions [28]. When the fraction
of $8\phi C_{16}$ increases in the mixture, the optimal salinity decreases as
expected, and for these surfactants, the solubilization parameter increases. This interesting result demonstrates that branching or
equivalently reducing Γ_s is not in all cases detrimental to surfactant
solvency. At issue is to what extent is Γ_s and $a_{\ell\ell}$ decreased and
what variable is changed to compensate for the branching.

Figure 7.20 shows that increased branching need not lead to greater solubilization when compensated by a hydrophilic change. Compared is the performance of two surfactants each having 15 carbon
atoms in the lipophile. The notation "bl" implies that the surfactants
are composed of linear lipophilic chains with variable points of attachment of the aromatic ring to the alkyl chain, but as indicated in
Fig. 7.20, with a high degree of second-carbon attachment. The terminal carbon attachment is excluded.

One of the surfactants studied is an alkyl benzene sulfonate and
all 15 carbon atoms are present in the alkyl chain, whereas the other
one is an orthoxylene sulfonate and two carbons in the form of meth-

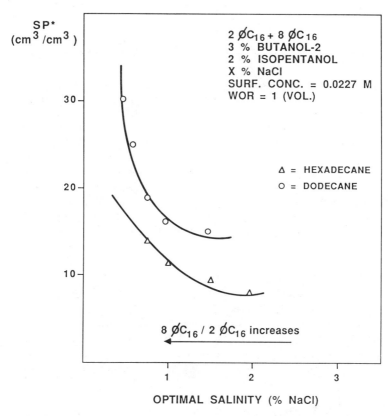

FIGURE 7.19 The effect on solubilization of compensating the change in lipophile structure by salinity. Increased proportions of $8\phi C_{16}$ correspond to lower salinities. The concentrations refer to the aqueous phase. (After Ref. 28.)

yl groups are attached directly to the aromatic ring. The remaining 13 are present as an alkyl chain.

The orthoxylene is clearly the more branched in the sense illustrated by Fig. 5.18. The driving force for adsorption into the C layer, namely the hydrophobic effect, is greatly diminished by distributing the lipophile onto various sites about the aromatic ring. In this case even though the branching is compensated by a decrease in the salinity, the solubilization parameter of the more branched surfactant decreases. This behavior is to be contrasted with the results shown by Fig. 7.19, which show an increasing solubilization with increased branching.

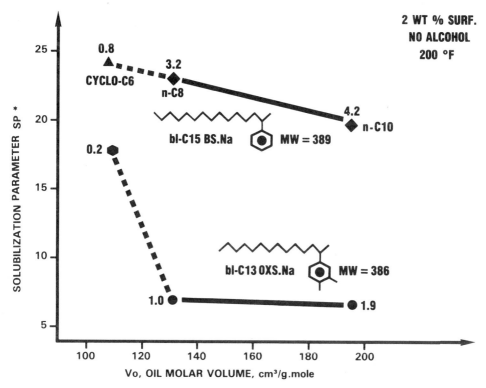

FIGURE 7.20   Effect of the lipophile structure on optimal phase behavior of three alkyl aryl sulfonates of similar molecular weight for
various hydrocarbons.   Cyclo-C6, nC8, and nC10 refer to cyclohexane, n-octane and n-decane, respectively.   The number next to each
point indicates the optimal salinity (% NaCl).   (After Ref. 12.)

The effect of branching on solubilization is in general complex
when compensated by a hydrophilic change.   Solubilization will, however decrease if Γ_s is reduced substantially.

The sensitivity of the phase behavior to the surfactant structure
is dramatically evidenced in Fig. 7.21, which illustrates the effect
of the point of sulfonation of 1-dodecyl p-ethylbenzene sulfonate
[29].   When isomerically pure 1-dodecyl p-ethylbenzene is sulfonated,
two species may be obtained, with the sulfonate group attached to
the aromatic ring in either the meta or ortho position.   Figure 7.21
shows the behavior obtained with mixtures of these two species.

The o-sulfonate, as seen in Fig. 7.21a, produces a lower optimal
salinity than the m-sulfonate.   This is interpreted as indicating that

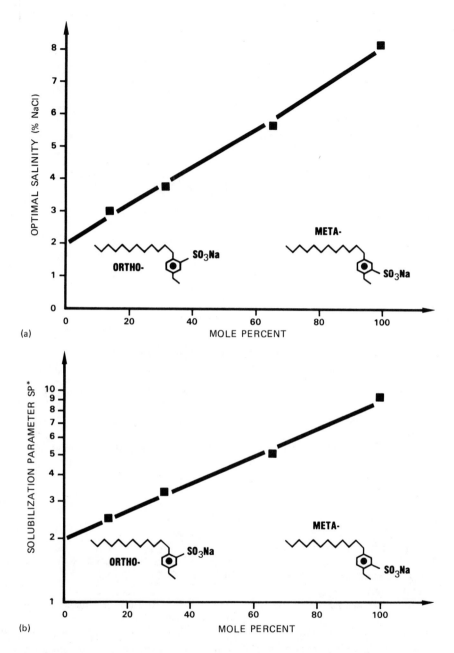

FIGURE 7.21 The effect of point of sulfonation for 1-dodecyl para-ethylbenzene sulfonate on (a) optimal salinity and (b) optimal solubilization parameter. Mixtures of the two species (ortho- and meta-) schematically shown in the figure have been investigated. (After Ref. 29.)

the ortho species has a more "branched" character than the meta
species, displays lower lateral interactions, and occupies a larger
area in the C region. As in Fig. 7.14, the latter effect is predominant and yields a decrease in solubilization despite the increase in
$(A_{cw} - A_{hh})$ obtained when the salinity decreases (Fig. 7.21b).

C. Influence of Alcohol

As seen in Section 5.IV.A, the influence of alcohol on the phase behavior of surfactant/water/hydrocarbon mixtures can be understood
by considering the interaction of the C layer with O (or W) to be
simply the sum of the interactions with O (or W) of the surfactant
and alcohol molecules contained in C. This same reasoning should
apply when considering the solvency in the presence of alcohol. According to Eqs. (5.9) and (5.10), A_{co} and A_{cw} are given by

$$A_{co} = \frac{x_a a^a_{co} + x_s a^s_{co}}{x_a S_a + x_s S_s} \qquad (7.7)$$

$$A_{cw} = \frac{x_a a^a_{cw} + x_s a^s_{cw}}{x_a S_a + x_s S_s} \qquad (7.8)$$

The effect of the concentration and the type of alcohol on solubilization can be discussed conveniently with the aid of Eqs. (7.7)
and (7.8), provided that simplifying assumptions are made: namely,
the changes in $A_{\ell\ell}$, A_{hh}, A_{oo}, and A_{ww} are negligible, and the
areas per surfactant molecule S_s and alcohol molecule S_a do not depend on their concentration within the C layer.

Depending on the values of a^s_{co}/a^a_{co} and S_s/S_a, A_{co}, as determined
by Eq. (7.7), can decrease, increase, or remain constant as the fraction of alcohol in the C layer increases. All of these different trends
are shown in Fig. 7.22.

The solubilization of oil (or water) at optimum is expected, under
the foregoing assumptions, to vary in proportion to A_{co} (or A_{cw}).
Thus the trends depicted in Fig. 7.22 should reflect the solubilization of different systems. Examples of the various possibilities are
given below.

1. Concentration

Most reported results show that adding alcohol to micellar systems
reduces the solubilization at optimum and correspondingly increases
the interfacial tension [2,5−7,10−12,18,19,23,28−31]. These studies generally refer to alcohols whose lipophilic moieties contain fewer

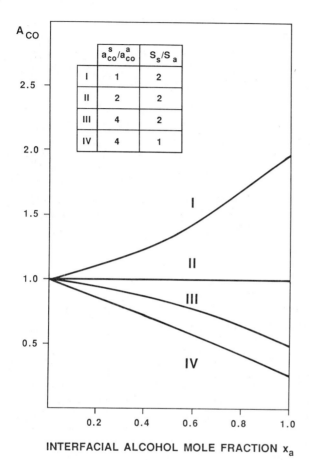

The table within the figure:

	a_{co}^s/a_{co}^a	S_s/S_a
I	1	2
II	2	2
III	4	2
IV	4	1

INTERFACIAL ALCOHOL MOLE FRACTION x_a

FIGURE 7.22 The effect of alcohol on A_{co} calculated from Eq. 7.7.
A_{co} is given in such units that $a_{co}^s/S_s = 1$. For $x_a = 1$, $A_{co} = a_{co}^a/S_a$.

than six carbon atoms, and therefore a_{co}^s/a_{co}^a is expected to be greater than unity. Curves exhibiting trends similar to III and IV shown
in Fig. 7.22 are therefore expected to apply.
   The trends shown in Fig. 7.23 are typical. The upper curve
(open circles) corresponds to alcohol-free systems, whereas the lower
one (closed circles) refers to systems containing pentanol. The optimal salinity decreases upon the addition of pentanol. This implies,
as is amply discussed in Section 5.IV.A, that the denominator of the
R-ratio has decreased more rapidly than the numerator and a decrease in salinity is required to restore optimum conditions.

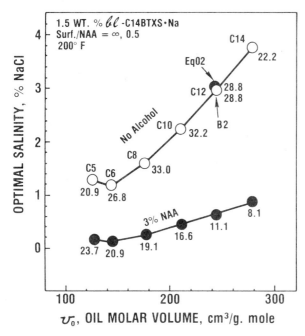

FIGURE 7.23 Effect of pentanol-1 (NAA) on optimal salinity and solubilization parameter (given by each point) for the series of n-alkanes.
The surfactant is a mixture of C14 benzene, toluene, and xylene sulfonates. (After Ref. 19.)

In most cases, although not all, solubilization has also decreased
in the presence of alcohol, as noted by the numbers adjacent to the
data points. This is the expected trend when the surfactant system
includes short-chain alcohols. Interestingly, a different trend is observed for the oil having the smallest molar volume (pentane), since
in Fig. 7.23 the solubilization parameter in the presence of alcohol
is shown to exceed that of the alcohol-free system. This contradictory behavior occurs because the approximations invoked when constructing Fig. 7.22 are not always valid. When the O region is composed of pentane, the surfactant lipophilic lateral interactions become
important when compared to oil/surfactant interaction and therefore
cannot be neglected. In this case increasing the fraction of alcohol
within the C layer decreases $A_{\ell\ell}$ faster than A_{co} and the numerator
increases rather than decreases, resulting in increased solubilization.

2. Alcohol Structure

For a given interfacial alcohol mole fraction and assuming constant molecular area, increasing the alcohol chain length primarily increases the interaction of the C layer with O. The solubilization at optimum is thus expected to increase whenever compensation is achieved by increasing the interaction with water through either a decrease in salinity or an increase in ethylene oxide number, for example. Correspondingly, the interfacial tension at optimum is expected to decrease. Healy et al. [1] and Salter [30] confirm these trends by replacing tertiary butyl alcohol with tertiary amyl alcohol. Figure 7.24 shows the complex effect on phase behavior when compensating a change in alcohol chain length by altering the ethylene oxide number of a nonionic surfactant keeping the surfactant/alcohol ratio constant [32]. The curve denoted "no alcohol" has been obtained without using alcohol, and for this case the ordinate refers to surfactant. To compare with results obtained when alcohol is present in the system, a crude correction has been applied to the no alcohol data. The surfactant concentration has been divided by 72% to reflect the fact that all of the other curves have been obtained with 72% surfactant and 28% alcohol. The corrected curve is designated in Fig. 7.24 as "no alcohol (corrected)."

To interpret the results shown in Fig. 7.24, it is convenient to begin by considering propanol and imagine an increase in the alcohol carbon number, assuming as a first approximation that the interfacial alcohol concentration does not vary greatly. The EON^* is expected to increase with increasing alcohol carbon number. Since both numerator and denominator increase in the R-ratio, $(S + A)^*$ should correspondingly decrease. This trend is confirmed by the experimental results plotted in Fig. 7.24 for alcohol carbon numbers less than 6. Above this value, EON^* begins to decrease unexpectedly, whereas $(S + A)^*$ continues to decrease. Consequently, at this point the initial assumption must be rejected and a varying alcohol interfacial concentration with alcohol carbon number must be considered. In fact, it is likely that increasing the alcohol chain length results in a larger partitioning into the O region and consequently a decreased interfacial density.

If the interfacial alcohol concentration decreases fast enough with alcohol chain length (and this trend is accelerated for the case shown in Fig. 7.24 since there is also a decrease in overall alcohol concentration as the alcohol chain length increases), it is possible that the overall hydrophile/lipophile balance of the C layer will begin to favor the W region despite the increased lipophilic affinity of higher-chain-length alcohol. Consequently, compensation on the water side of C must be achieved using a surfactant having a smaller EON. However, because of the increase in the interfacial surfactant concentration, the interaction energies continue to increase on both sides of the in-

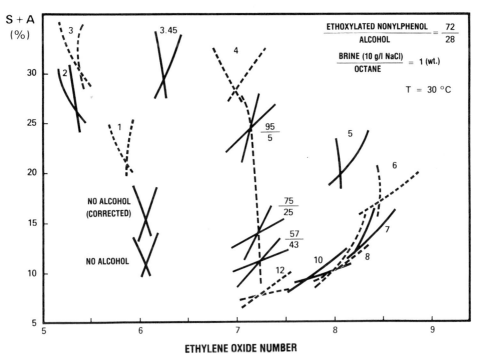

FIGURE 7.24 Effect of chain length of normal alcohols on phase behavior. Each integer number in the figure refers to the alcohol carbon number; 3.45 refers to a 55/45 molar mixture of propanol-1 and
butanol-1; 95/5, 75/25, and 57/43 refer to molar mixtures of butanol-1
and dodecanol-1. (After Ref. 32.)

terface, yielding a continuous decrease in $(S + A)^*$. The limit of
such a decrease is expected to be the value for an alcohol-free system since, for very long chain alcohols, the C region should be constituted almost entirely of surfactant molecules with almost the entire
alcohol inventory being in the oil.

Figure 7.24 shows that the alcohol-free solubilization is not the
limit since for alcohol homologs higher than heptanol, values of $(S +
A)^*$ that are significantly lower than those found for the "no-alcohol"
corrected system are obtained. It is conjectured that this enhanced
solvency may be related to the fact that the area of the alcohol molecule is smaller than the surfactant. In accordance with the results
plotted in Fig. 7.22, this may ultimately result in increased interaction energies per unit area of interface. This trend corresponds to
curve I in Fig. 7.22. The increased A_{co} in the presence of alcohol
corresponds to increased solubilization.

The results obtained with ethanol and methanol can be interpreted similarly. Assuming their concentration in the C region to be the same as propanol, exchanging propanol for ethanol should decrease the interaction energy with the O region, and to compensate for this, the EON should be correspondingly decreased. Consequently, $(S + A)^*$ should increase. The opposite trend is actually observed in Fig. 7.24, showing that both ethanol and methanol exhibit smaller interfacial concentrations than propanol. These alcohols are now dissolved primarily in the W region.

Figure 7.24 also shows results obtained with mixtures of alcohols. The phase map corresponding to a 55:45 molar mixture of propanol and butanol is positioned intermediate between the results obtained with the pure alcohols, and this seems to imply that mixtures of alcohols behave as pure compounds and can therefore be treated as alcohols of nonintegral carbon numbers. Mixtures of butanol and dodecanol, which are more disparate in their tendency to partition between the C and O regions, do not match that of the corresponding pure alcohols. It is likely that this difference is due primarily to the presence of both alcohols at the interface in proportions ruled by their respective partition coefficient; that is, there is no average partitioning. Similar behavior has been observed in different systems [32–34].

The values of EON* and $(S + A)^*$ are shown in Fig. 7.25 as a function of the alcohol carbon number. The alcohol-free system is characterized by EON* = 6.05. From this figure it can be seen that an alcohol having 3.3 carbon atoms requires the same EON* as the alcohol-free system. In this case both the surfactant and the alcohol have the same balance of hydrophilic/lipophilic interactions. The same conclusion can also be inferred from the results shown in Fig. 7.24, since an alcohol consisting of a mixture of propanol and butanol in proportions such that the average number of carbon is 3.3 gives a nonrotated phase map (see Sections 5.VII.C and 5.VII.H). This implies that the shape of the phase maps as a function of the alcohol chain length is worth noting. For the low-molecular-weight alcohols, the phase maps are rotated slightly in a direction opposite to that of heavy alcohols. The rotation of the phase maps increases as the alcohol carbon number is increased. The unbalanced character of alcohol is more pronounced with the long-chain compounds (favoring oil) than with the short-chain compounds (favoring water). For the alcohol-free system, no rotation is observed at the high surfactant concentrations used in the present case.

Based on the interpretation of the results shown in Fig. 7.24 given above, the alcohol having equal interaction energies with the O and W regions is the one that exhibits the greatest adsorption. That alcohol is the one having a chain length between three and four carbons. Figure 5.11 shows that the same alcohol was identified as being the equally balanced one based on an entirely different type of experiment.

The foregoing discussion can perhaps be better understood with the aid of a simple calculation based on Eq. (6.7). For a given total alcohol concentration, the fraction of alcohol in the C layer depends on the alcohol carbon number n, which can be qualitatively described by

$$x_a = \frac{Kn}{1 + Fn^2} \tag{7.9}$$

where K and F are constants. When n increases, x_a goes through a maximum and then tends to zero. This is qualitatively the behavior indicated by the trends shown by Fig. 7.25. The maximum occurs at n = 3.3. If the interaction of the alcohol with the oil is assumed proportional to n (Section 1.V.A), then

$$a_{co}^a = \beta n \tag{7.10}$$

where β is a constant. Replacing x_a and a_{co}^a in Eq. (6.7) using Eqs. (7.9) and (7.10) yields

$$EON^* = \frac{(a_{co}^s F + \beta K)n^2 - (a_{co}^s + a_{cw}^a)Kn + a_{co}^s}{\alpha(Fn^2 - Kn + 1)} \tag{7.11}$$

The results of a numerical calculation are reported in Fig. 7.25, using values of the parameters given in the figure caption.

Similarly, at optimum, the solubilization parameter can be assumed proportional to the numerator of R, namely to A_{co} given by Eq. (7.7). Therefore,

$$SP^* = g \frac{a_{co}^s}{S_a} \frac{x_a(\beta n - 1) + 1}{x_a(1 - p) + p} \tag{7.12}$$

where g is a constant and $p = S_s/S_a$. $(S + A)^*$ can then be approximated as

$$(S + A)^* = \frac{1}{2SP^* + 1} = \frac{x_a(1 - p) + p}{2g\left(\dfrac{a_{co}^s}{S_a}\right)[x_a(\beta n - 1) + 1] + x_a(1 - p) + p} \tag{7.13}$$

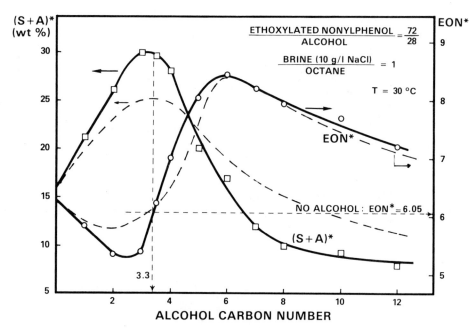

FIGURE 7.25 EON^* and $(S + A)^*$ as a function of alcohol carbon number.   The dotted curves are calculated according to Eqs. 7.11 and 7.13 above, where F = 0.035; K = 0.36; α = 1.26; β = 0.066; a_{cw}^a = 0.22; a_{co}^s = 7.6; g a_{co}^s/S_a = 5.25; p = 2.   (After Ref. 2.)

Numerical calculations are shown in Fig. 7.25.   Results similar to those reported in Fig. 7.24, and thus confirming the interpretations above, have been observed in different systems, where the variation of alcohol chain length was compensated by changing the surfactant chain length [32] or by changing the electrolyte concentration [2,6].

D.   Temperature

As discussed in Section 5.IV.B, for ionic amphiphiles increased temperatures generally result in slightly increased interactions (A_{cw} − A_{hh}), but the main effect is that the interactions per unit area of interface are decreased because the area occupied per surfactant molecule increases.   This is due primarily to a decrease in the hydrophobic effect described in Section 2.V, which reduces the adsorption driving force.   At sufficiently high temperatures the driving force greatly diminishes and solvency is correspondingly reduced.   As a consequence, the solubilization at optimum is expected to decrease.   Figure 7.26 provides an illustration.

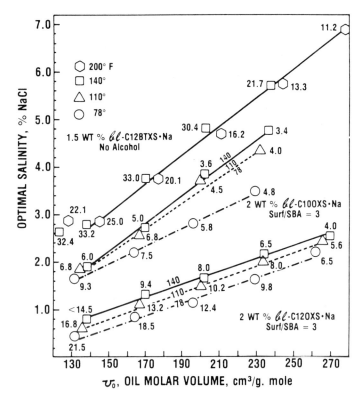

FIGURE 7.26   Effect of temperature on optimal salinity and optimal
solubilization parameter (indicated next to each point) for the series
of n-alkanes. bl-C12BTXSNa, bl-C10OXS Na, and bl-C12OXSNa are
alkylarylsulfonates. SBA stands for butanol-2. (After Ref. 19.)

The results obtained with the alcohol-free system shows that increasing the temperature from 140°F to 200°F does not change the
optimal salinity, but the solubilization parameter decreases substantially. For those systems containing alcohol, increasing the temperature yields, for a given alkane, large optimal salinities and substantial decreases in the solubilization parameter. Similar results were
reported earlier by Healy et al. [1].

In the presence of alcohol, Fig. 7.26 shows that at a given optimal salinity, compensating the increase in temperature by a decrease
in oil molar volume (i.e., by reducing the alkane carbon number)
also seems to result in decreased solubilization. The difference is
less pronounced than when an increase in temperature is compensated
by an increase in salinity since compensation is accomplished by increasing the cohesive energy in the numerator rather than decreasing

it in the denominator. The effect of temperature on optimal solubilization parameter is also evident in Fig. 6.3, where the hydrocarbon
used are the n̲-alkylbenzenes. The trends seen with alkyl benzenes
correspond to alkanes.

For nonionic surfactants, the main effect of temperature is to decrease the interaction between polyethylene oxide chain and water
(A_{cw}). However, the number of amphiphilic molecules per unit area
of interface (Section 5.IV.B) also decreases with a rise in temperature. Any compensation, whether hydrophilic or lipophilic, is thus
expected to yield a smaller solvency. The greatest decrease in solvency will be observed when an increase in temperature is compensated by reducing the interactions between the C layer and the O
region. In addition, the partitioning of ethoxylated surfactants in
favor of the O region may be enhanced by a rise in temperature, resulting in further decrease of the apparent solubilization parameter.

When compensation is achieved by increasing the interactions on
the aqueous side of the C region, decreased solubilization parameters
at optimum have also been reported [28,35,36]. A striking example
is shown in Fig. 7.27, where temperature is compensated by increasing the ethylene oxide number. Further illustration is provided by
Fig. 7.28, where temperature is compensated by reducing the electrolyte concentration at constant ethylene oxide number [35].

The excess oil phase has been analyzed to measure the surfactant
partitioning and the solubilization parameter has been calculated based
on that amount of surfactant which is present in the C layer. Despite this fact, the solubilization at optimum is seen to decrease,
even though the optimal salinity also decreases (as opposed to the
trend observed for anionic surfactants, Fig. 7.26) with increased
temperature. This must be related to the decrease in the number of
amphiphilic molecules per unit area of interface because of the decreased hydrophobic effect.

E.   Nature of the Polar Phase

Winsor [37] first reported that replacing the water with ethylene
glycol in a type II system (the surfactant being Aerosol OT) produces a type I system and proposed that this effect could be due
to the decrease in A_{ww} and/or in A_{cw} accompanying the change in
polar phase. In any event, if, starting from an optimized system,
such a change is compensated by modifying the surfactant head
(both by hydrophilic changes), the solubilizing power is expected
to stay constant provided, of course, that the number of surfactant
molecules per unit area of interface remains the same.

Figure 7.29 shows the results obtained when the surfactant is an
ethoxylated nonylphenol, the ethylene oxide number of which is varied to achieve the optimization [5] and when water is replaced by

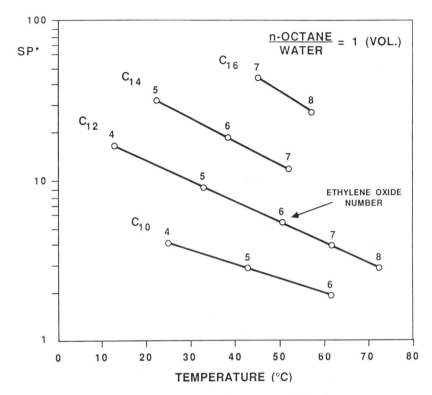

FIGURE 7.27 Effect of temperature on solubilization parameter at
optimum, when compensated by the ethylene oxide number. The
surfactants are pure monoisomeric ethoxylated normal alcohols. C_n
refers to the number of carbon atoms in alkyl chain. (After Ref.
36.)

a series of diols, namely, ethanediol (ethylene glycol or glycol), 1,3-
propanediol, and 1,4-butanediol, which are referred to as $C_2(OH)_2$,
$C_3(OH)_2$, and $C_4(OH)_2$, respectively.

When water is replaced by $C_2(OH)_2$, the optimal ethylene oxide
number EON^* increases slightly, whereas S^* increases markedly.
For $C_3(OH)_2$ and $C_4(OH)_2$, the expected trends are observed. EON^*
decreases to compensate, through A_{cw}, the decrease in A_{ww} and/or
in A_{cw}. Both changes being carried out on the same side of the interface, S^* remains constant.

The large increase in S^* when water is replaced by $C_2(OH)_2$ appears at first thought to be quite surprising since one might intui-
tively think that it would be easier to "mix" decane with glycol than

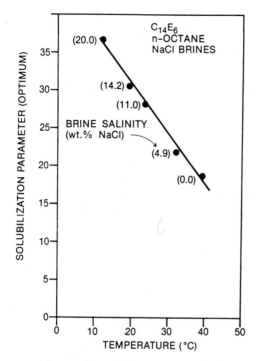

FIGURE 7.28 Effect of temperature on solubilization parameter at optimum, when compensated by salinity. $C_{14}E_6$ is the pure mono-isomeric tetradecanol containing 6 ethylene oxide units. (After Ref. 35.)

decane with water. Furthermore, analysis of the excess oil and gly-col phases of a type III system have shown that they contain very limited quantities of surfactant (<1%). Surfactant partitioning cannot thus explain the decrease in solubilization.

Solubilization is, as shown by Eq. (1.19), dependent on the sur-face packing of amphiphilic compounds and the primary driving force for surfactant adsorption is the hydrophobic effect. This is greatly reduced in glycols.

If the surface density of amphiphilic compounds is reduced, one would expect a corresponding decrease in lateral interaction between the molecules at the glycol/decane interface, which may explain why, according to Winsor [4], "in solubilized systems based on ethylene glycol, no liquid crystalline solutions have been found" (see Section 5.VI for further discussion).

As seen above, EON* increases slightly when water is replaced by ethylene glycol, which disagrees with Winsor's initial observations.

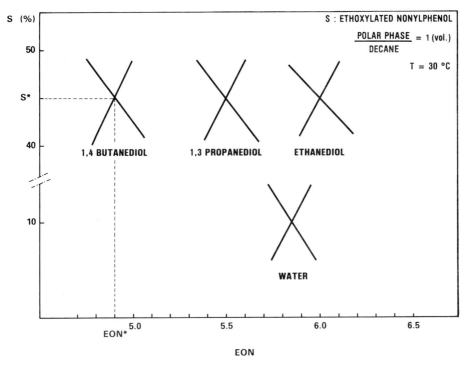

FIGURE 7.29   The effect on phase behavior of replacing water by
three diols (see text for notations).   (After Ref. 5.)

In fact, a change in the nature of the polar phase may affect various interaction terms of the R-ratio in a complex manner.   For example, it is likely that the interaction with the hydrophile A_{cw} is
smaller with glycol than with water.   This may compensate to some
extent for the decrease in A_{ww}.   As a consequence, an increase in
EON* may very well be observed.   (Different behavior could be observed with surfactants of different type: anionic, cationic, etc.).
    Comparing the changes that occur when $C_2(OH)_2$, $C_3(OH)_2$, and
$C_4(OH)_4$ constitute in turn the W region, it is seen that EON* decreases, indicating that A_{ww} (and/or A_{cw}) decreases in agreement
with Winsor's finding.   Since the change in the polar phase is compensated by a change in the hydrophile, S* is expected to remain
constant as long as the adsorption density does not change.   This
is the trend shown in Fig. 7.29.

F. Molecular Structure of the Oil

It is interesting to recall that the solubilization of n-alkanes is the
same irrespective of the molecular weight provided that the change
of one alkane for another is compensated by a lipophilic interaction
(Section 7.II.C). Increasing the surfactant lipophile to compensate
for an increase in ACN is an example of changes that were found to
leave the solubilization unchanged. This is a strong indication that
the area per amphiphile remains essentially constant over the range
of n-alkanes studied. This observation is in agreement with the
conclusions of Spegt et al. [38] drawn from x-ray diffraction studies of liquid-crystalline phases containing n-octadecane. On the
other hand when the n-octadecane is replaced by ethylbenzene, the
effective area per amphiphile increases with increased proportions of
ethylbenzene.

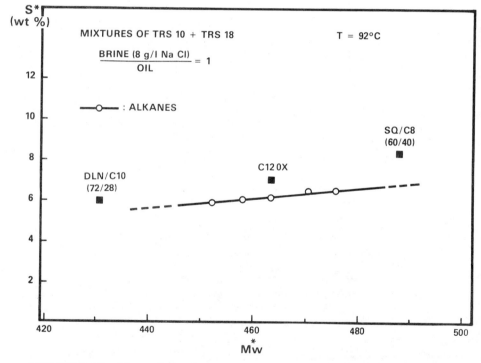

FIGURE 7.30 The effect of the type of hydrocarbon on solubilization
at optimum, in conditions of Fig. 7.7. DLN, SQ, C10, and C8 stand
for decalin, squalane, decane, and octane, respectively. Mixtures of
oils are in moles. C12OX is a commercial dodecyl orthoxylene, supplied by Esso Chimie. (After Ref. 16, replotted data.)

This single experiment indicates that the solubilization at optimum may very well depend on the molecular structure of the oil phase. It may be anticipated that the results found for n-alkanes depend on the ability of the members of this series to orient themselves perpendicular to the surface. Other structures may not be able to accommodate this type of packing. Figure 7.30 shows S^* for different oils, with the balance being affected by varying the average surfactant lipophile. This is done by blending two petroleum sulfonates together. Their average equivalent weight is designated by M_w^*. The S^* is seen to vary rather than remain constant as would be expected if the oils behaved like alkanes. This shows in a qualitative way that the solubilization may very well depend on the oil structure.

Similar results obtained using different surfactant, namely mixtures of synthetic orthoxylene sulfonates (sodium salt) C_nOXSNa with n = 12 and n = 24, are reported in Fig. 7.31. C_{12}OXSNa is synthesized by sulfonating the C_{12}OX oil. The behavior observed with the normal alkane series is also shown for comparison.

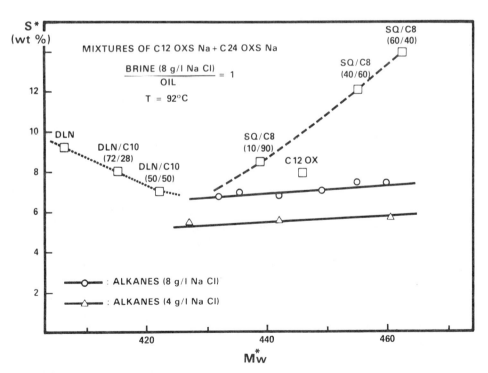

FIGURE 7.31 The effect of the type of hydrocarbon on solubilization at optimum in conditions of Fig. 6.8. See Fig. 7.30 for acronyms. (After Ref. 16, replotted data.)

From Figs. 7.30 and 7.31 it can be seen that contrary to the trends observed with the normal alkane series, S^* is found to depend on the nature of the oil. Furthermore, the difference between S^*(oil) and S^*(alkane) depends on which surfactant, petroleum sulfonate, or synthetic sulfonate is employed. Figures 7.30 and 7.31 show clearly that mixtures of oils cannot be fully modeled by a simple normal alkane (Section 6.II.F).

The difference between S^*(oil) and S^*(alkane) appears to be rather small when the surfactant is a petroleum sulfonate, but can be quite large using the C_nOXSNa sulfonates. A striking example is provided by the mixture of squalane and octane at the 60:40 molar ratio. For this oil the surfactant concentration at optimum is twice that observed with the corresponding normal alkane.

An interpretation of this result based on the R-ratio is proposed. A portion of the oil solubilized is located between the surfactant tails. Depending on the arrangement of oil molecules and surfactant lipophiles, the number Γ_s of amphiphilic molecules per unit area of interface may vary, yielding a change in solubilization. The results shown in Figs. 7.30 and 7.31 yield the conclusion that the TRS tails can more easily accommodate oil molecules than can the C_nOXSNa tails [16]. Thus for a given surfactant, when the solubilization of different types of oils is to be compared, the oil molar volume, which is indicative of the bulkiness of the molecule, may be used to help predict the relative behavior [19]. Figure 6.2 provides an example. At a given optimal salinity, the solubilization parameter decreases on passing from n-alkanes to alkylcyclohexanes to alkyl benzenes, that is, on increasing the oil molar volume.

VI. SOLUBILIZATION IN DILUTE MICELLAR SOLUTIONS

A. General

Solubilization of otherwise insoluble or sparingly soluble substances such as hydrocarbon in water with the aid of surface-active agents has long attracted interest [3,4,37,39−46]. A comprehensive review of the earlier work has been given by Klevens [47]. Most of this work has been carried out by measuring the amount of solute that can be added in an amphiphilic solution of given concentration before phase separation occurs, but often the nature of the phases precipitated at the saturation point was not identified.

Thus interpreting the results is not straightforward since different systems are not compared in similar states and, in addition, any change (other than overall increase in amphiphile concentration) which increases solubilization in type I systems has the inverse effect in type II systems.

As pointed out by Palit [48], the R-ratio provides a way of classifying the diversified phenomena encountered when considering sol-

ubilization and discussing them from a unified standpoint. In particular, a complete scan of a formulation variable may show a maximum
in the amount of solute solubilized, as shown in Fig. 7.32, where
the formulation variable is the nature of the nonpolar phase.

The paraffin-rich nonpolar phase corresponds to separation in a
type I system (R < 1), while the carbon tetrachloride−rich region
corresponds to a type II system (R > 1). A maximum apparent solubility of water is found in between these two regimes and is referred
to as R = 1. A number of similar results have been reported by Palit
et al. [48−50], allowing one to interpret the common observation that
solubilization by an amphiphile may be greater for mixed solvents
even when that amphiphile has little solubilizing power for either of
the individual solvents. In this case one of the solvents must yield
a type I system and the other a type II one.

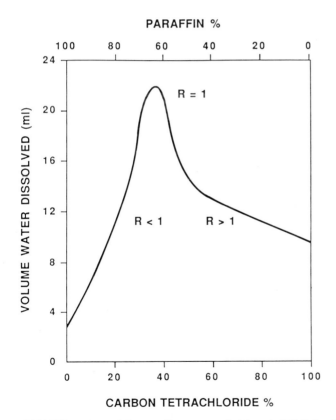

FIGURE 7.32 Volume of water soluble at 20°C in 100 ml of 25% (wt/
vol.) solutions of Aerosol OT (di-octyl sulfosuccinate) in carbon tet-
rachloride/paraffin mixtures. (After Ref. 46.)

The existence of a maximum in the solubilization curve such as that shown by Fig. 7.32 provides a reference state which allows a comparison of various systems in a meaningful way. Furthermore, this curve appears actually constituted by two demixing curves of a more complete phase map which would be obtained over the entire water/oil ratio range, as seen typically in Figs. 4.26 and 5.4. The maximum in solubilization is thus related to type III behavior, which justifies the value R = 1 indicated in Fig. 7.32 and establishes the relevance of considering the corresponding system at optimum.

B.   Complete Phase Maps

Shinoda and co-workers [51–57] have extensively investigated both the solubilization of hydrocarbon in aqueous surfactant solutions and the solubilization of water in organic surfactant solutions. They have

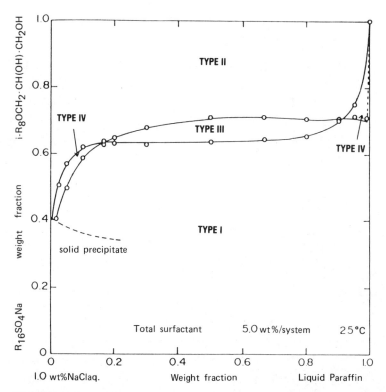

FIGURE 7.33   Phase diagram of 1.0% wt. NaCl aqueous/$C_{16}H_{33}SO_4Na$/ glycerol mono (2-ethylhexyl) ether/liquid paraffin at 25°C. The total surfactant concentration is 5% wt. (After Ref. 52.)

reported a number of phase maps such as those shown in Figs. 4.26 and 5.4. The type of system observed when the phase separation occurs is indicated on these diagrams using Winsor's nomenclature. The boundary of both single-phase regions (type IV) corresponding to the solubilization of oil in water and of water in oil exhibit a maximum. The single-phase regions are linked by a type III domain. In the examples provided by Figs. 4.26 and 5.4, the maximum solubilization of water in oil is seen to be less than the maximum solubilization of oil in water. Part of this difference is due to the greater partitioning of the nonionic surfactant into the oil (Section 5.VII.D). Thus the amount of amphiphile dissolved in the O region (and lost from the C layer) can be quite large at high oil content. As discussed in Section 5.II.D, the similarity between the overall features of the phase diagrams shown in Figs. 4.26 and 5.4 is notable. Another example is shown in Fig. 7.33 [52], where the variation of the balance of interactions of the C layer with O and W is achieved by mixing two different surfactants. Again, the same topology of the phase map is observed. Similar phase diagrams are also observed by varying the concentration of added salt [53], the hydrocarbon chain length of the oils [54], and so on.

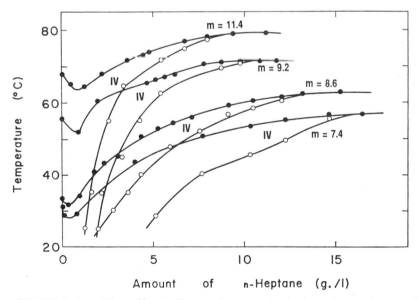

FIGURE 7.34 The effect of oxyethylene chain length on the solubilization of heptane in 1% aqueous solutions of polyoxyethylene (m) nonylphenyl ether. (After Ref. 54.)

C.   Solubilization of Hydrocarbons in Aqueous Surfactant Solutions

The effect of ethylene oxide number m on solubilization of heptane in water is shown by Fig. 7.34.   For a given surfactant the open circles represent the demarcation between type I and type IV systems, whereas the full circles divide the type II and type IV systems.   The curves shown in Fig. 7.34 are segments of complete phase diagrams such as those represented by Fig. 7.32.

When m increases, the optimal temperature for maximum solubilization increases and the amount of heptane that can be dissolved at optimum decreases.   Both of these observations are in agreement with the behavior reported above (Section 7.V.D).

Other parameters, such as the type of hydrocarbon [54], the electrolyte concentration [54], the surfactant lipophile length, the surfactant lipophile structure [55], and the concentration of cosurfactant [55,56], have been investigated in similar conditions.

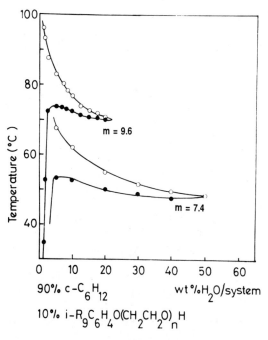

FIGURE 7.35   The solubilization of water in cyclohexane containing 10% wt polyoxyethylene (7.4 and 9.6) nonylphenyl ethers.   (After Ref. 57.)

D. Solubilization of Water in Hydrocarbon Surfactant Solutions

Similar to Fig. 7.34, the effect of the ethylene oxide number on the solubilization of water in a cyclohexane solution of amphiphile is shown in Fig. 7.35 [57].   Again the optimal temperature increases with the ethylene oxide number, and the solubilization decreases, as expected.

The solubilization of brine of various electrolyte concentrations in different oils belonging to the series n-alkanes is illustrated in Fig. 7.36 [58].   Here the surfactant used is anionic and changes in ACN can readily be compensated by a change in the salinity.   Note that the titration of alkane hydrocarbon solubility at various salt concen-

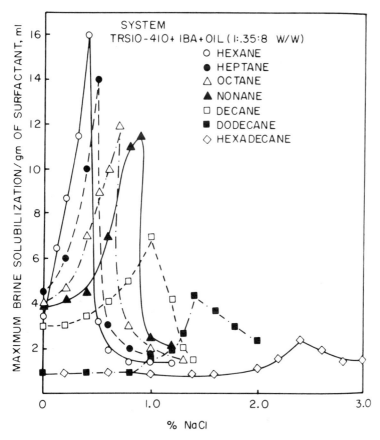

FIGURE 7.36   The brine solubilization capacity of n-alkane solutions of surfactant.   TRS 10-410 is a petroleum sulfonate supplied by Witco. IBA stands for isobutanol.   (After Ref. 58.)

trations reaches a maximum. While the authors did not report the nature of the phases that precipitate above the demixing curve, it seems likely that the maximum solubilization is obtained when the system is optimal. Thus for salinities less than the optimum, one expects to observe type I systems and for salinities greater than the optimum, type II systems are likely to exist. Here the optimum salinity is defined as that value yielding the maximum of solubilization.

The conjecture that the maximum corresponds to a type III system is strengthened by noting that as the ACN is increased, the optimum salinity increases correspondingly. This corresponds to the results expressed by Eq. (6.3), which closely predicts the increase of optimum salinity with increasing ACN shown by Fig. 7.36.

The decrease in solubilization at maximum shown in Fig. 7.36 is expected, since an increase in ACN which reduces the numerator of the R-ratio is compensated by an increase in salinity which reduces its numerator. Thus the results shown are in full accord with the expected trends.

It is perhaps worthwhile to stress the value of comparing systems at optimum when assessing surfactant solvency. For example, if all of the systems shown in Fig. 7.36 are compared at 1 wt % NaCl, it will be concluded that the solvency is rather poor but favors decane. Thus the use of this surfactant in applications that require higher solvency for heptane might be rejected without really knowing its full potential.

REFERENCES

1. R. N. Healy, R. L. Reed, and D. G. Stenmark, Soc. Pet. Eng. J., 16:147 (1976).

2. M. Bourrel and C. Chambu, Soc. Pet. Eng. J., 23:327 (1983).

3. J. W. McBain, Advances in Colloid Science, Vol. 1, Interscience, New York (1942).

4. P. A. Winsor, Solvent Properties of Amphiphilic Compounds, Butterworth, London (1954).

5. F. Verzaro, M. Bourrel, and C. Chambu, in Surfactants in Solution (K. L. Mittal and P. Bothorel, eds.), Vol. 6, p. 1137, Plenum Press, New York (1986).

6. M. C. Puerto and W. W. Gale, Soc. Pet. Eng. J., 17:193 (1977).

7. R. L. Reed and R. N. Healy, in Improved Oil Recovery by Surfactant and Polymer Flooding (D. O. Shah and R. S. Schechter, eds.), Academic Press, New York (1977).

8. W. C. Hsieh and D. O. Shah, paper SPE 6594 presented at the
    SPE Symposium on Oilfield and Geothermal Chemistry, La Jolla,
    Calif. (June 1977).

9. M. Bourrel, C. Chambu, R. S. Schechter, and W. H. Wade,
    Soc. Pet. Eng. J., 22:28 (1982).

10. Y. Barakat, L. N. Fortney, R. S. Schechter, W. H. Wade, and
    S. Yiv, Proceedings Second European Symposium on EOR, Tech-
    nip, Paris, p. 11 (Nov. 1982).

11. I. Carmona, R. S. Schechter, W. H. Wade, and U. Weerasooriya,
    paper SPE 11771 presented at the SPE Symposium on Oilfield and
    Geothermal Chemistry, Denver (June 1983).

12. M. C. Puerto and R. L. Reed, paper SPE 10678 presented at
    the 3rd SPE/DOE Symposium on Enhanced Oil Recovery, Tulsa
    (Apr. 1982).

13. M. Bourrel, D. Bernard, and A. Graciaa, Tenside Deterg.,
    21:311 (1984).

14. F. D. Martin and J. C. Oxley, paper SPE 13575 presented at
    the SPE Symposium on Oilfield and Geothermal Chemistry,
    Phoenix (Apr. 1985).

15. R. E. Salager and J. L. Salager, personal communication (1987).

16. M. Bourrel, F. Verzaro, and C. Chambu, SPE Reservoir Eng.,
    2:41 (1987).

17. Y. Barakat, L. N. Fortney, R. S. Schechter, W. H. Wade, S.
    Yiv, and A. Graciaa, J. Colloid Interface Sci., 92:561 (1983).

18. A. Graciaa, L. N. Fortney, R. S. Schechter, W. H. Wade, and
    S. Yiv, Soc. Pet. Eng. J., 22:743 (1982).

19. M. C. Puerto and R. L. Reed, Soc. Pet. Eng. J., 23:669 (1983).

20. J. L. Salager, M. Bourrel, R. S. Schechter, and W. H. Wade,
    Soc. Pet. Eng. J., 19:271 (1979).

21. R. C. Nelson, Soc. Pet. Eng. J., 23:501 (1983).

22. H. Kunieda and K. Shinoda, Bull. Chem. Soc. Jpn., 55:1777
    (1982).

23. M. Bourrel, C. Chambu, and F. Verzaro, Proceedings Second
    European Symposium on EOR, Technip, Paris, p. 39 (Nov. 1982).

24. E. H. Crook, D. B. Fordyce, and G. F. Trebbi, J. Phys.
    Chem., 67:1987 (1963).

25. P. K. Kilpatrick, C. A. Gorman, H. T. Davis, L. E. Scriven,
    and W. G. Miller, to be submitted to J. Phys. Chem.

26. P. K. Kilpatrick, H. T. Davis, W. G. Miller and L. E. Scriven, submitted for publication in J. Colloid Interface Sci. (cited in Ref. 25; not yet available).

27. I. Osborne-Lee, W. H. Wade, and R. S. Schechter, J. Colloid Interface Sci., 108:60 (1985).

28. L. N. Fortney, "Criteria for Structuring Surfactants to Maximize Solubilization of Oil and Water," M.A. thesis, The University of Texas at Austin (1981).

29. S. J. Salter, paper SPE 14106 presented at the SPE Meeting in People's Republic of China (1986).

30. S. J. Salter, paper SPE 6843 presented at the 52nd SPE Annual Technical Conference and Exhibition, Denver (Oct. 1977).

31. C. Lalanne-Cassou, I. Carmona, L. Fortney, A. Samii, R. S. Schechter, W. H. Wade, U. Weerasooriya, V. Weerasooriya, and S. Yiv, paper SPE 12035 presented at the 58th SPE Annual Technical Conference and Exhibition, San Francisco (Oct. 1983).

32. M. Bourrel, P. Gard, F. Verzaro, and J. Biais, in Interactions Solide-Liquide dans les Milieux Poreux, Collection Colloques et Séminaires No. 42, Technip, Paris, p. 303 (1985).

33. M. Baviere, R. S. Schechter, and W. H. Wade, J. Colloid Interface Sci., 81:266 (1981).

34. M. Baviere, Compte Rendu de Recherche, No. 26928, DGRST, Paris (1979).

35. L. A. Verkruyse and S. J. Salter, paper SPE 13574 presented at the SPE Symposium on Oilfield and Geothermal Chemistry, Phoenix (Apr. 1985).

36. S. J. Salter and L. A. Verkruyse, "Phase Behavior of Nonionic Surfactants," to be submitted.

37. P. A. Winsor, Trans. Faraday Soc., 44:451 (1948).

38. P. A. Spegt, A. E. Skoulios, and V. Luzzati, Acta Crystallogr., 14:866 (1961).

39. S. R. Palit, J. Colloid Sci., 44:455 (1949).

40. C. Engler and E. Dickoff, Arch. Parm., 230:561 (1892).

41. S. U. Pickering, J. Chem. Soc., 111:86 (1917).

42. E. Lester-Smith, J. Phys. Chem., 36:1401 (1932).

43. R. C. Pink, J. Chem. Soc., 53 (1939).

44. S. R. Palit and J. W. McBain, J. Soc. Chem. Ind. London, 66:3 (1947).

45. P. H. Richards and J. W. McBain, J. Am. Chem. Soc., 70:1338 (1948).

46. P. A. Winsor, Trans. Faraday Soc., 44:396 (1948).

47. H. B. Klevens, Chem. Rev., 47:1 (1950).

48. S. R. Palit and V. Venkateswarlu, Proc. R. Soc. London, A208:542 (1951).

49. S. R. Palit and V. Venkateswarlu, J. Chem. Soc., 2129 (1954).

50. S. R. Palit, V. A. Moghe, and B. Biswas, Trans. Faraday Soc., 55:463 (1959).

51. K. Shinoda and H. Kunieda, J. Colloid Interface Sci., 42:381 (1973).

52. K. Shinoda, H. Kunieda, T. Arai, and K. Saito, J. Phys. Chem., 88:5126 (1984).

53. K. Shinoda and H. Takeda, J. Colloid Interface Sci., 32:642 (1970).

54. H. Saito and K. Shinoda, J. Colloid Interface Sci., 24:10 (1967).

55. H. Sagitani, T. Suzuki, M. Nagai, and K. Shinoda, J. Colloid Interface Sci., 87:11 (1982).

56. K. Shinoda and H. Kunieda, in Microemulsions: Theory and Practice (L. M. Prince, ed.), Academic Press, New York (1977).

57. K. Shinoda and S. Friberg, Adv. Colloid Interface Sci., 4:281 (1975).

58. V. K. Pithapurwala and D. O. Shah, J. Am. Oil Chem. Soc., 61:1399 (1984).

8

Thermodynamics of Solubilized Systems

I. PREVIEW

The main thrust of this chapter is to better understand the R-ratio in a thermodynamic context. Research along these lines is continuing, although considerable progress has been recorded. It seems likely that future developments will provide new formulas which specifically relate the R-ratio to the molecular structure of all of the components composing a micellar solution. At the present time such equations do not exist. It is believed, however, that the fundamental framework established in this chapter will provide the basis for these anticipated results. Here the work of Gibbs is crucial and a connection between his conjecture that a free energy of bending an interface might, in some systems, be important in determining the state of thermodynamic equilibrium and the R-ratio is suggested. The behavior of micellar solutions cannot be modeled successfully without including a bending energy (or equivalently, the R-ratio). It is seen that without a term which imparts a preferred curvature to the C layer, micellar solutions are predicted to have an intrinsic symmetry which does not correspond to experiment.

It is shown, for example, that without the presence of a term akin to the R-ratio, the inversion of a micellar solution composed of a dispersion of submicroscopic drops always takes place near the point where the solution contains equal quantities of oil and water irrespective of the molecular structure of the surfactant or the oil, whether or not salt is added or whatever the temperature. This contradicts experimental results presented in Chaps. 5 and 6 showing that these variables exert a profound influence on the phase behavior. Therefore, any feasible thermodynamic theory must incorporate a symmetry-breaking bending energy, and of course a bending energy is the crux of the R-ratio.

The thermodynamic analysis also provides insight as to why the inversion from water to oil continuity as the oil/water ratio increases may take place even though the natural curvature (R-ratio) of the C layer favors the existence of oil drops. This question arose in Chap. 5 when it was seen that as the amount of water relative to the amount of oil is reduced, a micellar solution may revert from a water-continuous, S_1 system to an oil-continuous, S_2 system despite the fact that the R-ratio is less than unity and remains unchanged as the water/oil ratio is varied. It was noted that this transition is driven primarily by entropy. This point of view is justified in this chapter and it is shown that there exists a natural curvature which may differ, perhaps substantially, from that actual curvature which minimizes the free energy of the entire system.

While most of the discussion presented in this chapter considers the micellar solution to consist of a dispersion of spherical drops, other more phenomenological representations of the free energy must also necessarily include terms that contain the same information as given by the R-ratio if these representations are expected to model the phase behavior of micellar solutions. To make this clear, a model based on regular solution theory is examined and the symmetry breaking terms identified. These terms can, in a qualitative sense, be related directly to the R-ratio. This is interesting in that it shows that a quantity resembling the R-ratio must also be included in those empirical free-energy formulations intended to embody the entire phase diagram. The role of these R-ratio-type functions is to provide a means of forcing a I-III-II transition, that is, of causing the phase diagram to shift from a symmetric one (type III system) to others which are nonsymmetric. This is precisely the same role played by the R-ratio in the droplet models. Thus the R-ratio is an essential concept. It is not merely a handy qualitative tool which can be used to help visualize the trends that will occur when one of the formulation variables is changed, but it is a quantity that must appear in some way, perhaps disguised, in any feasible thermodynamic theory of micellar solutions.

The R-ratio or, equivalently, the bending energy is but one contributor to the free energy of a micellar system. As we shall see, other factors also influence the thermodynamic behavior. These include interfacial tension, the free energy of a dispersion of drops, adsorption into the C layer, and the water/oil ratio. These factors all play an important role governing the phase behavior of micellar systems, but to formulate their relationship a model is required.

Whenever the conjugate phases that form are very nearly critical, certain features of their behavior can be described without using a

model. This possibility arises because the concentration fluctuations
in near critical systems, that is, those on the verge of separating
into two phases having very nearly the same overall compositions,
are relatively large and the correlation lengths long ranged. In this
state molecular interactions are secondary and there are many properties common to all near-critical systems. In the next section these
general properties are considered and their relationship to microemulsion phase behavior noted.

II. ROLE OF CRITICAL PHENOMENA

A. Critical Scaling Laws: A Brief Review

In Chap. 4 it is noted that a three-phase region often originates at
a critical tie line which broadens into a tie triangle. The two phases
that evolve from one end of the critical tie line, as for example, the
lower and middle phases, have almost the same composition. These
represent near-critical states. Furthermore, under conditions for
which the phase compositions differ substantially, micellar solutions
still exhibit properties characteristic of near-critical behavior. This
important and remarkable behavior of micellar solutions occurs because
the chemical potentials of the constituent parts are not very sensitive
to the overall composition. Thus even if the proportions of water to
oil change substantially, the chemical potentials of individual water
and oil molecules do not experience large differences because micellar
solutions are composed of submicroscopic regions of water separated
from oil by the C layer. Since the fluid in these regions often closely
resemble bulk water or oil, their chemical potentials are slowly varying
functions of the concentration. Because of this, micellar phases apparently remote from critical compositions still display near-critical
properties. It is for this reason that critical behavior plays an important role in determining the properties of micellar solutions, and
its study will contribute to our understanding of them.
        Near-zero interfacial tension is one of the most notable [1,2]; how-
ever, there are a number of other properties which are also associated
with mechanisms characteristic of near-critical systems [3-7]. It is,
therefore, of interest to consider briefly this important subject.
        The theory of critical phenomena has undergone a spectacular de-
velopment over the past two decades. A description of the behavior
of thermodynamic properties in the neighborhood of the critical point
which includes the important contributions arising from the fluctuations has evolved [8]. This description identifies the similarity of
physical behavior in the critical region in terms of scaling theory [8],
which applies whenever the length scale of correlations is much greater
than the range of molecular interactions. Scaling theory effectively
relates physical properties of a system to the "distance" from a critical
point in terms of simple, nonintegral power laws. In this section a

brief review of scaling theory is provided and discrepancies between the classical theory and theories that take into account thermodynamic fluctuations are noted.

It is instructive to consider the behavior of a single-component system near the vapor-liquid critical point. If the Helmholtz free energy, the pressure, and the chemical potential are all assumed to be analytical functions in the neighborhood of the critical, they can be expanded into a Taylor series in terms of the canonical variables ρ and T about the critical density, ρ_c, and the critical temperature, T_c. Thus the Helmholtz free energy per unit of volume can be written as [9]

$$f = \sum \sum \frac{1}{m!n!} f_{mn}^c \, \delta\rho^m \, \delta T^n \qquad (8.1)$$

where $\delta\rho$ is $\rho - \rho_c$, δT is $T - T_c$, and

$$f_{mn}^c = \left(\frac{\partial^{m+n} f}{\partial^m \rho \, \partial^n T} \right)_c$$

Since for a single-component system

$$df = \mu d\rho - s \, dT \qquad (8.2)$$

it is seen that

$$f_{1,0} = \mu \qquad f_{2,0} = \frac{1}{\rho}\left(\frac{\partial p}{\partial \rho} \right)_T \qquad f_{3,0} = \frac{\partial}{\partial \rho}\left(\frac{1}{\rho} \, \frac{\partial p}{\partial \rho} \right)_T$$

Since at the critical point [9]

$$\frac{\partial p}{\partial \rho} = \frac{\partial^2 p}{\partial \rho^2} = 0 \quad \text{and} \quad \frac{\partial^3 p}{\partial \rho^3} > 0 \qquad (8.3)$$

then $f_{2,0}^c = f_{3,0}^c = 0$ and $f_{4,0}^c > 0$.

Because $\mu = \partial f / \partial \rho$, then by differentiating Eq. (8.1) the leading terms of the expansion are seen to be

$$\mu - \mu^c = \mu_{0,1}^c \delta T + \mu_{1,1}^c \, \delta\rho \, \delta T + \frac{1}{6} \mu_{3,0}^c (\delta\rho)^3 \qquad (8.4)$$

using the fact that $\mu_{1,0}^c = f_{2,0}^c$ and $\mu_{2,0}^c = f_{3,0}^c$ are both zero. Two equations of this kind can be written. The chemical potential of the liquid μ^L can be expresses in terms of $\delta\rho^L$ and δT, whereas the chemical potential of the vapor μ^V can be expanded in powers of $\delta\rho^V$ and δT. Since along the coexistence curve $\mu^L = \mu^V$, the two equations can be subtracted to yield [9]

$$\mu^c_{1,1}\,\delta T + \frac{1}{6}\,\mu^c_{3,0}[(\delta\rho^L)^2 + (\delta\rho^L\delta\rho^V) + (\delta\rho^V)^2] = 0 \tag{8.5}$$

From Eq. (8.5) it follows that $\delta\rho^L$ and $\delta\rho^V$ are of leading order $|\delta T|^{1/2}$. Thus the assumption of analyticity of the Helmoholtz free energy, the density and the chemical potential at the critical point, and the classical conditions expressed by Eq. (8.3) leads to [9]

$$\delta\rho^{L\ or\ V} = B^{\pm}_1\,|\delta T|^{1/2} + B^{\pm}_2\,|\delta T| + \cdots \tag{8.6}$$

where $B^{\pm}_1$ are system-dependent coefficients, with the plus sign denoting that coefficient which corresponds to the liquid density on the coexistence curve and the negative sign with the vapor density.

From Eq. (8.6) it is seen that

$$\frac{1}{2}\,(\delta\rho^L + \delta\rho^V) = \rho^c + B_1|\delta T|^\beta \tag{8.7}$$

and

$$\frac{1}{2}\,(\delta\rho^L - \delta\rho^V) = B_2|\delta T|^{1-2\alpha} \tag{8.8}$$

where from the classical results as expressed by Eq. (8.6) (Taylor series expansion), $\beta = 1/2$ and $\alpha = 0$.

The constants B_1, B_2, μ^c_{10}, and so on, all depend on the properties of the particular system under consideration, but β and α have the same values for all substances; that is, they are universal. They arise as exponents in the Taylor series expansion and should therefore apply to all substances.

Experiments show, however, the values $\alpha = 0$ and $\beta = 1/2$ which follow from classical thermodynamics to be incorrect. This discrepancy, which is significant, arises because near a critical point density fluctuations in each phase become large since there is very little free-energy difference between the phases. When two phases can transform back and forth spontaneously, they can no longer in fact be considered distinct phases. Classical thermodynamics does not consider fluctuations. Classical thermodynamics, for example, insists that the entropy of a closed system always increase. Fluctuations can in fact lead to a decrease in the entropy of a closed system. Thus to represent systems near their critical point adequately, statistical mechanics is a necessary starting point [8]. Estimates for α and β which include fluctuations are [1] $\alpha = 0.110 \pm 0.0045$ and $\beta = 0.325 \pm$

0.0015. These nonintegral, nonclassical values are seen to differ substantially from their classical counterpart; however, once fluctuations are included, it is assumed that these values should apply to all systems, including microemulsions.

B. Phase Behavior of Microemulsions

Fleming and Vinateri [4,5] have shown that the shape of the phase volume diagrams can be described in terms of critical scaling theory. A multicomponent system in a two-phase region near a critical point is considered. A path in composition space starting in the two-phase region terminating in the single-phase region and at constant temperature, pressure, and $N - 3$ of the N component chemical potentials is examined. Such a path admits only one degree of freedom, and this can be associated with a single chemical potential, μ. Two mole fractions C_1 and C_2 on the coexistence curve are considered. The third, C_3, is determined so that the chemical potentials of each of the remaining $N - 3$ components remain constant. Let C_1, C_2 represent a point on one branch of the coexistence curve and $\tilde{C}_1$, $\tilde{C}_2$ represent a point on the other branch conjugate to the first point. Figure 8.1 depicts the two-phase region showing C_1, C_2 and $\tilde{C}_1$, $\tilde{C}_2$ as endpoints of a tie line. It is known that for liquid-liquid systems, a concentration difference between the critical value and one on the coexistence curve is equivalent (in terms of scaling) to a density difference, and furthermore, that it is possible to replace one field variable by another [5]. A field variable is defined as one that takes on the same value in both phases at equilibrium. Thus temperature, pressure, and chemical potential are all examples of field variables.

Based on these principles it is possible to write the following [5] equations:

$$C_1 - \tilde{C}_1 = B_1 (\delta \mu^*)^\beta \tag{8.9}$$

and

$$C_2 - \tilde{C}_2 = B_2 (\delta \mu^*)^\beta \tag{8.10}$$

where $\delta \mu^* = (\mu - \mu^c)/ \mu^c$ and β is the nonintegral universal power discussed above. Here the liquid and vapor densities on the coexistence curve for a vapor-liquid system have been replaced by the variable mole fractions (C_1, C_2) and $(\tilde{C}_1, \tilde{C}_2)$, which are points on the coexistence curves for the multicomponent liquid-liquid system. δT^* has been replaced by $\delta \mu^*$ since both are field variables.

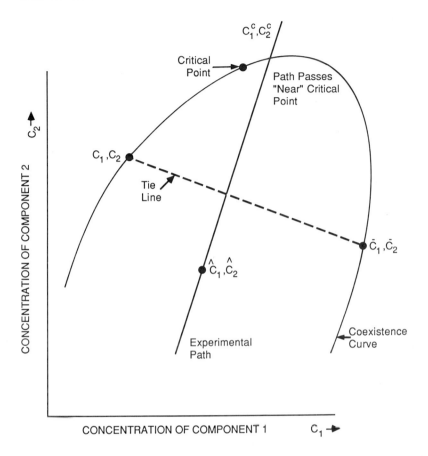

FIGURE 8.1 Schematic representation of two-dimensional composition space showing coexistence curve and experimental path for phase volume studies.

Similarly, Eq. (8.9) suggests that

$$C_1 + \tilde{C}_1 = 2C_1^c + 2A_1(\delta\mu^*)^{1-\alpha} \tag{8.11}$$

and

$$C_1 + \tilde{C}_2 = 2C_2^c + 2A_2(\delta\mu^*)^{1-\alpha} \tag{8.12}$$

where α is the critical exponent of the specific heat.

These four equations [Eqs. (8.9) to (8.12)] provide the input from critical theory. The constants B_1, B_2, A_1, and A_2 are system dependent and must be determined from experiment; however, α and β are universal constants. It is now necessary in the development of a relationship for the phase volumes to consider the way in which the experiment is to be conducted. There are, of course, an infinite number of paths along which the phase volumes can be measured. Equations (8.9) to (8.12) will apply along each of these paths as long as they are in the neighborhood of the critical point. Fleming and Vinatieri [5] hypothesize that the phase volume along a salinity scan can be considered to be equivalent to that found along the linear path shown in Fig. 8.1. The overall compositions for a sequence of increasing salinities then satisfy

$$\hat{C}_1 - C_1^c = \delta\hat{C}_1 + M(\hat{C}_2 - C_2^c) \tag{8.13}$$

M and $\delta\hat{C}_1$ are viewed as parameters which are fixed for a particular path, such as the one depicted by Fig. 8.2.

If the experimental path intersects a tie line as shown by Fig. 8.1, then the fraction of the phase F having compositions C_1 and C_2 is given by simple material balance as

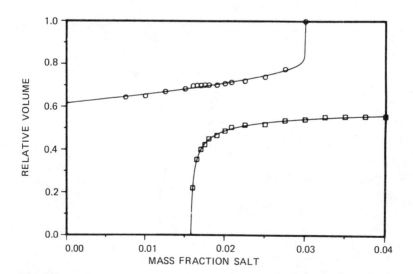

FIGURE 8.2 Phase−volume diagram for the system with 1-phenyltetra-decane. The surfactant consists of an extract of a petroleum sulfo-nate (Witco TRS-410), isobutyl alcohol and sodium chloride (Ref. 5).

$$F = \frac{\tilde{C}_1 - \hat{C}_1}{\tilde{C}_1 - C_1} \qquad (8.14)$$

In general the experiment involves measurement of F as a function of $\hat{C}_1 = \hat{C}$. It is these data which are to be correlated. It is possible to eliminate variables and obtain F as a function of $\hat{C}$ in the following form [4,5]:

$$F = F_0 \left\{ 1 - \left| \frac{\bar{C}^0 - C^0}{\hat{C} - C^0} \right|^{\beta/(1-\alpha)} [1 + \lambda(\hat{C} - C^0)] \right\} \qquad (8.15)$$

Here F is taken to be the volume fraction of the disappearing phase along the linear path described by the overall composition $\hat{C}$. In this expression F_0, C^0, $\bar{C}^0$, and λ are system-dependent quantities to be determined by curve fitting, α and β are the critical coefficients, and $\hat{C}$ represents the independent variable. Figure 8.2 shows a typical phase volume diagram. The diagram has two branches since there is a three-phase region for intermediate salinities. The solid curve represents a least-squares fit of Eq. (8.15) to the top phase volume with $F_0 = 0.206$, $C^0 = 0.003$, $C^0 - \bar{C}^0 = 1.5 \times 10^{-6}$, and $\lambda = 1023$ [5]. The fit to the bottom phase volume was obtained with $F_0 = 0.657$, $C^0 = 0.01571$, $C^0 - \bar{C}^0 = 0.01580$, and $\lambda = 3.95$ [5]. The physical interpretation of this diagram is that the system passes "near" an oil-microemulsion critical endpoint (two phases in the presence of a third) near 3 wt % salt and also near an aqueous-microemulsion critical endpoint near 1.58 wt % salt. The fit of the volume fraction data was obtained using the nonclassical values for β and α which are estimated to be 0.325 and 0.11, respectively.

C. Fluctuations

Micellar systems do exhibit unmistakable critical behavior [10–13] detectable by light-scattering experiments. Huang and Kim [13] have used photon correlation spectroscopy [14] to infer the correlation length in a microemulsion at a point near its critical. This procedure involves measurement of the intensity of light having a particular wavelength λ_ℓ at a particular scattering angle θ as a function of time. The correlation function $Cr(\tau)$ is defined as

$$Cr(\tau) = \lim_{T \to \infty} \frac{1}{T} \int_0^T I(t)I(t + \tau) \, dt \qquad (8.16)$$

This correlation function, which is discussed in Section 4.III.D, is calculated directly from the scattered light intensities, $I(t)$, measured

as a function of time. Since droplet motion is correlated over a time
small compared to a characteristic fluctuation time, the product of
the intensities is large for small values of τ, but at long times all
correlation is lost and Cr tends to approach the square of the average intensity.

The interpretation of Cr in terms of a correlation length which
plays a prominent role in critical phenomena requires the introduction
of a model. If, for example, the scattering centers are assumed to
be monodisperse and to have random movements, $C_r(\tau)$ can be expressed by an exponential function [13]

$$Cr = Cr^0 \exp(-2\tau\Gamma) + \langle I \rangle^2 \tag{8.17}$$

The characteristic decay time, $1/\Gamma$, is the average duration that the
scatters (regions of different composition) maintain a given spatial
configuration before being completely altered by Brownian motion.
Therefore, a Brownian diffusion coefficient can be inferred, and using
the Stokes-Einstein equation an effective hydrodynamic radius, ξ, can
in turn be calculated [13]. This hydrodynamic radius will be taken
to be the correlation length, which should scale as

$$\xi = \xi_0 |\delta\mu^*|^{-\nu} \tag{8.18}$$

or

$$\xi = \xi_0' |\delta T^*|^{-\nu} \tag{8.19}$$

with the value $\nu = 5/8$ [5]. Huang and Kim [13] found that

$$\xi = \xi_0'' |\delta T^*|^{-0.75} \tag{8.20}$$

which is very nearly the classical exponent. It is interesting that by
considering the alkane carbon number (ACN) to be a field variable,
they were also able to show that exactly the same exponent applies,
that is,

$$\xi = \xi_0''' |\delta ACN^*|^{-0.75} \tag{8.21}$$

where δACN^* has an obvious definition.

Experiments carried out by Dorshow et al. [15] yield a value of
ν which is much larger (≈ 1.13), but the reason for this deviation is
not understood.

The scattering experiments all indicate that the correlation length
becomes large as the microemulsion system approaches a critical point.
These critical points are found at compositions consistent with those
obtained from phase maps similar to those shown by Fig. 8.2.

D. Interfacial Tension

As the thickness of an interface increases, the interfacial tension
tends to vanish. Furthermore, the thickness of the interface and
hence the interfacial tension are related to the correlation length.
Scaling theory [16,17] yields

$$\gamma = \gamma_0 |\delta\mu^*|^{2\nu} = \gamma_0' |\delta C^*|^{2\nu/(1-2\alpha)} \tag{8.22}$$

Chatenay et al. [18] have reported interfacial tensions measured to
very small values (ca. 10^{-4} dyn/cm) and found reasonable agreement
between the universal scaling laws (based on density rather than concentration difference). Fleming and Vinatieri [5] have correlated interfacial measurements for the same system, shown in Fig. 8.2, and
found that Eq. (8.22) gives a good fit of the data. The comparison
is shown by Fig. 8.3. The solid lines are the best fit of Eq. (8.22)
to the interfacial tension measured on the two branches.

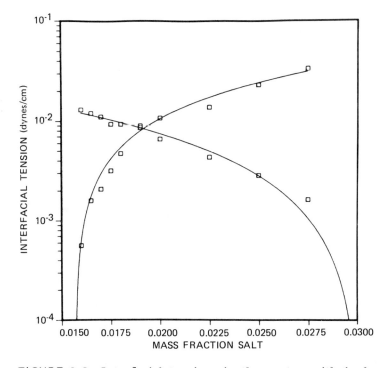

FIGURE 8.3 Interfacial tensions in the system with 1-phenyltetrade-
cane. Same surfactant system as Fig. 8.2 (Ref. 5).

The value of γ_0' required to fit the oil-microemulsion interfacial tensions of Fig. 8.3 is almost an order of magnitude smaller than the one required to fit the water-microemulsion tension data. This may imply that the surfactant structure is more active at the oil/microemulsion interface than at the water/microemulsion interface. This viewpoint has not been pursued to ascertain, for example, that surfactants having the same values of γ_0' at the two interfaces are in some sense better balanced.

III. THE THERMODYNAMIC DROPLET MODEL

A. General

The theory of critical phenomena provides valuable insight into the reasons for the high degree of universality assigned to some of the purely empirical correlations displayed in Chap. 4. The remarkable relationship between phase volume and interfacial tension is one example of such a correlation [19]. Unfortunately, those system-dependent parameters which appear in the equations that result from a consideration of critical phenomena cannot now be calculated because the theory has not progressed to the point where the properties of a micellar solution near a critical point are predictable.

Presently, the system-dependent properties must be measured. They cannot be calculated. Thus to gain insight into the relationship between the micellar structure and the physical properties of a micellar solution, it is necessary to attempt to construct models of the micellar solution and then to explore the thermodynamic consequences of the model. A number of attempts have been made to develop such models. Many of these will be discussed here. Furthermore, the primary goal will be to relate qualitatively the R-ratio to quantities that have a thermodynamic interpretation. As noted above, it is not now possible to cite specific formulas for the R-ratio.

Most thermodynamic analyses of micellar solutions have as their starting point a model system consisting of large submicroscopic water-like and oil-like regions separated by an interfacial region. The shapes of these regions have been taken to be spherical, polyhedral, or planar. In some cases the sizes of the regions have been permitted to vary, while in other studies more or less homogeneous-sized regions (i.e., constant drop sizes) have been considered. These models are all approximations. What is known is that the solubilized micellar systems are often isotropic, unlike liquid-crystalline systems, which are structured at the molecular level and require a different thermodynamic treatment. In this section we are concerned with the isotropic solubilized micellar systems and will focus primarily, although not entirely, on models that consider spherical drops.

The approach that is taken is first to consider a single spherical droplet in a micellar solution, neglecting the presence of other droplets. The interior of the single droplet will be assumed to have the properties of a bulk phase, so the analysis will apply to systems with substantial solubilization but will probably not accurately model micelles or inverted micelles containing only small quantities of solubilizate. The analysis of a single droplet leads to an expression for the change in interfacial free energy which contains a bending energy. This term is not new. It was introduced by J. W. Gibbs [20] in his seminal work dealing with the thermodynamics of interfaces. The bending energy term in the differential free energy will be shown to play an essential role which cannot be neglected for small droplets like swollen micelles, even though for most systems characterized by larger radii it can and has been ignored. The bending energy will be associated with the R-ratio. Arguments will be presented which suggest that the R-ratio includes the same information contained in the bending energy. This is the extent to which the theory has been advanced to date. It seems likely to the authors that future developments will quantify the relationships, which are only suggested here.

Given the free energy of the interface of an isolated drop, one can formulate the free energy of a collection of such drops by including the free energy of their interaction. This analysis yields a drop radius that minimizes the total free energy of the micellar solution. This radius depends on the free energy of the interfacial region of a single drop and also on the free energy of the entire dispersion of drops. The radius that minimizes only the free energy of the interfacial region will, in general, differ from that which minimizes the free energy of the entire dispersion. The radius that minimizes the interfacial free energy can be considered to represent a natural curvature of the interfacial region. It seems reasonable to equate this natural curvature to that curvature dictated by the R-ratio. This is indeed the approach taken here. The first step is, therefore, to consider the thermodynamics of a single isolated drop.

B. Thermodynamics of Spherical Interfaces

Gibbs [20] proposed that interfacial regions separating two bulk phases possess two-dimensional homogeneity. He postulated that if a base surface is located within the interfacial region in such a way that the state of matter is everywhere constant on the surface, then surfaces parallel to this base surface will also represent surfaces of constant state. For spherical droplets such as the one shown in Fig. 8.4, the parallel surfaces, each representing a constant state, will be spheres. It is assumed that despite the relatively small size of a micellar droplet, the pressure within the interior of the drop is homogeneous and isotropic. Thus the interior region is thought to be a bulk phase. This assumption is to be questioned and will certainly not apply to

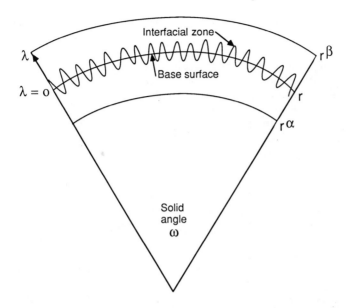

FIGURE 8.4   The interfacial zone of a spherical drop.

all swollen micelles. Indeed, only those micelles containing relatively large quantities of solubilizate are to be considered. For micelles containing only small quantities of solubilizate, the concept of interfacial tension may itself be questioned and perhaps of limited value in thermodynamic arguments. The latter systems are excluded from the analysis to follow.

The segment of spherical drop depicted by Fig. 8.4 contains phase α in its interior and phase β in its exterior. r^α and r^β are positioned so that these surfaces are contained within the region that is isotropic and homogeneous. The sphere of radius r is contained within the interfacial region, but its precise position is not yet defined. The system depicted by Fig. 8.4 is a conical segment and thus

$$A = \omega r^2 \qquad v^\alpha = \frac{1}{3}\omega[r^3 - (r^\alpha)^3] \qquad v^\beta = \frac{1}{3}\omega[(r^\beta)^3 - r^3] \qquad (8.23)$$

where A, v^α, and v^β are the areas, the volume of phase α, and the volume of phase β, respectively.

In addition to the entropy and composition, four geometric variables are therefore required to specify the state of the system. Choose v^α, v^β, r, and A to be these variables. Thus the internal energy is

$$dU = T\ dS + \Sigma \mu_i\ dn_i - p^\alpha\ dv^\alpha - p^\beta\ dv^\beta + \gamma\ dA + C\ dr$$

$$(8.24)$$

The free energy is therefore

$$dF = -S\ dT + \Sigma\ \mu_i\ dn_i - p^\alpha\ dv^\alpha - p^\beta\ dv^\beta + \gamma\ dA + C\ dr$$

$$(8.25)$$

Note that Eqs. (8.24) and (8.25) both admit the possibility that the energy of the system can depend on the drop radius. One seldom finds this term retained in thermodynamic analyses since, according to Gibbs, the radius of the dividing surface, r, can be selected so that C vanishes.

If the size of the system is increased by increasing ω, then v^α, v^β, A, n_i, and F all increase proportionately. Thus F is an Euler function of the first order in ω and therefore, at fixed T and r,

$$F = \Sigma\ \mu_i n_i - p^\alpha v^\alpha - p^\beta v^\beta + A\gamma \tag{8.26}$$

This expression for the free energy does not contain explicitly either T or r. This is true because these are intensive variables and do not change as the system is enlarged by increasing ω.

The position of the dividing surface defined by the radius r has not been precisely defined. One can therefore, imagine it being displaced a small distance without changing any variable except the volumes v^α and v^β. Such a change, which will be denoted by [dr], obviously does not alter the free energy, nor does it change any of the n_i or the chemical potentials, μ_i. T also remains constant. Thus for a change [dr], Eq. (8.25) becomes

$$\frac{dF}{[dr]} = 0 = -(p^\alpha - p^\beta)\omega r^2 + 2\omega r\gamma + C \tag{8.27}$$

Similarly, differentiate Eq. (8.26) to find

$$\frac{dF}{[dr]} = 0 = -(p^\alpha - p^\beta)\omega r^2 + 2\omega r\gamma + r^2 \frac{d\gamma}{[dr]} \tag{8.28}$$

From Eq. (8.27) a generalized version of the Laplace equation is obtained:

$$p^\alpha - p^\beta = \frac{2\gamma}{r} + \frac{C}{\omega r^2} \qquad (8.29)$$

Buff [21] has written this result and Melrose [22] has presented an extended version for nonspherical surfaces. Also, Murphy [23] has considered a force balance at a drop interface and found a result corresponding to Eq. (8.29). He defined a bending moment H that is equal to $-C/2\omega$. Murphy's notation will be adopted and H/r^2 is then a force per unit of area.

By comparing Eqs. (8.28) and (8.29), a significant result is found:

$$H = -\frac{1}{2}r^2 \frac{d\gamma}{[dr]} \qquad (8.30)$$

The bending moment is therefore related to the change in surface tension as the position of the dividing surface is varied, all other factors being held fixed. Both the bending moment and the surface tension depend on the position of the dividing surface. Thus, in considering the generalized Laplace equation (8.29), it should be clearly understood that the pressure difference between phases α and β is independent of the position of the dividing surface. Satisfying Eq. (8.30) ensures the independence of these pressures.

If f^α, s^α, and c_i^α are the free energy, entropy, and molar densities of phase α, respectively, then

$$dF^\alpha = -p^\alpha \, dv^\alpha - S^\alpha \, dT + \Sigma \, \mu_i \, dn_i^\alpha \qquad (8.31)$$

where

$$F^\alpha = f^\alpha v^\alpha \qquad S^\alpha = s^\alpha v^\alpha \qquad n_i^\alpha = v^\alpha c_i^\alpha$$

Similarly,

$$dF^\beta = -p^\beta \, dv^\beta - S^\beta \, dT + \Sigma \, \mu_i \, dn_i^\beta \qquad (8.32)$$

Sum these two free energies and subtract the resultant expression from Eq. (8.25) to obtain an expression for the surface free energy:

$$dF^S = -S^S \, dT + \Sigma \, \mu_i \, dn_i^S + \gamma dA + C \, dr \qquad (8.33)$$

where $F^S = F - F^\alpha - F^\beta$, and so on. Equation (8.33) can be integrated because F^S is an Euler function of order 1 in the variables A and n_i^S. Thus

$$F^S = \gamma A + \Sigma \ \mu_i n_i^S \qquad\qquad (8.34)$$

Differentiating Eq. (8.34) and subtracting from Eq. (8.33) gives the Gibbs-Duhem equation.

$$0 = -A \ d\gamma + C \ dr - \Sigma \ n_i^S \ d\mu_i - S^S \ dT \qquad\qquad (8.35)$$

or dividing by A,

$$d\gamma = -\frac{2H}{r^2} \ dr - \Sigma \ \Gamma_i \ d\mu_i - s^S \ dT \qquad\qquad (8.36)$$

This is the Gibbs adsorption equation where Γ_i is the surface excess of i and s^S refers to the entropy per unit area.

Before preceeding to present an analysis of the thermodynamics of micellar solutions, it is useful to investigate the information that can be gleaned from the generalized Gibbs adsorption equation with regard to the bending moment. Note that the interfacial tension $\gamma(T, \Gamma_i, r)$ is a function of state. Therefore,

$$\frac{\partial^2 \gamma}{\partial \mu_i \ \partial \mu_j} = \frac{\partial^2 \gamma}{\partial \mu_j \ \partial \mu_i} \qquad\qquad (8.37)$$

and

$$\frac{\partial^2 \gamma}{\partial \mu_i \ \partial r} = \frac{\partial^2 \gamma}{\partial r \ \partial \mu_i} \qquad\qquad (8.38)$$

From Eq. (8.37) the following conditions are to be imposed on the adsorptions:

$$\frac{\partial \Gamma_j}{\partial \mu_i} = \frac{\partial \Gamma_i}{\partial \mu_j} \qquad\qquad (8.39)$$

Furthermore, Eq. (8.38) coupled with Eq. (8.36) yields the interesting result

$$\frac{\partial \Gamma_i}{\partial r} = \frac{2}{r^2} \ \frac{\partial H}{\partial \mu_i} \qquad\qquad (8.40)$$

These equations impose conditions on the functional forms of $\Gamma_i(T,\mu_j,r)$ and $H(T,\mu_j,r)$ but do not specify them.

To illustrate this point further, consider an approximation to be used in subsequent sections. For ideal bulk solutions, a change in the chemical potential of component i at constant temperature is given by

$$d\mu_i = kTd \, \ln(c_i^\beta) \tag{8.41}$$

where c_i^β represents the molar density of component i in phase β. Note that i can represent s (surfactant), A (alcohol), o (oil), or w (water). For ideal bulk phases Eq. (8.39) requires that

$$\frac{\partial}{\partial c_A^\beta} \left(\frac{\Gamma_s}{c_s^\beta} \right) = \frac{\partial}{\partial c_s^\beta} \left(\frac{\Gamma_A}{c_A^\beta} \right) \tag{8.42}$$

This is a well-known result which applies to any multicomponent isotherm. For example, the competitive Langmuir isotherms satisfy this restriction. Thus

$$\Gamma_s = \frac{\alpha_s N_A c_s^\beta}{1 + \alpha_s c_s^\beta + \alpha_A c_A^\beta} \tag{8.43}$$

and

$$\Gamma_A = \frac{\alpha_A N_A c_A^\beta}{1 + \alpha_s c_s^\beta + \alpha_A c_A^\beta} \tag{8.44}$$

where α_A, α_s, and N_A are constants satisfying Eq. (8.42).

Equation (8.40) leads to the conditions

$$kT \frac{\partial}{\partial r} \left(\frac{\Gamma_A}{c_A^\beta} \right) = \frac{2}{r^2} \frac{\partial H}{\partial c_A^\beta} \tag{8.45}$$

and

$$kT \frac{\partial}{\partial r} \left(\frac{\Gamma_s}{c_s^\beta} \right) = \frac{2}{r^2} \frac{\partial H}{\partial c_s^\beta} \tag{8.46}$$

The bending moment H is, therefore, a measure of curvature-dependent adsorption, and curvature-dependent adsorption in turn signals the existence of a bending energy. These are features that dictate the phase behavior of micellar systems. If they are ignored or neglected, then, as we will see in the following section, there is not an intrinsic bending of the C-layer toward either the O or W regions. Thus all micellar systems will exhibit an inherent symmetry which is far from realistic. In order to confer type I or type II phase behavior, a bending moment must be included as a part of the thermodynamic analysis; moreover, since H does impart a preferred curvature to the C-layer, an intimate relationship must exist between H and the R-ratio, which also, as stressed in Chaps. 5 and 6, imposes a preferred interfacial curvature.

Since the curvature-dependent free energy per unit of area, $-2H/r^2$ dr, seldom appears in thermodynamic analyses of macroscopic interfaces, it is instructive to consider a case for which a bending energy can be calculated. Consider an elastic rectangular plate that is clamped along two of its sides and is subjected to a uniformly distributed transverse load. The deflection of the plate can, at some considerable distance from the clamped edge, be considered to be cylindrical with the axes of the cylinder parallel to the clamped edges of the plate. The rectangular plate is shown in Fig. 8.5.

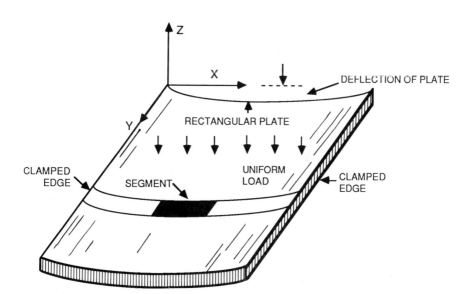

FIGURE 8.5 A rectangular plate which is clamped along its lateral edges is uniformly loaded.

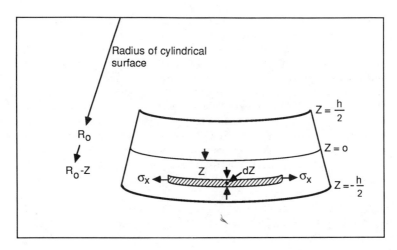

FIGURE 8.6 A cross-section showing a small segment of the uniformly loaded rectangular plate. The material is both under tension and compression depending on Z.

The fact that the deflection can be considered to be cylindrical represents a considerable simplification and Fig. 8.6 shows an elemental segment of the rectangular plate. The radius R_0 represents that cylinder for which a filament of the metal is neither extended nor compressed. For $R < R_0$, the metal is compressed, while for $R > R_0$ it is extended. Thus the strain (extension per unit length), ε_x, can be represented (see Fig. 8.6) as

$$\varepsilon_x = \frac{R(z) - R_0}{R_0} \simeq \frac{z}{R_0} \qquad\qquad (8.47)$$

There is assumed to be no strain in the y direction, so that $\varepsilon_y \simeq 0$. Making use of Hooke's law, the strains in terms of the normal stresses σ_x and σ_y are given by

$$\varepsilon_x = \frac{\sigma_x}{E} - \frac{\nu\sigma_y}{E} \qquad\qquad (8.48)$$

and

$$\varepsilon_y = \frac{\sigma_y}{E} - \frac{\nu\sigma_x}{E} = 0 \qquad\qquad (8.49)$$

where E is Young's modulus and ν is Poisson's ratio describing the elastic properties of the plate.

The free energy is the isothermal work required to bend the surface from its planar configuration ($R_0 \to \infty$) to its cylindrical form. This work, which is done by the transverse load exerted on the plate, is stored as elastic energy in the deformed plate, which can be calculated as follows. The x component of force acting on the shaded area shown in Fig. 8.6 is $\sigma_x\, dz$, and the strain created by a small increase in load is $d\varepsilon_x$.

From Eqs. (8.47), (8.48), and (8.49) it can be shown that

$$\sigma_x\, dz = - \frac{E}{(1 - \nu^2)} \frac{z}{R_0}\, dz \tag{8.50}$$

and

$$d\varepsilon_x = \frac{z}{R_0^2}\, dR_0 \tag{8.51}$$

It therefore follows that

$$f_{elastic} = - \int_\infty^{R_0} \int_{-h/2}^{h/2} \frac{z^2 E}{(1 - \nu^2)R_0^3}\, dz\, dR_0 \tag{8.52}$$

Thus

$$\frac{\partial f_{elastic}}{\partial R_0} = - \frac{Eh^3}{12(1 - \nu^2)} \frac{1}{R_0^3}\, dR_0 \tag{8.53}$$

The quantity $Eh^3/12(1 - \nu^2)$ is called in mechanics the flexural rigidity of a plate [24]. Equation (8.53) does not, of course, correspond precisely to the bending moment of a spherical surface, nevertheless, the form of the equation is instructive. The flexural rigidity increases with Young's modulus and with the thickness of the plate. It is anticipated that similar increases should increase H. Furthermore, the effect vanishes for a plane surface ($R_0 \to \infty$) and it would similarly be expected for spherical interfaces that $H/r^2 \to 0$ as r increases. Thus if the surface has a natural curvature for which H vanishes, the change in $f_{elasticity}$ with radius will, at that radius of curvature, vanish. For the case of a rectangular plate, the unstressed state corresponds to a plane. Any bending of the plate

from a planar configuration will give rise to a positive increase in
the free energy per unit of area. One might expect similar behavior
with respect to distorting a spherical surface from its natural curvature.

H can be given a precise definition in terms of stresses that
exist in a spherical interface. It can be shown [21,22] that

$$H = - \int_{-\infty}^{\infty} (p_t - p_{\alpha\beta}) \lambda \, d\lambda \qquad\qquad (8.54)$$

where

$$p_{\alpha\beta} = \begin{cases} p^{\alpha} & \text{for } \lambda < 0 \\ p^{\beta} & \text{for } \lambda > 0 \end{cases}$$

λ is a radial distance measured from the reference surface, so that H
is a moment of the excess pressure (stress) defined as the tangential
pressure, p_t, less the pressure that would exist within the interface
if it changed abruptly from p^{α} to p^{β} at the reference surface of
radius r. Equation (8.54) is a very satisfying result because clearly
H represents a moment (force multiplied by a lever arm) taken about
the reference surface. This is similar to Eq. (8.52), which also represents a moment. H can, therefore, be visualized as a measure of
the elastic energy stored in an interface which is deformed from its
natural curvature.

Equation (8.54) provides a means of calculating H. Since the
local pressure is determined given the intermolecular forces and the
distribution of molecules [25], H can, in principle, be determined
based on statistical mechanical methods. This calculation has not
been carried for complex systems such as micellar solutions. But
Miller and Neogi [26] have utilized Eq. (8.54) in an approximate
analysis that will be described in a subsequent section.

It is also useful to note that the surface tension is given by
[21,22]

$$\gamma = - \int_{-\infty}^{\infty} (p_t - p_{\alpha\beta}) \left(1 + \frac{2\lambda}{r}\right) d\lambda \qquad\qquad (8.55)$$

which when differentiated with respect to [dr] and multiplied by
$-r^2/2$, yields H as defined by Eq. (8.54). Thus Eqs. (8.54) and
(8.55) satisfy Eq. (8.30). This expression will help to understand
the factors that contribute to H.

Gibbs [20] observed the possibility of selecting a position of the dividing surface for which H vanishes; however, as noted by Miller and Neogi [26], this dividing surface, the surface of tension, may, for micellar droplets, be positioned at a radius which is distant from either the hydrodynamic radius or the Gibbs surface ($\Gamma_{H2O} = 0$), and therefore it is not reasonable to assume that all three surfaces correspond. Generally, one must retain the bending moment as a part of the thermodynamic analysis. This is done in the following section.

C. Solutions of Spherical Swollen Micelles

In this section a micellar solution composed of oil, water, surfactant, and perhaps other components is considered. This solution is a single isotropic phase having a submicrostructure which is assumed to consist of a dispersion of spheres of constant radius r. The spheres are sufficiently large so that the thermodynamic equations developed in Section 8.III.B for a spherical interface apply. A dividing surface is positioned at the oil/water interface so that $\Gamma_{water} = 0$. In practical calculations given here it will be assumed that this same dividing surface yields $\Gamma_{oil} = 0$, although this is certainly not the case in general and for precise treatments such an approximation is unacceptable. It is clear that this dividing surface does not necessarily correspond to the surface of tension. A spherical model for micellar solutions has been used by Rehbinder [27], Ruckenstein and Chi [28], Reiss [29], and Ruckenstein and Krishan [30]; however, these analyses have taken the surface of tension to correspond to a dividing surface at the oil/water interface, and they will be extended here to include the bending energy. Other spherical models taking into account the bending energy have been reported by Miller and Neogi [26], Ruckenstein [31], and Lam et al. [32].

Once a dividing surface is positioned, the volumes of the continuous and the discontinuous phases are well defined. Let ϕ be the volume of the discontinuous phase per unit of volume. Thus $1 - \phi$ is the continuous phase volume fraction. The surface area per unit of volume is $3\phi/r$, and c_i^β and c_i^α are the concentrations of component i in the continuous and discontinuous phases, respectively. Thus if the c_i are the total moles of component i per unit volume of micellar solution,

$$c_i = \frac{3\phi\Gamma_i}{r} + \phi c_i^\alpha + (1 - \phi)c_i^\beta \qquad (8.55)$$

For this system the Gibbs-Duhem equation applies to both the discontinuous phase,

$$\Sigma \ c_i^{\alpha} \ \delta\mu_i \ - \ \delta p^{\alpha} = 0 \tag{8.57}$$

and the continuous phase,

$$\Sigma \ c_i^{\beta} \ \delta\mu_i \ - \ \delta p^{\beta} = 0 \tag{8.58}$$

Summing Eqs. (8.57), (8.58), and (8.36) yields

$$\Sigma \ c_i \ \delta\mu_i = (1 - \phi) \ \delta p^{\beta} + \phi \ \delta p^{\alpha} - \frac{3\phi}{r} \ \delta\gamma - \frac{6\phi H}{r^3} \ \delta r \tag{8.59}$$

Using the generalized Laplace equation (8.29), p^{α} can be eliminated as follows:

$$\Sigma \ c_i \ \delta\mu_i = \delta p^{\beta} - \left(\frac{2\gamma\phi}{r^2} + \frac{2H\phi}{r^3} \right) \delta r - \frac{\phi}{r} \ \delta\gamma - \frac{2\phi}{r^2} \ \delta H \tag{8.60}$$

There are three contributions to the free energy: the free energy of the interface, the free energies of the bulk phases, and the free energy of the drops. The first two contributions have been discussed in Section 8.III.B. The free energy of the drops includes both an interaction energy and an entropic contribution. The formulation of this free energy, which is an important contribution to the total, has been considered by a number of authors [28,33]. Denote it Δf_e. The free energy per unit of volume is therefore

$$f = \frac{3\phi}{r} \ (\gamma + \Sigma \ \Gamma_i\mu_i) + \Sigma \left(c_i - \frac{3\phi}{r} \ \Gamma_i \right)\mu_i - \phi p^{\alpha} - (1 - \phi)p^{\beta} + \Delta f_e \tag{8.61}$$

The first term is the surface free energy [see Eq. (8.34)], the following terms represent the Helmholtz free energies of the two bulk phases, and the last term accounts for the entropy of the drops. Equation (8.61) reduces to the following form with the aid of the Laplace equation (8.29):

$$f = \frac{\gamma\phi}{r} + \frac{2\phi H}{r^2} + \Sigma \ c_i\mu_i - p_{\beta} + \Delta f_e \tag{8.62}$$

This is the free energy of the microemulsion system expressed per unit of volume. A necessary condition for thermodynamic equilibrium is

$$\left(\frac{\delta f}{\delta r}\right)_{c_i, p_\beta, T} = 0 \tag{8.63}$$

Thus

$$\frac{\delta f}{\delta r} = -\frac{\phi\gamma}{r^2} + \frac{\phi}{r}\frac{\delta\gamma}{\delta r} - \frac{4\phi H}{r^3} + \frac{2\phi}{r^2}\frac{\delta H}{\delta r} + \Sigma\, c_i\,\frac{\delta\mu_i}{\delta r} + \frac{\delta\Delta f_e}{\delta r} \tag{8.64}$$

This condition can be simplified with the aid of Eq. (8.60):

$$\frac{\delta f}{\delta r} = -\frac{3\phi\gamma}{r^2} - \frac{6\phi H}{r^3} + \frac{\delta\Delta f_e}{\delta r} = 0 \tag{8.65}$$

Equation (8.65) must apply at thermodynamic equilibrium independent of the number of components, whether or not the surfactant is charged, independent of the partitioning of the components between the oil and aqueous phases, and for all positions of the dividing surface. This result shows that the radius of microemulsion droplet which satisfies the minimum of free energy represents a balance of three terms. Two of these terms relate to the state of the interface and the last term relates to the overall microemulsion. If the drops are noninteracting, Δf_e is composed entirely of the entropic terms and this term will then increase as the number of drops increase for a fixed dispersed phase volume. Miller and Neogi [26] have developed a condition similar to Eq. (8.65).

D. Symmetric Micellar Solutions

In this section the bending moment is set equal to zero. The consequences of this with respect to a particularly simple system will then be explored and it will be seen that the thermodynamic predictions are then unsatisfactory. The difficulties that arise with regard to the simple system are profound and persist even for more realistic models.

In the absence of bending energy ($C = 0$), Eq. (8.36) yields

$$-d\gamma = \Sigma\Gamma_i\, d\mu_i \tag{8.66}$$

at constant temperature. If Γ_{water} and Γ_{oil} both vanish for a given choice of the dividing surface (an approximation that would have to be relaxed in any serious consideration), then considering a system containing both alcohol and another more typical amphiphile, Eq. (8.66) can be written as

$$-d\gamma = \Gamma_s\, d\mu_s + \Gamma_A\, d\mu_A \tag{8.67}$$

where to be definite, Γ_S and Γ_A are defined by Eqs. (8.43) and
(8.44). While these equations may not precisely define the adsorption that takes place in any real system, it is certain that the trends
they represent are qualitatively correct. One may therefore use
this simple model to explore further the implications of the thermodynamic analysis presented in the preceding section.

For ideal solutions, Eq. (8.67) can be integrated to yield

$$\gamma = \gamma_0 - N_A kT \ln(1 + \alpha_A c_A^W + \alpha_S c_S^W) \qquad (8.68)$$

where γ_0 is the interfacial tension of the oil/water system in the absence of either alcohol or surfactant and the constants α_A and α_S
are the Langmuir parameters evaluated with respect to aqueous-
phase concentrations. For this reason c_A^W and c_S^W are the concentrations of alcohol and surfactant, respectively, in the aqueous phase,
whether it is the continuous or the discontinuous phase.

If the bending energy vanishes, then Eq. (8.65) reduces to
two terms. Since the change in the droplet free energy with respect
to droplet radius is often small (this term is explored in a subsequent section), a condition for equilibrium is that $\gamma \approx 0$. This implies that

$$\alpha_A c_A^W + \alpha_S c_S^W = Q \qquad (8.69)$$

where

$$Q = \exp\left(\frac{\gamma_0}{N_A kT}\right) - 1$$

This expression relates the aqueous-phase concentrations which yield
ultralow interfacial tensions and hence stabilize the micellar solution.

The aqueous-phase compositions, which are seen by Eq. (8.68)
to be important, depend on the droplet radius. As the droplet
radius decreases, the interfacial area increases and a larger proportion of both the alcohol and the surfactant will be found at the interface. The relationship between the drop radius and the solution composition is found by material balance. To simplify the calculation but
at the same time preserve the essence of the process, it will be assumed that both the oil and the aqueous phase are ideal solutions,
so that

$$\frac{c_S^0}{c_S^W} = \bar{H}_S \qquad (8.70)$$

and

$$\frac{c_A^0}{c_A^W} = \bar{H}_A \tag{8.71}$$

where $\bar{H}_S$ and $\bar{H}_A$ are distribution coefficients. Therefore, a material balance becomes for the surfactant when the aqueous phase is the continuous phase ($c_S^\beta = c_S^W$ and $c_A^\beta = c_A^W$)

$$c_S = \frac{3\phi}{r} \frac{N_A \alpha_S c_S^W}{1 + \alpha_S c_S^W + \alpha_A c_A^W} + \phi c_S^W \bar{H}_S + (1 - \phi)c_S^W \tag{8.72}$$

where ϕ is the volume fraction of oil. The first term represents that amount in the interfacial layer and the second and third terms account for that amount in the O and W regions, respectively. A similar balance for alcohol yields

$$c_A = \frac{3\phi}{r} \frac{N_A \alpha_A c_A^W}{1 + \alpha_S c_S^W + \alpha_A c_A^W} + \phi c_A^W \bar{H}_A + (1 - \phi)c_A^W \tag{8.73}$$

Equations (8.72) and (8.73) make explicit the dependence of c_A^W and c_S^W on r for the water-continuous system. Similar equations can be written for the oil-continuous case. Equations (8.72) and (8.73) can be solved together with Eq. (8.69) to obtain

$$c_S^W = \frac{c_S}{F_S/r + B_S} \tag{8.74}$$

$$c_A^W = \frac{c_A}{F_A/r + B_A} \tag{8.75}$$

and

$$B_A B_S \left(Q - \frac{\alpha_A c_A}{B_A} - \frac{\alpha_S c_S}{B_S} \right) r^2 + [Q(B_A F_S + B_S F_A) - \alpha_A c_A F_S$$

$$- \alpha_S c_S F_A]r + QF_S F_A = 0 \tag{8.76}$$

where

$$B_A = \phi\bar{H}_A + (1 - \phi) \qquad F_A = \frac{3\phi N_A \alpha_A}{Q + 1}$$

$$B_s = \phi\bar{H}_s + (1 - \phi) \qquad F_s = \frac{3\phi N_A \alpha_s}{Q + 1}$$

Equation (8.76) is quadratic in r. The positive solution of this equation is R_O^*, the oil droplet radius, which minimizes the free energy when water is the continuous phase. There is a similar expression that can be written when water is the discontinuous phase and another drop radius R_W^* minimizes the free energy of that system.

Physical considerations dictate that there can be one and only one real, positive solution of this quadratic equation. This requirement can be expressed by the inequality

$$Q - \frac{\alpha_A c_A}{B_A} - \frac{\alpha_s c_s}{B_s} < 0 \qquad\qquad (8.77)$$

This equation expresses the fact that there must be a minimum amount of surfactant and alcohol present in the system before a stable micellar system is achieved, where stability is defined as the existence of a finite-size droplet that minimizes the free energy.

The minimum alcohol and surfactant requirement is consistent with the condition for obtaining ultralow interfacial tensions, as expressed by Eq. (8.69). It should be noted that the surfactant concentration in the aqueous phase will at all times be less than the CMC; otherwise, a micelle will form, solubilize either oil or water depending on which phase is continuous, and therefore be counted as a submicroscopic droplet. Thus c_s^W is limited by the CMC. There is no corresponding limit on the alcohol concentration. Thus one role of alcohol is to help reduce the interfacial tension to zero in those cases where the maximum surfactant concentration does not permit sufficient surfactant adsorption to accomplish this task. Clearly, the minimum alcohol requirement is a function of the parameters α_A, α_s, $\bar{H}_A$, or $\bar{H}_s$ and will vary from system to system. There are, therefore, two radii that minimize the free energy: R_O^* corresponds to the case for which the aqueous phase is continuous and R_W^* corresponds to the case for which the oil phase is continuous. There are two minimum free energies, f_W^* and f_O^*, corresponding to water-continuous and oil-continuous systems, respectively. The smaller of the two free energies represents a global minimum free energy. This situation is depicted in Fig. 8.7. The overall volumes of oil and water together with

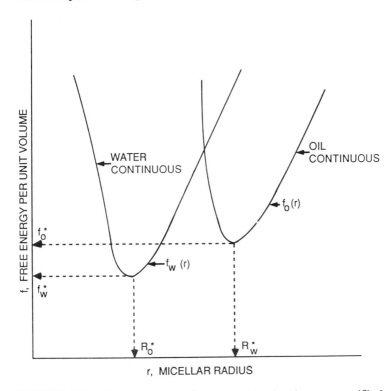

FIGURE 8.7   Free energies for a system having a specified overall composition.

the total moles of alcohol and surfactant are specified.   The micro-structure can, according to the present model, exist in one of two forms.   The aqueous phase can be the continuous phase or it can be the discontinuous phase.   The free energy corresponding to these two different systems having the same overall composition will exhibit different dependence on radius.   Both are shown in Fig. 8.7.   Corresponding to R_O^* is f_W^* and corresponding to R_W^* is f_O^*.   For the particular case depicted, f_W^* is the global minimum and hence for the given overall composition, water-continuous systems would be expected.

The free energy per unit of volume is essentially given by

$$f^* = c_A \mu_A + c_s \mu_s - p^\beta \tag{8.78}$$

since in Eq. (8.62), γ, H, and Δf_e are presumed small.   For the water-continuous system

$$f_w^* = c_A(\mu_A^* + kT \ln c_A^W) + c_s(\mu_s^* + kT \ln c_s^W) - p^\beta \tag{8.79}$$

For the oil-continuous system having the same overall composition,

$$f_o^* = c_A(\mu_A^* + kT \ln \bar{c}_A^W) + c_s(\mu_s^* + kT \ln \bar{c}_s^W) - p^\beta \tag{8.80}$$

where $\bar{c}_A^W$ and $\bar{c}_s^W$ are the aqueous-phase compositions for the oil-continuous system. Since the comparison is made at the same overall composition and pressure, c_s, c_A, and p^β are the same in Eqs. (8.79) and (8.80). Thus

$$f_w^* - f_o^* = c_A kT \ln \frac{c_A^W}{\bar{c}_A^W} + c_s kT \ln \frac{c_s^W}{\bar{c}_s^W} \tag{8.81}$$

If $f_w^* - f_o^* > 0$, the oil-continuous system will be most probable. If, on the other hand, $f_w^* \approx f_o^*$, the system will fluctuate between an oil-continuous and a water-continuous state. This might be thought of here as exemplifying type III behavior. The system for which $f_w^* \gg f_o^*$ can be qualitatively regarded as a type II system.

    It is, therefore, interesting to investigate the free energy further. The ratios $c_A^W/\bar{c}_A^W$ and $c_s^W/\bar{c}_s^W$ are of prime importance. These are seen to be given by

$$\frac{c_s^W}{\bar{c}_s^W} = \frac{F_s^O/R_w^* + B_s^O}{F_s^W/R_o^* + B_s^W} \tag{8.82}$$

and

$$\frac{c_A^W}{\bar{c}_A^W} = \frac{F_A^O/R_w^* + B_A^O}{F_A^W/R_o^* + B_A^W} \tag{8.83}$$

based on Eqs. (8.74) and (8.75). The superscript o designates that oil is the continuous phase and ϕ becomes the volume fraction of water. Thus

$$F_s^O = \frac{3\phi_w \alpha_s N_A}{Q + 1} \tag{8.84}$$

and

$$F_s^w = \frac{3(1 - \phi_w)\alpha_s N_A}{Q + 1} \qquad (8.85)$$

where ϕ_w is the volume fraction of water. Substituting these values of F_s and similar equations for F_A into Eq. (8.75), one can readily show that the following relationship exists for all volume fractions:

$$\frac{1 - \phi_w}{R_o^*} = \frac{\phi_w}{R_w^*} \qquad (8.86)$$

This result means that the micellar system is arranged so that the surface area per unit volume is the same irrespective of which phase is continuous. For such arrangements, the surface concentrations of surfactant and alcohol remain constant. This is the hypothesis proposed by Schulman et al. [34,35] to explain the structure of solubilized systems. There is, however, a profound difficulty with this result. Based on Eqs. (8.82), (8.83), and (8.86), it can be seen that

$$c_A^w = \bar{c}_A^w \qquad (8.87)$$

and

$$c_s^w = \bar{c}_s^w \qquad (8.88)$$

for all compositions. This then leads to the condition

$$f_0^* = f_w^* \qquad (8.89)$$

for all compositions. In other words, conditions do not exist for which water-continuous or oil-continuous systems are favored. Apart from possible contributions stemming from Δf_e, which have been neglected in this development, there is no mechanism that can destroy the symmetry of the system. Water-continuous and oil-continuous systems are, according to this analysis, equally likely. This symmetry is not, however, consistent with the observation. At equal volumes of oil and water some micellar solutions exhibit an electrical conductivity which indicates that oil is essentially the continuous

phase, whereas for others, water is definitely the continuous phase
(see Section 4.II.D). Thus a bias must be introduced in which one
continuous phase is preferred relative to the other. The free energy
of the dispersion of drops does not provide this bias where volumes
of oil and water are equal. It does, however, favor the dispersion
of the smallest phase volume into the larger one. Thus if the phase
volume of oil is smaller than that of water, Δf_e will tend to dictate a
water-continuous system.

This bias is most likely to arise through the curvature-dependent
contribution. If H does not vanish and depends, as it should, on
the composition of the interfacial layer, the adsorption must also depend on the droplet radius and furthermore depend on which phase
is continuous. This represents a symmetry-breaking mechanism that
can bring the theory in correspondence to the experimental observations. Thus to achieve realistic results one must consider the bending contribution to the free energy.

E.   Curvature-Dependent Adsorption and the
Bending Moment

In Section 8.II.D it has been shown that when the adsorption is symmetric, that is, when the amount of amphiphile per unit area of C
layer is a function of the aqueous phase (or oil phase) composition
but does not depend on the curvature of the interface, the thermodynamic predictions reflect an intrinsic symmetry not consistent with
experimental observations. The thermodynamic analysis must therefore include unrealistic assumptions.

Even when the influence of double-layer overlap and more realistic adsorption isotherms [30] are taken into account, the symmetry
of the thermodynamic results is not markedly changed. Furthermore,
the problem does not reside with the assumption of spherical drops.
Even more sophisticated models for the submicroscopic structure of a
microemulsion yield essentially the same symmetries found for spherical drops. This intrinsic symmetry has been discussed at length
by de Gennes and Taupin [37], and they have established the need
for a curvature-dependent term to be added to the free energy even
for sophisticated models such as that used by Talmon and Prager
[36]. Widom [38] has included a curvature-dependent energy in
his analysis.

To break the symmetry and develop more realistic thermodynamic
predictions, the bending energy neglected in the preceding section
must therefore be included. Miller and Neogi [26] have considered
the simplified model of a droplet in a micellar solution shown in Fig.
8.8 for a water-continuous system. It is assumed that the surface
layer has a uniform thickness δ and the tension or cohesive interaction energy between the C layer and the W region is γ_{WS}, while
γ_{os} represents the C layer interaction with the oil phase. In ac-

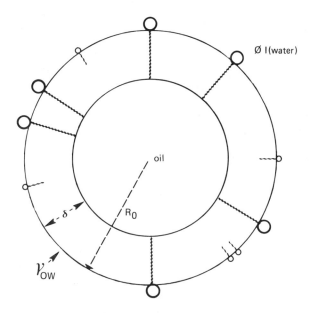

(a) Ø/W DROP MODEL

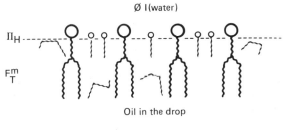

(b) LATTICE MODEL

FIGURE 8.8 (a) The spherical oil-in-water interface and (b) its equivalent lattice structure. (After Ref. 39.)

cordance with the definition of interfacial tension of a spherical surface as expressed by Eq. (8.55), we can write approximately[†]

$$\gamma = \gamma_{ws}\left(1 + \frac{\delta}{r}\right) + \gamma_{os}\left(1 - \frac{\delta}{r}\right) \tag{8.90}$$

Thus in this approximation the interfacial tension is the sum of the
inner and outer tensions modified by a small correction to take into
account the curvature. To a similar approximation based on Eq.
(8.54), we write

$$H = \frac{\delta}{2}(\gamma_{ws} - \gamma_{os}) \tag{8.91}$$

To examine the nature of the bending energy further, suppose
that there exists an interfacial area per molecule S_m^* and a radius of
curvature R_N for which $\gamma = 2\gamma^*$ and $H = 0$. This radius corresponds
to the natural curvature of the interface where there is no tendency
for bending; that is, $\gamma_{ws} = \gamma_{os} = \gamma^*$. R_N also represents the radius
of the surface of tension since H vanishes. Now suppose [26] that
γ_{ws} and γ_{os} individually satisfy an expression similar to the Gibbs
adsorption Eq. (8.36) and thus an expansion of γ_{ws} and γ_{os} can be
developed in terms of S_m, the interfacial area per surfactant molecule
(note that $\Gamma_s = 1/S_m$), and r, the droplet radius. For values of S_m
and r that are slightly different from S_m^* and R_N, Eq. (8.36) suggests
the following equations:

$$\gamma_{ws} = \gamma^* + K_{ws}\frac{S_m - S_m^*}{S_m^*} + d_{ws}\left(\frac{1}{r} - \frac{1}{R_N}\right) \tag{8.92}$$

and

$$\gamma_{os} = \gamma^* + K_{os}\frac{S_m - S_m^*}{S_m^*} - d_{os}\left(\frac{1}{r} - \frac{1}{R_N}\right) \tag{8.93}$$

[†]It is not possible to develop Eqs. (8.90) and (8.91) rigorously starting with Eqs. (8.54) and (8.55). Thus these equations should be
taken as qualitative recognizing that other, perhaps more accurate,
results may evolve when detailed analyses of the complex interfacial
molecular orientations are considered.

Here K_{WS} and K_{OS} are the interfacial compressional moduli (reciprocal interfacial compressibilities) between the C layer and the W and
O regions. In the analysis to follow, K_{WS} and K_{OS} are assumed constant over the range of S_m of interest. The expansion coefficients
d_{WS} and d_{OS} are also taken to be constants.

For the special case $K = K_{WS} = K_{OS}$ and assuming that the main
effects of a decrease in drop radius at constant mean film area S_m
are an expansion of the outer surface of the film and the corresponding compression of the inner surface, one can show [26] that

$$H = \frac{K\delta^2}{1 + \delta/R_N}\left(\frac{1}{r} - \frac{1}{R_N}\right) \qquad (8.94)$$

This expression corresponds to a contribution added to the interfacial free energy by Helfrich [40], Robbins [41], de Gennes and
Taupin [37], and Widom [38] so as to break the thermodynamic symmetry. Equation (8.94) cannot, however, be quite correct if K and
R_N are assumed independent of the concentration of surfactant, as it
fails to satisfy Eq. (8.46). A slightly modified form,

$$H = c_S^W K_r kT\left(\frac{1}{r} - \frac{1}{R_N}\right) \qquad (8.95)$$

does satisfy the integrability condition if surfactant, but not alcohol,
is present in the system.[†] Here c_S^W is the aqueous-phase surfactant
concentration. Perhaps it would be better to replace c_S^W by Γ_S, the
amount of surfactant actually present at the interface. However,
Γ_S and C_S^W are related by an isotherm, so that there is a corresponding change in Γ_S whenever c_S^W is changed. Equation (8.95) is selected for its convenience.

Equation (8.95) has an interesting and suggestive structure. It
insists, for example, that the droplets have a preferred or natural
curvature R_N. This is concordant with the concept embodied by the
R-ratio. R_N is, furthermore, determined by the interaction energies
that exist between the C layer and the W and O regions, just as the
R-ratio is determined by these same quantities. There is then a
clear link between the natural curvature and the R-ratio.

[†]The authors wish to note that Eq. (8.95) should be regarded as a
strictly empirical one and its primary attraction is that it is simple.
We use this expression here to determine the influence of H on
thermodynamic results but do not claim that it corresponds to any
real system.

Note that the sign which is to be assigned to R_N requires careful consideration, and to define it properly one must associate with a particular R_N that region O or W which is discontinuous. Winsor type I phase behavior is generally associated with those micellar solutions thought to be water continuous and hence to consist of oil drops dispersed in an aqueous phase. For type I systems one asserts, based on $R < 1$, that the interface tends to be convex with respect to the O region. Thus for type I systems R_N is positive for oil drops. This means that the interface prefers this arrangement. However, other considerations, such as droplet entropy, enter into the ultimate determination of the final equilibrium droplet radius or even, in fact, which region will ultimately be continuous. As noted in Section 4.II.D, Winsor type I systems can exhibit either water-continuous or oil-continuous behavior, depending on the volumetric water/oil ratio. This can be seen by examining the phase behavior depicted in Fig. 4.33. The same phase behavior is also shown by Fig. 8.9a. Over a wide range of water/oil ratios, overall compositions falling within the two-phase region divide into an excess oil phase in equilibrium with a micellar phase. This is type I behavior. However, in the neighborhhod of point A in Fig. 8.9a within the single-phase region, inverted micelles are known to exist and the system is, for small water volumes, oil continuous even though the natural bending of the interface and the R-ratio express a preference for water-continuous systems.

When r represents a droplet radius of curvature opposite to that which is natural, then R_N is negative and H expressed by Eq. (8.95) is positive for all values of r. This means that its contribution to the free energy will be to tend to decrease with increasing r. This is seen based on Eq. (8.65). Winsor type I systems are therefore those for which R_N is positive for oil drops in water and is a negative number of the same magnitude for water drops in oil.

Both the R-ratio and R_N are determined by the composition of the interfacial layer and the interaction between the constituents residing in this layer and the adjacent regions. They should be independent of the water/oil ratio unless variation of this ratio produces corresponding changes in the composition of the interfacial layer. Table 8.1 provides a summary defining the sign of R_N and also compares R_N to the R-ratio. The qualitative relationship between the two is evident. Both define the bending tendency of the C layer. As stressed in Chap. 7, the R-ratio also provides information regarding the solvency of amphiphilic compounds. It is seen that the magnitude of the denominator (numerator) is an important measure. The other parameter in Eq. (8.95), K_r, provides this same measure. It will be shown that the solvency increases with increasing K_r. Of course, as noted in Chap. 7, there exists a limitation to the benefit that may be derived by continuously increasing both the numerator and the denominator of the R-ratio to improve solvency. As the

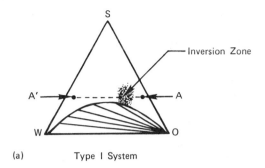

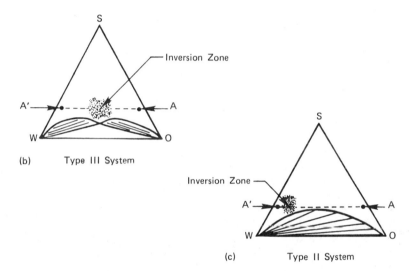

FIGURE 8.9 Diagram showing zones in which an inversion from water to oil continuous systems may be expected.

TABLE 8.1 Comparison of the R-Ratio with the Natural Curvature

| Winsor classification | R-ratio | R_N | |
		Water drops	Oil drops
Type I	<< 1	Negative	Positive
Type III	1	≈ ∞	≈ ∞
Type II	>> 1	Positive	Negative

numerator or the denominator increases, liquid-crystalline rather
than isotropic solutions tend to form. Similarly, large values of K_r
tend to yield structures which are rigid, presumably modeling liquid-
crystalline structures.

To summarize briefly, the existence of a bending moment breaks
the symmetry which otherwise exists. One representation of H that
has been used, Eq. (8.94), includes two parameters which are
closely related to the R-ratio. Because of this relationship, it is
appropriate to investigate further the significance of the bending
moment. If the aqueous surfactant solution is considered to be ideal,
then Eq. (8.46) applies and upon substituting Eq. (8.95) into this
expression, the following interesting result is obtained:

$$\frac{\Gamma_s - \Gamma_s^{(\infty)}}{c_s^w} = \frac{K_r}{r} \left(\frac{2}{R_N} - \frac{1}{r} \right) \tag{8.96}$$

where Γ_s is the adsorption at a plane interface. This expression
shows that the curvature-dependent adsorption is greater than at a
plane interface for $r > R_N > 0$. Otherwise, it is smaller. The system
tries to adapt to an unnatural radius $r \neq R_N$ by either adding or ex-
pelling surfactant from the interfacial region. Because the adsorption is curvature dependent, the thermodynamic symmetry found in
the preceding section is now broken and the system does have a
preference for a particular surface curvature.

There have been few attempts to fix H based on molecular models.
By far the most comprehensive study to date has been reported by
Huh [42]. He has considered that the electrical double layer formed
within the aqueous regions near the C layer plays an important role
in establishing the natural curvature of the interface. When the salinity is low, repulsion between the ionic heads of the surfactant is
strong because the shielding effect provided by the salt is small.
The interface then tends to curve away from the brine, making the
electrical double layer diffuse. Conversely, an increase in the salinity increases the shielding and tends to compress the double layer
by bending the interface toward the oil side.

On the oil side of the interface, Huh considers that the surfactant lipophiles residing there have less configurational freedom than
do those solely in solution. If the interface is curved toward the
brine, the loss of configurational freedom is less severe, resulting
in an increase of entropy and decrease of system free energy. Thus
the surfactant lipophiles will also tend to repel one another, forcing
the interface to curve toward the W region. In addition, the free
energy is affected by enthalpy of mixing between the lipophile and
oil. The natural radius is a compromise between these forces acting

on the C layer. Huh [42] has obtained an expression for the bending moment which incorporates these factors.

Using arguments similar to those invoked by Huh, Mukherjee
and co-workers [39] have examined the tendency of a surfactant
film to assume its natural radius of curvature R_N. Figure 8.8 shows
the model used by Mukherjee et al. [39]. The lipophiles occupy the
region between two concentric spheres. The curvature dependence
of the interfacial free energy is due primarily to the differential
change in area per surfactant within the region. Consistent with
this view, the spherical interface is replaced by I lattice planes
(Fig. 8.8) with unequal numbers of lattice sites in each plane. Here
I is the number of segments in the surfactant chain. The number
of lattice points (N_i^m) in the ith plane is

$$N_i^m = \frac{4\pi r_i^2}{S_m} \tag{8.97}$$

where r_i is the radius of the sphere corresponding to the ith plane
and S_m is the area occupied by a lattice site. The numbers of surfactant molecules n_s^m and cosurfactants n_A^m per drop are

$$n_s^m = \frac{4\pi R_N^2}{S_m} \tag{8.98}$$

and

$$n_A^m = \frac{n_s^m}{\varepsilon} \tag{8.99}$$

where ε is the assumed ratio of surfactant to cosurfactant in the
film.

The configurational free energy F_T^m of the tail region in the mean
field approximation can be written as

$$F_T^m = E_T^m - kT \ln \Omega_T^m \tag{8.100}$$

where F_T^m is the internal energy of the interface and Ω_T^m is the number
of ways of arranging the molecules in the region.

Mukherjee et al. [39] have approximated Ω_T. The units of the surfactant's lipophilic tail, the alcohol tail, and the oil (CH_2 or CH_3) were

all assumed to be equivalent, and the alcohol and surfactant tails were restricted so that successive units of a chain occupy lattice sites in successive layers. The combinational problem even for this simple situation is complex and can only be solved approximately.

Mukherjee et al. assert that the drop size obtained by minimizing the free energy is the natural radius since it represents the one that the system would assume in the absence of all other constraints. Based on these arguments, Mukherjee et al. found trends that correspond to observation. In their analysis the length of the lipophile, the molecular weight of the alkane oil, the type of alcohol, and the surfactant/alcohol ratio were all varied. In certain regions of the parameter space, oil-in-water microemulsions having natural radii less than about 100 Å were found, whereas in other regions, water-in-oil microemulsions are the only system possible. In certain regions corresponding to larger natural radii, both types of microemulsions can exist. These regions might be imagined to be those for which the transition from one type of microemulsion to another takes place.

The approach used by Mukherjee et al. [39] appears to be a promising one and a similar technique has been used by Lemaire et al. [43]. The trends closely corresponding to those observed are predicted, and this suggests that further work is warranted, especially if one incorporates the approach set forth by Huh [42].

Suezaki [44] has presented a statistical mechanical analysis which includes the short-range interactions and a bending stiffness energy that resembles the bending energy of a thin elastic sheet. His results include a parameter like K_r, but this parameter requires experimental determination. It is not given molecular significance. Experimental evaluation of K_r may be possible at least in those cases for which long-range structure exists [45-47]. Other possible methods for determining both K_r and R_N are described in Section 8.III.

F. Droplet Free Energy

1. Noninteracting Drops

The calculations displayed in Section 8.III.D neglect the quantity Δf_e. In this section some of the formulas that have been used to represent this contribution are presented. The simplest case is that of noninteracting drops. The free energy is then entirely composed of entropy. An approximation to the entropy of a dispersion of noninteracting drops has been obtained by Ruckenstein and Chi [28] using a lattice model. The total number of lattice sites per N_t per unit of volume is given by

$$N_t = mn_e + N_c \tag{8.101}$$

where m is the number of drops per unit of volume ($m = 3\phi/4\pi r^3$), n_e the number of lattice sites occupied by one drop, and N_c the

number of sites occupied by the molecules of the continuous phase
assuming that one molecule occupies one site. The number of sites
occupied by one drop is therefore

$$n_e = \frac{\phi}{mv_c} \qquad (8.102)$$

where v_c is the volume of a single continuous-phase molecule.

Denoting as Ω the number of possible lattice configurations that
a mixture formed by N_c molecules and m drops can assume, the entropy is

$$\Delta S = k \ln \Omega \qquad (8.103)$$

Therefore, the number of possible configurations must be counted.
This is a difficult combinatorial problem if one allows for the geometry of a drop, but a crude approximation is possible. The center
of the first drop can be located at any of the N_t sites, the center
of the second drop at the remaining $N_t - \alpha_2$ sites, and the center of the
jth drop at the remaining $N_t - \alpha_j$ sites. Therefore,

$$\Omega = \frac{N_t(N_t - \alpha_2)(N_t - \alpha_3) \cdots (N_t - \alpha_m)}{m!} \qquad (8.104)$$

The factorial m! is introduced because the drops are indistinguishable. It is extremely difficult to calculate α_j and this has not rigorously been done. One can, however, obtain an upper bound for Ω
if one assumes that the drops can be deformed into any shape. In
this case the number of sites α_j unavailable to the jth drop is simply
the number of sites already occupied by the j − 1 drops,

$$\alpha_j = (j - 1)n_e \qquad (8.105)$$

Hence

$$\Omega = \frac{n_e^m(N_t/n_e)}{(N_t/n_e - m)! \; m!} \qquad (8.106)$$

and

$$\Delta f_e = -mkT \left(-\frac{1}{\phi} \ln(1 - \phi) + \ln \frac{1 - \phi}{mv_c} \right) \qquad (8.107)$$

Equation (8.103) becomes, on noting that $m = 3\phi/4\pi r^3$,

$$\Delta f_e = -\frac{3\phi kT}{4\pi r^2}\left[-\frac{1}{\phi}\ln(1-\phi) + \ln\frac{4\pi r(1-\phi)}{3\phi v_c}\right] \qquad (8.108)$$

From this expression $\delta\Delta f_e/\delta r$ can be calculated. It is easily seen that Δf_e can make a significant, in fact sometimes dominant, contribution to the terms $(f_w^* - f_0^*)/c_s kT$. However, when $\phi = 0.5$, there are no terms that break the symmetry. Thus, even upon adding the droplet entropy, the need for a symmetry-breaking curvature-dependent term remains.

2. Hard-Sphere Model

Overbeek [33] has used a hard-sphere model to represent the free energy of a dispersion of droplets. Miller [48] has written this model in the form

$$\Delta f_e = +\frac{3\phi}{4\pi r^3}kT\left[\ln\phi - 1 + \phi\frac{4-3\phi}{(1-\phi)^2} + \ln\frac{3v_c}{4\pi r^3}\right] \qquad (8.109)$$

It is seen that the same variables appear in this expression as in Eq. (8.108). In fact, the hard-sphere model gives values of the droplet free energy which are smaller (larger negative values) than those given by the approximate lattice model. Nevertheless, both show the same trends and involve the same variables.

3. Van der Waals Interactions

Huh [49] has included a Van der Waals attractive potential together with the hard-sphere model. Huh's analysis takes the system to be a collection of spheres of fixed radius. The introduction of an attractive interaction energy between droplets is a necessary step for understanding the critical behavior of micellar solutions, which was his goal. Near a critical endpoint, two phases, both micellar solutions and having very nearly the same composition, coexist. This type of phase separation arises because of an interaction between droplets [50].

Huh has shown that the free energy of a collection of hard spheres with Van der Waals interactions can be expressed as follows:

$$\Delta f_e = \frac{3kT}{4\pi(1+\lambda)^3 r^3}\left[\ln\frac{3v_c}{4\pi r^3(1+\lambda)^3} + \ln\eta - 1 + \frac{4-3\eta}{(1-\eta)^2}\right.$$
$$\left. - \varepsilon_{11}(\lambda)\right] \qquad (8.110)$$

In this analysis the minimum possible separation between drops is defined as 2b, where

$$b = \frac{A_1}{\kappa} \qquad \text{for water-continuous systems} \qquad (8.111)$$

or

$$b = A_2 \, \alpha \sqrt{m} \, \ell \qquad \text{for oil-continuous systems} \qquad (8.112)$$

Here $1/\kappa$ is the Debye length, α is Flory's expansion parameter, m the number of segments in the lipophilic chain, and ℓ the segment length. A_1 and A_2 are constants. The parameter η is related to the dispersed phase volume by the equation

$$\eta = (1 + \lambda)^3 \phi \qquad (8.113)$$

where $\lambda = b/r$.

Note that as λ and $\varepsilon_{11}(\lambda)$ both tend to zero, Eq. (8.110) reduces to Eq. (8.109), the hard-sphere model. The attractive Van der Waals potential has been written by Huh [49] as

$$\varepsilon_{11}(\lambda) = \frac{-2A_h}{3kT(1 + \lambda)^3} \left[\frac{1}{4} \ln(1 + \frac{2}{\lambda}) + (1 + \lambda) \right.$$
$$\left. + (1 + \lambda)^3 \ln \frac{2\lambda + \lambda^2}{(1 + \lambda)^2} \right] \qquad (8.114)$$

where $A_h = A_{ww} + A_{oo} - 2A_{wo}$ and A_{ij} is the Hamaker constant between the i and j phases.

Huh has shown that if $\varepsilon_{11}(\lambda)$ exceeds a critical value, there exist two values of η that satisfy the conditions for phase equilibrium. These two values are shown in Fig. 8.10. For the critical value of ε_{11}, it is seen that $\eta_c = 0.1304$. Thus the critical composition is predicted to be about 13% by volume droplets.

Equation (8.114) shows that one can increase $\varepsilon_{11}(\lambda)$ by either increasing A_h or by decreasing λ. The relationship between these variables is shown in Fig. 8.11. In this plot $a_h = A_h/A_{ww}$. For a given a_h, phase separation can occur if λ is small enough so that $\varepsilon_{11} > \varepsilon_c$. For large values of λ (larger b or smaller droplet radii), the solubilized micellar solution is stable.

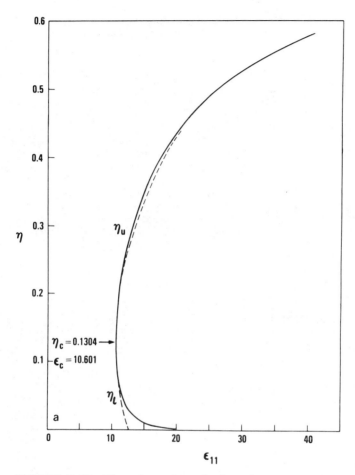

FIGURE 8.10 The phase boundaries shown as a function of the attractive potential. (After Ref. 49.)

With the addition of the attractive terms to Δf_e, a thermodynamic theory is, qualitatively at least, capable of predicting the critical behavior of micellar solutions; however, because the analyses assume the system to consist of uniform spherical drops, other more complex types of behavior observed in water/surfactant/hydrocarbon mixtures, such as the formation of liquid crystals or gels, is not predicted.

Calje et al. [51] have observed that the value of the Hamaker constant which must be used in Eq. (8.114) to represent the phase behavior is large, perhaps too large to be within the range expected, and therefore other models have been proposed [52]. For the micellar

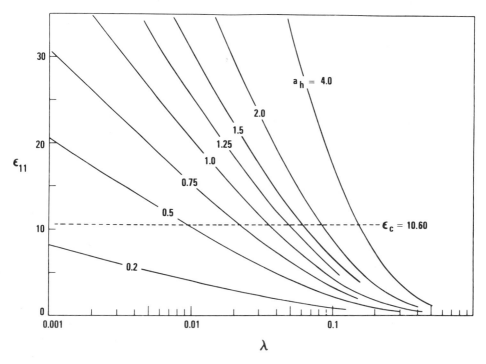

FIGURE 8.11 The relationship between the attractive potential and
the distance (dimensionless) of closest approach showing the domain
for which phase separation is possible. (After Ref. 49.)

systems that can be diluted, the second virial coefficient can be evaluated using light-scattering techniques (see Section 4.III.D) and models
of the interaction energy tested at least in the dilute region. The experiments show that the second virial coefficient varies over a wide
range of values depending on both the composition of the continuous
and the interfacial phases. The most sensitive parameter was the alcohol used [52]. To account for the full range of values observed,
Lemaire et al. [53] proposed that for oil-continuous systems there is
an interpenetration of one droplet into the interfacial layer of another.
This interpenetration is interpreted as being a mixing of the lipophilic chains with consequent expulsion of that oil associated with
the surfactant lipophile.

G. The Influence of the Bending Moment on Drop Size

Equation (8.95) provides a simple relationship between the c_S^W surfactant aqueous-phase composition and the bending moment for a given

droplet radius. Consider, together with this expression, a simple adsorption model for a plane interface

$$\Gamma_s^{(\infty)} = \frac{\alpha N_A c_s^W}{1 + \alpha c_s^W} \tag{8.115}$$

which gives from Eq. (8.96),

$$\Gamma_s = \frac{\alpha N_A c_s^W}{1 + \alpha c_s^W} + K_r c_s^W \left(\frac{2}{rR_N} - \frac{1}{r^2} \right) \tag{8.116}$$

Substituting Eqs. (8.95) and (8.115) into Eq. (8.36) yields the following expression for the curvature-dependent interfacial tension:

$$\gamma = \gamma_0 - N_A kT \ln(1 + \alpha c_s^W) + c_s^W \frac{kTK_r}{r} \left(\frac{1}{r} - \frac{2}{R_N} \right) \tag{8.117}$$

To determine the equilibrium drop radius, Eq. (8.65) must be solved together with a material balance equation that distributes the total surfactant inventory into the appropriate regions such that the surfactant chemical potential is everywhere constant. The surfactant present is distributed between the W and O regions and in the C-layer. The following equation expresses that simple balance:

$$c_s = \phi_w c_s^W + \phi_o \bar{H}_s c_s^W + \frac{3\phi}{r} \left(\frac{\alpha N_A c_s^W}{1 + \alpha c_s^W} + \frac{c_s^W K_r}{r^2} \frac{2r - R_N}{R_N} \right) \tag{8.118}$$

Equations (8.65) and (8.118) can be solved simultaneously, with γ and H being given by Eqs. (8.117) and (8.95), respectively. The results for one particular example are shown in Figs. 8.12 and 8.13. The drop free energy has been included in the calculation by assuming the hard-sphere model [Eq. (8.109)], to apply [32]. It is seen that for a given K_r and dispersed phase volume, ϕ, the drop radius depends on the natural curvature, R_N. Shown in Fig. 8.12 are the radii for the same system except that in one case R_N is positive and in the other negative. This would correspond, for example (see Table 8.1), to a type I system at points A and A' on the phase diagrams depicted in Fig. 8.9. At point A, with oil drops in water, the natural curvature is positive and the actual radius is somewhat

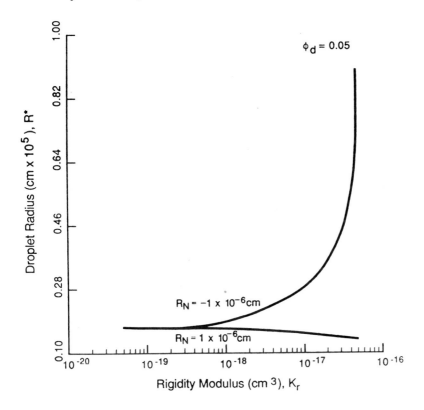

FIGURE 8.12 The equilibrium droplet radius for the parameters
$V_C = 75 \ \text{Å}^3$, $T = 25°C$, $Q = \gamma_0/N_A kT = 0.725$, $c_s = 10^{-4}$ mole/cm^3,
$\alpha = 1 \times 10^4$ cm^3/mole and $\bar{H}_s = 1.0$. The natural curvatures are expressed in cm. (After Ref. 32.)

larger than the natural radius, the equilibrium being dictated by a
complex compromise between the various terms entering Eq. (8.65).

The drop radius corresponding to point A' Fig. 8.9 is, on the
other hand, substantially larger than that corresponding to A because
at this point the interface is convex toward the W region, although
the natural tendency is opposed to this curvature. This conflict is
denoted by a negative value of R_N. As seen in Fig. 8.12, a negative
R_N causes the equilibrium radius at a given volume of fraction of dispersed phase to be larger than it is when R_N is positive. The difference is, furthermore, greater with increasing K_r.

For those values of K_r large enough to cause the equilibrium
radius to depend on R_N, adsorption also depends on R_N. This dependence is shown in Fig. 8.13. The adsorption density generally

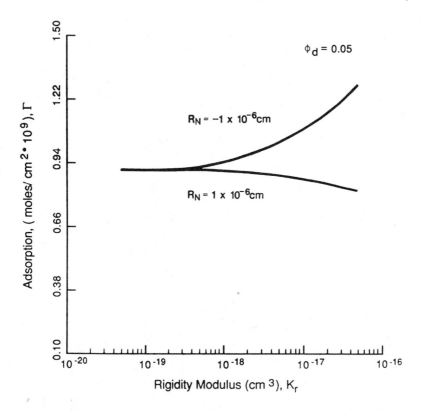

FIGURE 8.13 The curvature dependence of adsorption. The same parameters are used as Fig. 8.12. (After Ref. 32.)

decreases when the surface is deformed into an unnatural curvature. A consideration of the adsorption reveals the broken symmetry of the thermodynamic results and demonstrates the possibility of representing the various types of phase behavior based on a thermodynamic model provided that the bending energy is included. In the following section the ramifications of this simple thermodynamic model for the bending energy are developed further.

H. Thermodynamics of Phase Separation

In this section we consider the supplementary conditions that apply when an excess phase exists in equilibrium with a micellar solution composed of discrete droplets having an interior composition equal to that of the excess phase. Here again we adopt an idealized model which considers the interior of the swollen micelles to be a bulk phase, that is, one in which the pressure is isotropic. In this case we impose a further condition not considered previously. It is supposed that since the composition of the interior bulk phase is the same as that of the excess phase, then equilibrium requires that the internal pressure, p^α, correspond to the pressure exerted on the system, which is also, of course, the pressure of the excess phase. This equality ensures that the chemical potential of the dispersed phase is the same as that of the excess phase since they have the same composition and exist at the same pressure. It is probably worth noting that the need to require p^α to be equal to p_{system} arises only because we have insisted that the composition of the excess phase and the interior of the drops be the same. In reality, small differences in the composition may exist between the two phases, and one must equate chemical potentials rather than pressures.

The important feature of this analysis is that the pressure external to the drops, p^β, is not the same as the system pressure. It is less. This situation arises because the pressure exerted on the excess phase includes the sum of pressures exerted by the continuous-phase molecules plus that exerted by the drops, which are themselves in Brownian motion [$p_{system} - p^\beta$ may be thought of as an osmotic pressure (42)]. Thus p^β is not p_{system}. The system pressure is found by the thermodynamic equation

$$p_{system} = -\left(\frac{\partial F}{\partial V}\right)_{T,n_i} \qquad (8.119)$$

Here F is the Helmholtz free energy.

Based on Eq, (8.62), we see that F = Vf becomes

$$F = \frac{V_d \gamma}{r} + \frac{2V_d H}{r^2} + \Sigma \, n_i \mu_i - Vp^\beta + V \, \Delta f_e \qquad (8.120)$$

Where V_d is the dispersed phase volume and n_i represents the total moles of component i. If we assume that Δf_e depends on r and ϕ, then

$$\frac{\partial \Delta f_e}{\partial V} = \frac{\partial \Delta f_e}{\partial r} \frac{\partial r}{\partial V} + \frac{\partial \Delta f_e}{\partial \phi} \frac{\partial \phi}{\partial V} \qquad (8.121)$$

which can be written, imposing Eq. (8.65), as

$$\frac{\partial \Delta f_e}{\partial V} = \left(\frac{3\phi\gamma}{r^2} + \frac{6\phi H}{r^3}\right) \frac{\partial r}{\partial V} + \frac{\partial \Delta f_e}{\partial \phi} \frac{\partial \phi}{\partial r} \qquad (8.122)$$

Equation (8.65) still applies. It can be regarded as an internal condition for thermodynamic equilibrium.

Differentiating Eq. (8.120), substituting Eq. (122) and using
Eq. (8.60) leads to the following result:

$$-p_{system} = -p^\beta + \left(\frac{\gamma}{r} + \frac{2H}{r^2}\right) \frac{\partial V_d}{\partial V} + \Delta f_e + V \frac{\partial \, \Delta f_e}{\partial \phi} \frac{\partial \phi}{\partial V} = -p^\alpha$$

$$(8.123)$$

This equation represents a new condition which will come into play
only when an excess phase having the system pressure and the internal phase composition exist. Otherwise, p^α is not necessarily
equal to the pressure of the internal phase.

The difference in pressure is, in any event, given by the Laplace
equation (8.29). Thus the pressure difference $p^\alpha - p^\beta$ can be eliminated from Eq. (8.123) to yield a new condition to be imposed when
a second excess phase exists as follows:

$$\frac{\gamma}{r}(2 + \phi) - \frac{2H}{r^2}(1 - \phi) + \left(\frac{\gamma}{r} + \frac{2H}{r^2} + \frac{\partial \, \Delta f_e}{\partial \phi}\right) \frac{\partial \phi}{\partial \ln V} = 0 \qquad (8.124)$$

To use this expression, some statement as to the relative compressibilities of the continuous and dispersed phases is required. If the
compressibilities are roughly the same, a change in total volume resulting from an increase in pressure will not result in a change in
the fraction of the volume occupied by the droplets. Thus for equal
compressibilities, we have $\partial \phi / \partial V \simeq 0$.

If we apply this condition, Eq. (8.124) reduces to a particularly
simple form. This becomes an additional equilibrium condition to be
imposed together with Eq. (8.65). This additional constraint applies
if a separate excess phase is to exist. If the dispersed phase volume is fixed and the surfactant concentration allowed to vary, conditions for which Eqs. (8.124) and (8.65) are simultaneously satisfied
may be found. A solution so obtained represents a point on the demixing curve. Thus a critical surfactant concentration exists. It
represents that concentration of surfactant needed to solubilize the
dispersed phase volume ϕ. If greater surfactant concentrations than
the critical one are used, p^α is less than p_{system} and no excess
phase can coexist with the micellar solution. For surfactant concentrations less than critical value, the micellar solution is unstable and
a portion of the dispersed phase will coalesce and form a separate
phase. The critical surfactant concentration is therefore a measure

of the solvency of a particular surfactant and depends on the parameters K_r and R_N in the simple model used here.

Thus for the system described in Section 8.III.G, if Eq. (8.124)
is additionally imposed, the results shown by Figs. 8.14 and 8.15 are
found. The importance of these curves relates to their correspondence to the expected trends based on consideration of the R-ratio as
well as the experimental results seen in Chap. 7.

Figure 8.14 shows the critical surfactant concentration decreases
as R_N, the natural radius defined by Eq. (8.95), is increased. Of
course, as R_N is increased, the system tends to become type III.

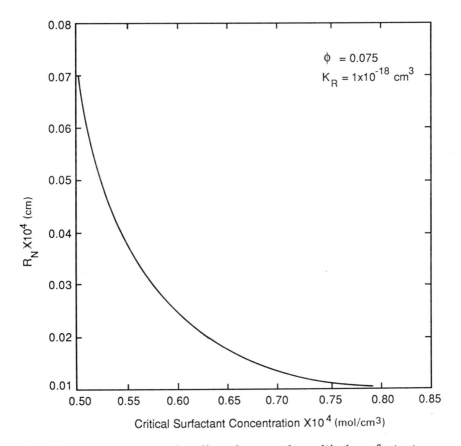

FIGURE 8.14 The natural radius changes the critical surfactant con-
centration. The case shown here is for $K_r = 10^{-18}$ cm^3 and $\phi = 0.075$.
The other parameters are as shown in the legend of Fig. 8.12. (After
Falk, unpublished results.)

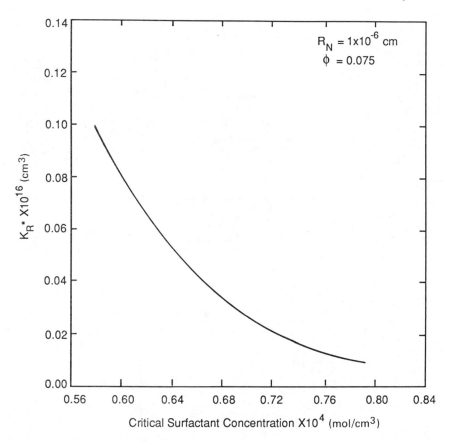

FIGURE 8.15 The modulus of rigidity influences the critical surfac-
tant concentration. The case shown here is for $R_N = 10^{-6}$ cm and
$\phi = 0.075$. The other parameters are the same as used in Fig. 8.12.
(After Falk, unpublished results.)

This is therefore precisely the behavior expected. The surfactant
solvency improves as R_N increases. Less surfactant is required to
maintain a micellar solution having a dispersed phase volume fraction
0.075.

Similarly, if the energy of the interaction with both the O region
and the W region is increased, we also expect the solvency to increase.
An increasing K_r at fixed R_N should model the increasing interactions.
Figure 8.15 shows that as K_r increases, the critical surfactant concentration decreases. The behavior dictated by the simple relationship
expressed by Eq. (8.95) is therefore largely in agreement with the
trends expected based on a consideration of the R-ratio. It should

be reiterated, however, that Eq. (8.95) is not theoretically based. Even though the expected trends are found, Eq. (8.95) is not expected to correspond to any real system.

The equations for phase separation shown here are similar to those derived by Ruckenstein [54]. The difference is that here we insist that the difference between p^α and p^β be given precisely by the Laplace equation as does Huh [42]. Ruckenstein amends the Laplace equation to include terms contributed by free energy of the droplets. The validity of this approach relative to that of Ruckenstein has not yet been resolved.

I. Phase Inversion

Figure 8.9 shows several ternary diagrams representing idealized phase behavior (see Section 5.V.B). Also depicted is a path AA' representing a locus of points above the demixing curve. Each point on this path contains the same fraction of surfactant but different proportions of oil and water. It is certain that in the neighborhood of point A', where the proportion of oil is small, the micellar solution is water continuous irrespective of the type of system, I, II, or III, which is considered. Similarly, near point A, all three systems shown are oil continuous. Therefore, it is also certain that at some composition or sequence of compositions between A' and A, an inversion must take place. Water-continuous S_1 systems must become S_2 oil continuous ones. This transition is not necessarily, or perhaps ever, abrupt. As stressed in Chap. 4, all experiments that relate to isotropic micellar solutions indicate a gradual transition.

It is speculated here that three different system types will exhibit the S_1 to S_2 transition at different compositions along the AA' path. If, for example, the system is Winsor type I, the inversion is expected to take place over a range of concentrations that is displaced toward the surfactant-oil leg of the ternary triangle. This behavior is anticipated because the natural curvature, R_N, of type I systems contributes a term to the free energy favoring water continuity. For this system, the R-ratio defines a C layer that is convex toward the O region. The precise position and the area of this transition zone must depend strongly on the values of K_r and R_N.

Similarly, for type II systems the transition zone (see Fig. 8.9c) must necessarily be displaced toward the water-surfactant leg of the ternary diagram. Furthermore, for type III systems with R_N large (see Table 8.1), the transition should be more or less centered in the ternary diagram. In this case the thermodynamic model is again symmetric; however, the range of compositions along AA' over which the transformation takes place will still depend on K_r.

Since the transition zones are susceptible to measurement, as for example, by a study of the electrical conductivity, it is of interest to attempt to establish a relationship between the position of this zone and the parameters K_r and R_N. Since, as it has been argued in Section 8.III.E, K_r and R_N also relate to the R-ratio, this will also provide a means of qualitatively assessing the R-ratio.

Since the transition from S_1 to S_2 is gradual, in a micellar solution having an intermediate composition along AA' there must coexist zones which are oil continuous (i.e., where the C-layer is convex to the W region) and at the same time other zones which are water continuous. As a model, one can visualize the micellar solution to be an ensemble of small volumes some in the S_1 state and others in the S_2 state. Near point A' shown in Fig. 8.9, almost all of these small subvolumes are in the S_1 state, whereas in the neighborhood of A, S_2 behavior predominates. A micellar solution is dynamic, not static. The interfaces are flexible and constantly being warped by thermal fluctuations. Except perhaps near A or A', it is not proper to regard a micellar solution as being strictly water or oil continuous. The actual system is a mosaic of small volumes in which the curvature of the C layer varies from volume to volume.

To formulate the free energy of a system composed of regions that at any instant in time are both water and oil continuous, let us first consider the free energy of a water-continuous micellar solution expressed per oil droplet. Thus starting with Eq. (8.62), we write

$$g_W^* = \frac{g_W}{n_o^*} = \frac{f_W + p^\beta}{n_o^*} = \frac{\gamma_o \phi_o / R_o^* + 2\phi_o H_o / R_o^{*2} + \Sigma\, c_i \mu_i^W + \Delta f_e(\phi_o, R_o^*)}{n_o^*}$$

$$(8.125)$$

where R_o^* is the oil drop radius satisfying Eq. (8.65) and n_o^* is the number of oil drops per unit of volume found from the simple expression $n_o^* = 3\phi_o / 4\pi R_o^{*3}$. The γ_o and H_o are the interfacial tension and bending moment, respectively, evaluated at the surface of an oil drop of radius R_o^*. Clearly, g_W^* is the free energy per oil drop of a micellar solution containing c_i moles of component i and assembled so that the aqueous phase is the continuous one. We recognized that using a spherical model to represent systems for which ϕ_o is large gives conceptual and theoretical problems. Certainly, ϕ equal to 76% is the maximum dispersed phase volume for packing spheres that are uniform in diameter. Our calculations should be limited to values of ϕ less than this maximum one. However, this limitation is not imposed here and in some cases a small but finite probability of existence is assigned to systems having somewhat larger dispersed phase volumes than the maximum possible geometrical value. This does not, however, represent a serious difficulty. We should recognize that visualizing the micellar droplets as spherical when the phase volumes are large represents a more serious problem, and other models, such as those introduced by Talmon and Prager [36] and Widom [38], may yield more realistic results. The merits of this approach are its simplicity and its ready interpretability. The quantity g_w^* can also be considered to be the chemical potential of an oil drop in a micellar solution. This viewpoint will prove to be quite useful.

The same constituents, c_i, taken in precisely the same proportions, can be reassembled as an oil-continuous micellar solution. In this case the free energy per water drop is

$$g_o^* = \frac{\gamma_w \phi_w / R_w^* + 2\phi_w H_w / R_w^{*2} + \Sigma\, c_i \mu_i^o + \Delta f_o(\phi_w, R_w^*)}{n_w^*} \qquad (8.126)$$

The terms here have obvious definitions when this equation is compared to the one valid for oil drops in water [Eq. (8.125)]. It is perhaps worth noting that even though the same number of moles of a component, say surfactant, may be present per unit of volume, the chemical potential of that component will not necessarily be the same within the oil-continuous and water-continuous zones. A difference can arise because the adsorption is curvature dependent. Therefore, the distribution of surfactant between water, oil and interface may differ depending on the curvature of the interface. Thus, in general, $\mu_i^w \neq \mu_i^o$.

The free energies g_w^* and g_o^* may be regarded as being the free energy of two definite chemical compounds, albeit very complex ones. Furthermore, since each "compound" is composed of precisely the same molecular constituents—oil, water, and surfactant—one of them can be obtained from the other by reassembling the constituents. Thus a system composed of one water drop can be reassembled into a system composed of b oil drops, where $b = n_o^*/n_w^*$. Thus b is a stoichiometric coefficient represented by the chemical reaction

$$A \rightleftharpoons bB \qquad (8.127)$$

where A denotes an oil-continuous system composed of one water drop and B designates a water-continuous system composed of a single oil drop. Equation (8.127) simply states that A can be spontaneously and reversibly converted into B, and vice versa.

Certainly, one of the systems, either A or B, will have the lower free energy; that is, either g_o^* or bg_w^* will be smaller than the other. Just as with chemical reactions, this does not imply only that state having the smaller free energy will be the one observed. In the chemical reaction between two compounds, one of them will, in general, have the smaller free energy of formation. Nevertheless, the reacting system will at equilibrium be a mixture. That compound having the smaller free energy of formation is not observed to the exclusion of the other. Both will be observed to coexist in the same vessel.

Some regions will contain A, while others may contain B. A final
reaction mixture is observed because an additional free energy of
mixing arises and this must be included in the total system free energy. The same is true for micellar solutions. Thus

$$g = (n_o g_w^* + n_w g_o^*) + \Delta f_{mix} \qquad (8.128)$$

Here n_o and n_w are the respective numbers of oil and water drops
per unit of volume. The term in parentheses is the free energy contributed by the presence of each of the components participating in
the reaction. This term is the "standard free energy" and Δf_{mix} is
the additional free energy resulting from the presence of coexisting
water- and oil-continuous regions. It is important to note that Δf_{mix}
is a term additional to the free energy of a dispersion of drops Δf_e.
The calculation of this additional free energy is an interesting and
challenging problem. A first approximation is to use the Flory-
Huggins formulation of the entropy. This appears to be appropriate
since the volume per oil drop may differ significantly from the volume
per water drop. Thus to an approximation

$$\frac{\Delta f_{mix}}{kT} = n_o \ln \alpha_w + n_w \ln \alpha_o \qquad (8.129)$$

where α_w is the volume fraction of those regions that are water-
continuous and $\alpha_o = 1 - \alpha_w$. Clearly, $\alpha_w = n_o/n_o^*$ and $\alpha_o = n_w/n_w^*$.
The volume fraction α_w must be distinguished from ϕ_w, the volume
fraction of water. If a considerable fraction of the micellar system
consists of oil-continuous regimes, ϕ_w may be large while α_w is small.

Substituting Eq. (8.129) into Eq. (8.128) and minimizing g with
respect to n_w yields the familiar equation for chemical reactions,

$$\frac{\alpha_w^b}{1 - \alpha_w} = \exp\left(\frac{g_o^* - b g_w^*}{kT}\right) \qquad (8.130)$$

This equation shows that the "product" of the chemical reaction expressed by Eq. (8.127) taken to the stoichiometric power b divided
by the reactant concentration (here volume fraction) must equal an
equilibrium constant derived from the standard free energies. In
this case the standard free energies relate to the entire micellar system, including bending free energies, interfacial tensions, chemical
potentials, and droplet free energies, as represented by Eqs. (8.125)
and (8.126).

We note here that the Flory-Huggins entropy is not likely to
represent Δf_{mix} precisely because the juxtaposition of water- and

oil-continuous regions will be opposed by an additional free energy proportional to the surface area of contact. This may, however, be a small contribution. Despite this obvious deficiency Eq. (8.130) has been used successfully to predict the electrical conductivities of micellar solutions [55]. Furthermore, Eq. (8.130) predicts that the transition from water to oil-continuity is gradual, not abrupt. This corresponds with observation.

The influence of R_N and K_r on the transition can now be explored. Figure 8.16 shows the fraction of the micellar solution which is water continuous as a function of the oil volume fraction, ϕ_0. Since $R_N > 0$ the intrinsic interfacial curvature favors oil drops.

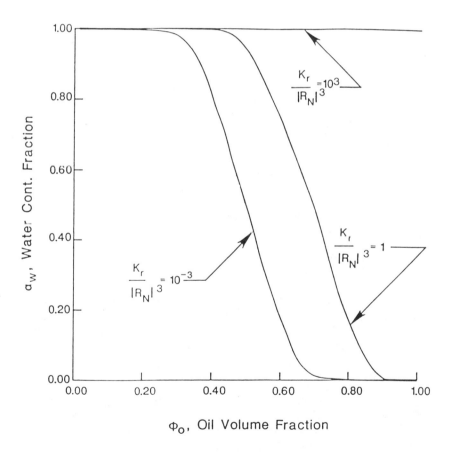

FIGURE 8.16 The volume fraction of water continuous regions in a micellar solution increases gradually as the volume fraction of water increases. (After Falk, unpublished results.)

The results shown are therefore as expected. For small values of K_r, the first perceptible regions of oil continuity appear at volume fractions of oil which are less than 0.4 and the micellar solution does not attain essentially complete oil continuity until ϕ_0 is increased to about 0.7. This illustrates the gradual nature of the transition.

Increasing K_r amplifies the importance of the interfacial curvature relative to other terms in the free energy. Since for the case illustrated by Fig. 8.16, oil drops are preferred, the transition to oil continuity is delayed to larger values of ϕ_0 when K_r is increased. Indeed, for the larger value of K_r shown in Fig. 8.16, the water-continuous fraction does not decrease to 50% ($\alpha_w = 0.5$) until the volume fraction of oil is increased to about 65%. Further increases in K_r will result in a predicted water continuity well into the region where it would not be possible to pack spherical oil drops.

For larger K_r the system would be expected to be liquid crystalline (rigid interfaces) rather than isotropic. The crystalline structure is promoted by a rigid interface and the calculations suggest water continuity to be retained for values of ϕ_0 near unity. When such rigidity is predicted, the model results do not correspond to real systems; however, the trends shown are instructive.

The transitions indicated by the positions of the inversion zones shown in Fig. 8.9 are now seen to be dictated by the interfacial bending energy. This quantity is, of course, related to the same factors that determine the R-ratio. Thus the intrinsic curvature arising from a consideration of those cohesive energies that enter into the calculation of the R-ratio determine the position and the extent of the transition zones.

These transition zones can be measured because electrical conductivities are determined mainly, although not entirely, by the fraction α_w [55]. For example, the electrical conductivities measured from A' to A along the path shown on the pseudoternary diagrams (Fig. 4.33) are plotted in Fig. 4.34. The gradual transition from water continuity (high conductivity) to a decane-continuous system is evident. The curves are similar in shape to those shown in Fig. 8.16. Indeed, the solid lines shown in Fig. 4.34 are theoretical curves calculated based on Eq. (4.130). The fit is good. Refer to Lam and Schechter [55] for further details.

J. Summary

It is evident that the difficult task of quantifying the R-ratio in thermodynamic terms is presently not complete. No single formula for calculating R has been presented. The thrust of this presentation has been to demonstrate that a thermodynamic droplet model of a micellar solution will not yield valid predictions unless terms relating to the natural bending tendency of the C layer are included. This demonstrates that a concept like the R-ratio is absolutely necessary if one is to understand the behavior of systems that include amphiphilic compounds.

The other feature which appears is that because of a number of factors, including the interfacial tension and the droplet free energy, the actual curvature of the C layer may differ markedly from the natural curvature.   Thus a system which, based on the R-ratio, is type I will under certain conditions exhibit type II behavior.   This difference simply means that other thermodynamic factors determining the phase behavior overcome the dictates of R-ratio.   This explains the inversions that necessarily take place when traversing any one of the three AA' paths shown in Fig. 8.9.   Based on the R-ratio alone, one cannot, therefore, predict the complete phase behavior of micellar solutions.   Other factors, including droplet-droplet interactions that lead to critical points, for example, are also of prime importance. The R-ratio does, however, remain an essential feature dictating the properties of micellar solutions.

IV.   FREE-ENERGY EQUATIONS OF STATE

Rossen et al. [56] and Kilpatrick et al. [57] have treated micellar solutions as a mixture represented by a modified version of the Flory-Huggins equation of state.   This approach provides a systematic scheme for representing the phase behavior in a quaternary system given the phase equilibrium on the ternary faces.   Furthermore, precise statements can be made with respect to the critical behavior of this model system.

It is interesting to note that to model correctly systems that include surfactants, Rossen et al. [56] found it necessary to add empirically terms which contain features that are similar to those imposed by the R-ratio.   It is this aspect that is to be pursued here.

The dimensionless free energy per unit volume is

$$g = \frac{G}{kTV} \qquad (8.131)$$

where G is the free energy of the entire mixture, which may include more than one phase and V is the total volume of the system.

Rossen et al. [56] assume that g is composed of three terms:

$$g = g_1 + g_2 + g_3 \qquad (8.132)$$

where

$$g_1 = \phi_w \ln \phi_w + \phi_o \ln \phi_o + \varepsilon_s \phi_s \ln \phi_s$$

$$g_2 = \alpha_{wo} \phi_w \phi_o \exp(-\beta_s \phi_s)$$

$$g_3 = +\alpha_{co} \phi_s \phi_o + \alpha_{cw} \phi_s \phi_w$$

g_1 represents the entropic contribution, where ε_s is in general a
small number to reflect the fact that the surfactant is located primarily at the interface and is not randomly disposed. g_2 represents
the enthalpic term, where α_{wo} is the binary interaction between the
O and W regions.   This interaction is reduced by the presence of
the C layer.   Thus g_2 is formulated so that the tendency for oil and
water to phase separate which is embodied in the large value of α_{wo}
is reduced by an intervening C layer that shields the interaction.
The parameter β_s represents that shielding.   Large values of β_s will
effectively reduce the contribution α_{wo}.

The free energy g_3 is the one that is of primary interest here
because the interactions between the C layer and the O and W regions
are explicitly written.   Thus $-\alpha_{co}$ is the interaction enthalpy between
the O region and the C layer.   The parameter $-\alpha_{cw}$ is that between
the C layer and the W region.   The R-ratio is therefore approximately
$|\alpha_{so}|/|\alpha_{sw}|$.   This ratio should be compared to Eq. (1.17).

The free energy function represents a complete description of
the phase behavior.   Since [56]

$$\mu_i = kT \left(g - \sum_{\substack{j \neq i}}^{n} \frac{\partial g}{\partial \phi_j} \phi_j \right) \tag{8.133}$$

so that

$$\mu_w = kT \left(g - \frac{\partial g}{\partial \phi_o} \phi_o - \frac{\partial g}{\partial \phi_w} \phi_w \right) \tag{8.134}$$

The composition of phases α and β in equilibrium are given by

$$\mu_i^\alpha = \mu_i^\beta \quad i = 1, \, 2, \, \ldots \tag{8.135}$$

The locus of points in composition space determined by Eq. (8.135)
represents the binodal curve.   Similarly, the compositions of three-
phase equilibria are given by

$$\mu_i^\alpha = \mu_i^\beta = \mu_i^\delta \quad i = 1, \, 2, \, \ldots \tag{8.136}$$

The calculations based on these conditions are complex. Three-phase equilibria may arise in one of two ways: either a critical tie-line opens at one endpoint to form a three-phase triangle, or a tricritical point opens into a small three-phase triangle. Kilpatrick et al. [57] have considered the conditions that must be satisfied at these critical points and have developed numerical methods for isolating them as well as the equilibrium phase compositions. Figure 8.17 shows the result of some of their calculations.

It is appropriate to consider some of the parameters in more detail. The quantity ε_S represents the tendency for the surfactant to form sheets or layers. A value of 0.04 is used in calculating the phase behavior shown in Fig. 8.17. It was selected to yield a low base of the three-phase triangle. Thus ε_S is related to the composition of surfactant in the excess phases.

The value of α_{wo} relates to the mutual solubility of oil in water. As shown in Fig. 8.17, the value selected, $\alpha_{wo} = 4.8$, permits little solubility of oil in water or water in oil.

Figure 8.17 shows that $\alpha_{ws} = \alpha_{os}$ yields an optimum middle phase system. The surfactant-rich third phase contains equal volumes of oil and water. This symmetry is expected. This point can be emphasized by rearranging g_3 as follows [58]:

$$g_3 = \frac{\phi_s}{2} \, [(\alpha_{so} + \alpha_{sw})(\phi_o + \phi_w) + (\alpha_{so} - \alpha_{sw})(\phi_o - \phi_w)]$$

$$(8.137)$$

The quantities g_1 and g_2 are symmetric with respect to ϕ_o and ϕ_w and hence cannot provide the I-III-II transition. g_3 is not, however, symmetric provided that $\alpha_{so} \neq \alpha_{sw}$. Thus the symmetry-breaking term is

$$\alpha_{so} - \alpha_{sw} = \Delta \qquad (8.138)$$

If Δ vanishes, the system is symmetric and the optimum is for WOR = 1. This is shown by Fig. 8.17a. If, on the other hand, $\Delta > 0$, g is no longer symmetric. In this case the minimum of free energy occurs at a single point $\phi_w - \phi_o > 0$ and the system tends to be type I. The R-ratio is calculated to be less than unity (actually it vanishes) for this case. Figure 8.17b shows that the composition of the surfactant-rich apex of the phase diagram has shifted toward the brine leg of the ternary diagram, indicating the expected tendency for type I behavior.

Figures 8.17c and d show that the trend is continuous as α_{ws} decreases. Thus the system of free-energy expressions represented by Eq. (8.132) provides a symmetry-breaking term by setting $\alpha_{so} \neq \alpha_{ws}$. There is a second interesting feature of these same equations.

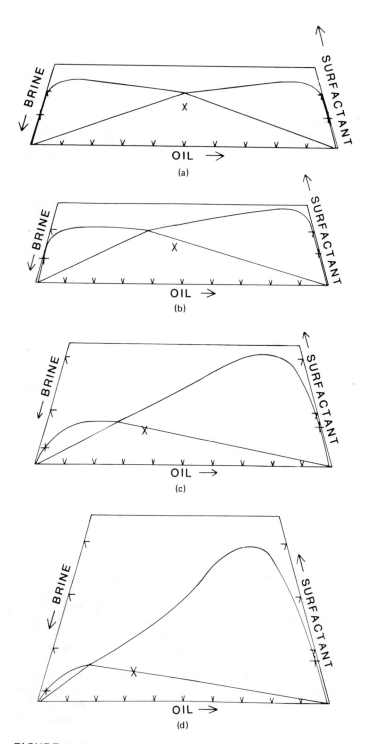

FIGURE 8.17

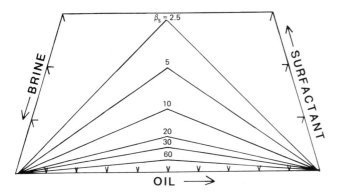

FIGURE 8.18 Sequence of symmetric amphiphile–oil–water three-phase tie-triangles for $\alpha_{WO} = 4.8$, $\alpha_{WS} = \alpha_{OS} = 0$, $\varepsilon_S = 0.04$, and values of β_S varying from 2.5 to 60. (After Ref. 57.)

As β_S, the shielding of the oil from the water, is increased, the height of the three-phase vertex decreases. This is shown in Fig. 8.18. Thus increasing β_S increases the solvency of the amphiphile. One can imagine that this shielding is improved as both the hydrophilic and lipophilic interactions of the amphiphile with the O and W regions increases. The model of Kilpatrick et al. contains all the features embodied by the R-ratio. Without including a term that breaks the symmetry and one that increases the shielding, this model would be unable to give realistic predictions.

Our contention here is quite simple. We have shown that all models which provide realistic predictions of the phase behavior of micellar solutions must input, as a minimum, the same information as contained in the R-ratio. The authors do not know how future research will develop and cannot state that the R-ratio per se will be the primary quantity of interest, but we can state that even as more fundamental approaches develop, symmetry-breaking and solvency-enhancing terms will appear and these will relate to the interaction of the C layer with the O and W regions.

FIGURE 8.17 Ternary phase diagram of surfactant oil-water system. Parameter values $\alpha_{WO} = 4.8$, $\alpha_{OS} = 0$, $\alpha_S = 0.04$, and $\beta_S = 12.5$. For Fig. 8.17, (a): $\alpha_{WS} = 0$; (b): $\alpha_{WS} = -2$; (c): $\alpha_{WS} = -4$; and (d): $\alpha_{WS} = -6$. (After Ref. 57.)

REFERENCES

1. H. E. Stanley, Introduction to Phase Transitions and Critical Phenomena, Oxford University Press, New York, (1971).

2. J. V. Sengers and J. M. H. Levelt-Sengers, "Concepts and Methods for Describing Critical Phenomena in Fluids," NASA Contractors Report 149665 (Sept. 1977).

3. H. T. Davis and L. E. Scriven, paper SPE 9278 presented at the 55th SPE Annual Technical Conference and Exhibition, Dallas (Sept. 1980).

4. P. D. Fleming and J. E. Vinatieri, AIChE J., 25: 493 (1979).

5. P. D. Fleming and J. E. Vinatieri, J. Colloid Interface Sci, 81: 319 (1981).

6. A. M. Bellocq, D. Bourbon, B. Lemanceau, and G. Fourche, J. Colloid Interface Sci, 89: 427 (1982).

7. A. M. Cazabat, D. Langevin, J. Meunier, and A. Pouchelon, Adv. Colloid Interface Sci., 16: 175 (1982).

8. S. K. Ma, Modern Theory of Critical Phenomena, Benjamin-Cummings, Menlo Park, Calif. (1976).

9. J. S. Rowlinson and F. L. Swinton, Liquids and Liquid Mixtures, 3rd ed., Butterworth, London (1982).

10. J. S. Huang and M. W. Kim, in Scattering Techniques Applied to Supramolecular Systems (S. H. Chen, B. Chur, and R. Nossal, eds.), Plenum Press, New York, p. 890 (1981).

11. J. S. Huang and M. W. Kim, Phys. Rev. Lett., 47: 1462 (1981).

12. M. Corti and V. Degiorgio, Phys. Rev. Lett., 45: 1045 (1980).

13. J. S. Huang and M. W. Kim, Soc. Pet. Eng. J., 24: 197 (Apr. (1984).

14. B. J. Berne and R. Pecora, Dynamic Light Scattering with Applications to Chemistry, Biology, and Physics, Wiley, New York (1976).

15. R. Dorshow, J. Briggs, C. A. Burton, and D. F. Nicoli, J. Phys. Chem., 86: 2388 (1982).

16. P. D. Fleming, J. E. Vinatieri, and G. R. Glinsman, J. Phys. Chem., 84: 1526 (1980).

17. H. T. Davis and L. E. Scriven, Adv. Chem. Phys., 49: 357 (1982).

18. D. Chatenay, O. Abillon, J. Meunier, D. Langevin, and A. M. Cazabat, in Macro- and Microemulsions: Theory and Practice (D. O. Shah, ed.), American Chemical Society, Washington, D.C., p. 119 (1985).

19. R. L. Reed and R. N. Healy, in Improved Oil Recovery by Surfactant and Polymer Flooding (D. O. Shah and R. S. Schechter, eds.), Academic Press, New York, p. 383 (1977).

20. J. W. Gibbs, "On the Equilibrium of Heterogeneous Substances," in The Scientific Papers of J. Williard Gibbs, Vol 1, Longmans, New York, (1906).

21. F. P. Buff, J. Chem. Phys., 19: 1591 (1951).

22. J. C. Melrose, Ind. Eng. Chem., 60: 53 (1968).

23. C. L. Murphy, "The Thermodynamics of Low Tension and Highly Curved Interfaces," Ph.D. thesis, University of Minnesota (1966).

24. S. Timoshenko, Theory of Plates and Shells, McGraw-Hill, New York, p. 1 (1940).

25. H. S. Green, The Molecular Theory of Fluids, North-Holland, Amsterdam, p. 186 (1952).

26. C. A. Miller and P. Neogi, AIChE J., 26: 212 (1980).

27. P. A. Rehbinder, Proc. 2nd Int. Congr. Surf. Act., 1: 476 (1957).

28. E. Ruckenstein and J. C. Chi, J. Chem. Soc. Faraday Trans. 2, 71: 1690 (1975).

29. H. Reiss, J. Colloid Interface Sci, 53: 61 (1975).

30. E. Ruckenstein and R. Krishnan, J. Colloid Interface Sci., 76: 201 (1980).

31. E. Ruckenstein, Chem. Phys. Let., 98: 573 (1983).

32. A. Lam, N. A. Falk, and R. S. Schechter, "The Thermodynamics of Microemulsions," J. Colloid Interface Sci., 120 (1), 30 (1987).

33. J. T. G. Overbeek, Faraday Discuss. Chem. Soc., 65: 7 (1978).

34. J. H. Schulman, W. Stockenius, and L. M. Prince, J. Phys. Chem., 63: 1677 (1959).

35. J. H. Schulman and J. B. Montagne, Ann. N.Y. Acad. Sci., 92: 366 (1961).

36. Y. Talmon and S. Prager, J. Chem. Phys., 69: 2984 (1978).

37. P. G. de Gennes and C. Taupin, J. Phys. Chem., 86: 2294 (1982).

38. B. Widom, J. Chem. Phys., 81: 1030 (1984).

39. S. Mukherjee, C. A. Miller, and T. Fort, J. Colloid Interface Sci., 91: 233 (1983).

40. W. Helfrich, Z. Naturforsch., C28: 693 (1973).

41. M. L. Robbins, in Micellization, Solubilization, and Microemulsions, Vol. 2 (K. L. Mittal, ed.) Plenum Press, New York, p. 273 (1977).

42. C. Huh, Soc. Pet. Eng. J., 23: 829 (1983).

43. B. Lemaire, P. Bothorel, and D. Roux, J. Phys. Chem., 87: 1023 (1983).

44. Y. Suezaki, J. Disp. Sci. Tech., 4: 371 (1983).

45. L. Auvray, J. P. Cotton, R. Ober, and C. Taupin, J. Phys. Chem., 88: 4586 (1984).

46. J. M. Di Meglio, M. Dvolaitzky, and C. Taupin, J. Phys. Chem., 89: 871 (1985).

47. J. M. Di Meglio, L. Paz, M. Dvolaitzky, and C. Taupin, J. Phys. Chem., 88: 6036 (1984).

48. C. A. Miller, J. Disp. Sci. Tech., 6(2): 159 (1985).

49. C. Huh, J. Colloid Interface Sci., 97: 201 (1984).

50. C. A. Miller, R. Hwan, W. J. Benton, and T. Fort, J. Colloid Interface Sci., 61: 554 (1977).

51. A. A. Calje, W. G. M. Agterof, and A. Vrij, in Micellization, Solubilization, Microemulsions, Vol 2 (K. L. Mittal, ed.), Plenum Press, New York, p. 779 (1977).

52. S. Brunetti, D. Roux, A. M. Bellocq, G. Fourche, and P. Bothorel, J. Phys. Chem., 87: 1028 (1983).

53. (a) B. Lemaire and P. Bothorel, Macromolecules, 13: 311 (1980); (b) B. Lemaire, P. Bothorel, and D. Roux, J. Phys. Chem., 87: 1023 (1983).

54. E. Ruckenstein, Fluid Phase Equilibria, 20: 189 (1985).

55. A. Lam and R. S. Schechter, "A Study of Diffusion and Conduction in Microemulsions," J. Colloid Interface Sci., 120 (1), 42 (1987).

56. W. R. Rossen, R. G. Brown, H. T. Davis, S. Prager, and L. F. Scriven, Soc. Pet. Eng. J., 22, 947 (1982).

57. P. K. Kilpatrick, L. E. Scriven, and H. T. Davis, 25, 330 Soc. Pet. Eng. J., (1985).

58. J. Biais and J. L. Trovilly, to be published.

Author Index

Underscored numbers give the page on which
the complete reference is cited.

Subject Index

Absorbance, 112
Activity of Ions, 64
Adsorption, 229-233, 249-264,
   353, 412-413, 427-435, 441-
   443
Aggregates (see Micelles)
Aggregation
   of alcohols, 233, 250
   numbers, 35, 69, 98, 112
   process, 8, 33, 69, 81, 115
Alcohol
   as an amphiphile, 229
   effect on lateral interactions
      of, 239-247
   effect on shape of phase maps
      of, 232-233, 254, 270, 376
   effect on solubilization of,
      371-378
   effect on I to II transition of,
      229-233, 293-296, 314-320
   partitioning of, 249-256, 295-
      296, 300, 316, 374-378
Alkane Carbon Number
   effect on I to II transition of,
      223-225
   effect on mesophases of, 239-
      241
   optimum (see Optimum)

Amphiphiles
   definition of, 1
   molecular area of amphiphiles,
      effect of alcohol on, 231-233
      effect of branching on, 228-
         229, 289, 365-371
      effect of hydrocarbon type
         on, 384-386
      effect of hydrophile on, 362-
         365
      effect of polar phase on, 380-
         383
      effect of temperature on,
         233-234, 378-380
      effect on solubilization of,
         349, 361-386
Amphiphilic Compounds (see
   Amphiphiles)
Antonov's Rule, 167
Association, 6 (see also Aggrega-
   tion)

Bending (see Free Energy of)
Bending Moment, 411-417, 429-
   435, 440-443, 445-453
Bicontinuity, 12, 140
Branching (Surfactant), 32, 97,

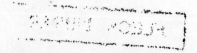